SPACE SCIENCE, EXPLORATION AND POLICIES

BLACK HOLES

EVOLUTION, THEORY AND THERMODYNAMICS

SPACE SCIENCE, EXPLORATION AND POLICIES

Additional books in this series can be found on Nova's website under the Series tab.

Additional e-books in this series can be found on Nova's website under the e-book tab.

PHYSICS RESEARCH AND TECHNOLOGY

Additional books in this series can be found on Nova's website under the Series tab.

Additional e-books in this series can be found on Nova's website under the e-book tab.

SPACE SCIENCE, EXPLORATION AND POLICIES

BLACK HOLES

EVOLUTION, THEORY AND THERMODYNAMICS

ALOIS J. BAUER
AND
DIETHELM G. EIFFEL
EDITORS

Nova Science Publishers, Inc.
New York

For permission to use material from this book please contact us:
Telephone 631-231-7269; Fax 631-231-8175
Web Site: http://www.novapublishers.com

NOTICE TO THE READER

Additional color graphics may be available in the e-book version of this book.

LIBRARY OF CONGRESS CATALOGING-IN-PUBLICATION DATA

Black holes : evolution, theory, and thermodynamics / [edited by] Alois J. Bauer and Diethelm G. Eiffel.

p. cm.

Includes bibliographical references and index.

ISBN 978-1-61942-929-1 (hardcover)

1. Black holes (Astronomy) 2. Gravity. I. Bauer, Alois J. II. Eiffel, Diethelm G.

QB843.B55B5865 2011

523.8'875--dc23

2012001088

Published by Nova Science Publishers, Inc. † New York

CONTENTS

PREFACE

In this book, the authors present current research in the study of the evolution, theory and thermodynamics of black holes. Topics discussed include the effective theory of a matter field near the event horizon of a black hole; dilatonic black holes in 4D Gauss-Bonnet gravity; the relationship between the tunneling rate of a rotating black hole and the first law of black hole thermodynamics; alternative theories of gravity without black holes; exotic smoothness structures for the application of standard Einstein equations to black hole structures; and lower dimensional quantum black holes.

Chapter 1 discusses the effective theory of a matter field near the event horizon of a black hole. Through the progress in the theory of relativity various solutions of higher dimensional black holes have been found. There is a variety of black hole solutions in higher dimensional spacetime because the restriction associated with topologies of the event horizon is relaxed. As complicated as the metric of spacetime is, the action of a matter field is also involved. However, it is shown that the effective theory of a matter field is quite simple in the region near the horizon when using the technique of the dimensional reduction near the horizon. This dimensional reduction technique was applicable for most black holes known, until recently. In this chapter, the authors show that a complicated metric of a black hole effectively becomes a 2-dimensional spherically symmetric metric by using the technique; and, they review the usability of the effective metric. For example, the derivation of Hawking radiation is simplified in elaborate black hole backgrounds.

In the last ten years, increasing attention has been devoted to Extended Theories of Gravity with the aim to understand several cosmological and astrophysical issues such as the today observed accelerated expansion of the universe and the presence of Dark Matter in self-gravitating structures. Some of these models assume modifications of General Relativity by adding higher order terms of curvature invariants like the Ricci scalar R, the Ricci tensor $R_{\mu\nu}$ and the Riemann tensors $R_{\mu\nu\lambda\sigma}$, or the presence of suitable scalar fields like the former Brans-Dicke theory. It is therefore natural to ask for black hole solutions in this context since, on the one hand, black holes signatures may be the test-bed to compare new models to the Einstein gravity; on the other hand, they may lead to rule out models which disagree with observations. Although black holes are one of the most striking predictions of General Relativity, they remain one of its least tested concepts. Electromagnetic observations allow indirectly to infer their existence, but direct evidences remains elusive. In the next decade, data coming from very long-baseline interferometry and gravitational wave detectors should allow to image and study black holes in detail. Such observations will test General Relativity in the non-linear and strong-field regimes where data are currently lacking. Testing strong-field features of General Relativity is of utmost importance to physics and astrophysics as a whole. This is because the black holes solutions, such as the Schwarzschild and Kerr metrics, enter several calculations, including accretion disk structure, gravitational lensing, cosmology and gravitational waves theory. These black hole solutions could indicate strong-field departures from General Relativity with deep implications for the still unknown fundamental theory of gravity. Beside the physical interest, black hole solutions represent a very active area for mathematical physics investigations. In Chapter 2 the authors review the problem of black holes in a particular class of Extended Theories of Gravity, the so called $f(R)$-gravity, discussing some resolution techniques, obtaining exact solutions and comparing results with standard General Relativity. Furthermore, the authors discuss the problems of hydrostatic equilibrium and stellar structure in the context of $f(R)$-gravity showing that new features could emerge. The observation of such features could both explain the physics of exotic self-gravitating objects and constitute a signature for Extended Theories of Gravity.

Chapter 3 summarizes different aspects of dilatonic black hole in 4D Gauss-Bonnet gravity that were hardly investigated not so long ago. Such study has been become real because of a new numerical integration method, based on the analysis of the main system determinant. The most interesting result is the existence of a minimal black hole in Gauss-Bonnet gravity, and, so, a new approach to the last stages of black hole evaporation is discussed. Such a black hole does not evaporate comletely, but only to some relic remnant. If this remnant is stable, being the endpoint of the evaporation, it can represent the relic remnant of the black holes formed during the initial stages of our Universe formation. The coupling with early Universe parameters is also discussed.

In Chapter 4 the authors study the relationship between the tunneling rate of a rotating black hole and the first law of black hole thermodynamics. The authors show that the tunneling rate of a rotating black hole $e^{-2\mathrm{Im}\mathcal{I}}$ can be set up from the emission rate $e^{\Delta S}$ and the first law of black hole thermodynamics. Besides, the authors give an analysis of the one loop corrected tunneling rate of a four-dimensional Kerr black hole from the corrected Bekenstein–Hawking entropy of the Kerr black hole.

Do observations of black hole candidates rule out alternative theories of gravity without horizon formation? This depends on the existence, viability and reasonableness of alternative theories of gravity without black holes.

In Chapter 5, a theory of gravity without black hole horizon formation is presented. The gravitational collapse stops shortly before horizon formation and leaves a stable frozen star. In the limit $\Xi, \Upsilon \to 0$ the Einstein equations of GR are recovered, and the frozen stars become observationally indistinguishable from GR black holes. The theory therefore provides a counterexample to recent claims that observational evidence from black hole candidates "all but requires the existence of a horizon".

The theory presented here shares its equations with RTG. Nonetheless, as is shown, there remain important conceptual and physical differences. In particular, some serious problems of RTG are not present in the theory proposed here. So it can be argued that the theory is a physically viable and conceptually sound alternative to GR.

In Chapter 6 the authors review some implications of the existence of exotic smoothness (differentiable) structures on otherwise topologically trivial spaces, such as $\mathbb{R}^4$ or $\mathbb{R}^2 \times S^2$, for the application of standard Einstein equations to black hole structures. These results in differential topology have shown that in considering the Einstein or any other equations globally we have always implicitly made a choice of one from a possible infinity of smoothness structures in continuing solutions from one coordinate patch to another. Since there are no known physical reasons for restricting studies to this one choice it is important to consider implications of choosing others. Although there do not yet exist finite presentations of non-trivially smooth coordinate atlases, we should consider possible implications of globally extending what we obtain locally in one coordinate patch to solutions of the Einstein equations. This may be physically important since in the Kruskal presentation of black hole solutions we only have direct physical information in our local coordinates outside the horizons. The authors discuss implications of going beyond the standard smooth extension globally to the topological $R_2 \times S_2$' Kruskal space, and point out that predictions of what happens beyond our local, physically observed, coordinate patch to the full topological Kruskal space may not be justified by any known physics. In particular, the authors have no *a priori* physical reason for using only the standard smoothness on topological $\mathbb{R}^2 \times S^2$ when predicting the nature of singularities within the horizon for solutions to the vacuum Einstein equations.

There are theoretical reasons to think that space-time is lower dimensional at very short distances. If this is the case, quantum black holes produced at the LHC or in cosmic rays scattering live in lower dimensions. Chapter 7 discusses production cross section and signatures for the corresponding quantum black holes at the LHC within several different models of low-dimensional space-time.

In Chapter 8 the authors review the current theoretical state of the art of small black holes at the LHC. The authors discuss the production mechanism for small non thermal black holes at the LHC and discuss new signatures due to a possible discrete mass spectrum of these black holes.

In the context of the scenario with Extra Dimensions (either Large or slightly Warped), the authors consider in Chapter 9 the prospect of the creation of mini black holes during high-energy particle collisions at ground-based accelerators. The authors then focus on the main observable signal of their creation, that is the emission of Hawking radiation in the form of either scalars, fermions, gauge bosons or gravitons during both the spherically-symmetric ("Schwarzschild") and axially-symmetric ("spin-down") phases. The authors look for footprints in the radiation spectra of brane-localised Standard Model particles that could help us determine fundamental parameters of the theory, such as the number of additional spacelike dimensions that may exist in nature and the angular momentum of the black hole. The angular distribution of the emitted particles on our brane is studied, too, as an additional source of information and a distinctive feature of the Hawking radiation emitted by a rotating black hole. The authors also attempt to shed light on the question of the energy balance between the brane and the bulk emitted by the black hole: they consider the emission of all particles allowed to live on the brane and in the bulk and the authors compute their relevant emissivities. Finally, the authors discuss the effect of the mass on the scalar radiation spectra of a spinning black hole both in the bulk and on the brane as an indicative example of the role of the mass of the emitted particles.

In: Black Holes: Evolution, Theory and Thermodynamics ISBN: 978-1-61942-929-1
Editors: A. J. Bauer and D. G. Eiffel, pp. 1-39

Chapter 1

EFFECTIVE THEORY OF MATTER FIELD NEAR THE EVENT HORIZON

Koichiro Umetsu*
Maskawa Institute for Science and Culture,
Kyoto Sangyo University, Kyoto, Japan

Abstract

We discuss the effective theory of a matter fiel near the event horizon of a black hole. Through the progress in the theory of relativity various solutions of higher dimensional black holes have been found. There is a variety of black hole solutions in higher dimensional spacetime because the restriction associated with topologies of the event horizon is relaxed. As complicated as the metric of spacetime is, the action of a matter fiel is also involved. However, it is shown that the effective theory of a matter fiel is quite simple in the region near the horizon when using the technique of the dimensional reduction near the horizon. This dimensional reduction technique was applicable for most black holes known, until recently. In this chapter, we show that a complicated metric of a black hole effectively becomes a 2-dimensional spherically symmetric metric by using the technique; and, we review the usability of the effective metric. For example, the derivation of Hawking radiation is simplifie in elaborate black hole backgrounds.

1. Introduction

The existence of black holes has been predicted as a consequence of the general theory of relativity, namely, as special solutions of the basic Einstein equation. According to the Einstein equation, the spacetime is curved by the effects of gravity [1]. The spacetime curved by a very strong gravity can form a closed region from which nothing, not even photons can escape. The closed region is the black hole.

It is considered that black holes are formed as a result of the gravitational collapse of a star with a very large mass [2]. The original stars that will form the black hole have

*E-mail address: umetsu@cc.kyoto-su.ac.jp

various physical quantities and properties. As soon as a black hole is formed by the gravitational collapse, the state of the black hole becomes a stationary state. It is known that the stationary state is characterized by only three physical parameters; namely, the mass, the angular momentum and the electrical charge. This means that a black hole does not retain the various information of the original star except for these three parameters. In particular, it is known that a black hole can uniquely be decided by the mass, the angular momentum and the charge in 4-dimensional spacetime. This consequence is called the black hole uniqueness theorem [3,4], or the no-hair theorem [5].

To understand characteristics of black holes, it is very useful to understand the black hole physics in the context of generalized thermodynamics. By digging deep into the properties of black holes, it was clarif ed that black hole physics has analogous relationships from the zeroth law to the third law of thermodynamics [6]. Note that the analogous quantities between of thermodynamics and black hole physics are respectively "Energy ⇔ Mass," "Temperature ⇔ Surface gravity," and "Entropy ⇔ Surface area."

On the basis of these charactertistic resemblances, Bekenstein proposed that a black hole has its entropy from the viewpoint of information theory [7]. According to information theory, a loss of information means the increase of entropy [8]. Now, to consider what occurs when matter falls into a black hole. We can know information about the matter before it falls into the black hole. Yet, we cannot determine the state of the matter becomes after it falls into the black hole, namely, all information of the matter is lost. Bekenstein considered that the information loss is connected with the increase in entropy of the black hole and proposed a black hole has its entropy. The increase of black hole entropy is associated with the increase of black hole area by the matter that falls into the black hole. Bekenstein suggested that a black hole has its entropy while this comment stopped short of remarking that a black hole has its temperature. The reason is caused by the classical def nition of a black hole, namely, "a closed region from which nothing, not even photons can escape." If a black hole has a temperature, one must explain a mechanism which a black hole radiates matter according to circumstances. Therefore Bekenstein did not suggest that a black hole has a temperature.

Hawking searched a mechanism of black hole radiation following this discussion by Bekenstein. Then Hawking succeeded in showing the radiation from a black hole by taking quantum effect into account [9]. In this way, Hawking concretely calculated the expectation value of the particle number by using the Bogoliubov transformations. In comparison with the obtained result and a blackbody spectrum, it was found that a black hole has a particular temperature and continuously radiates as a blackbody. The temperature of black holes and the radiation from black holes are respectively called Hawking temperature and Hawking radiation. Hawking temperature is given by

$$T_{\rm BH} = \frac{\hbar\kappa}{2\pi c k_{\rm B}}, \tag{1.1}$$

where $\hbar$ is the Planck constant (Dirac's constant), c is the speed of light in vacuum, $k_{\rm B}$ is the Boltzmann constant and κ is the surface gravity on the horizon of a black hole.

The complete correspondence between black hole physics and thermodynamics was obtained because Hawking explained the mechanism of the black hole radiation. In the wake of the black hole temperature, the black hole entropy is also determined from the f rst law of thermodynamics if one believes the black hole thermodynamics. The black hole entropy is given by

$$S_{\rm BH} = \frac{k_{\rm B}c^3}{4\hbar G} A_{\rm BH}, \tag{1.2}$$

where G is the gravitational constant and $A_{\rm BH}$ is the area of a black hole. As a result of the black hole thermodynamics, we f nd that most of black holes which exist in the universe at present become absorption-dominated. Namely, the temperature for typical black holes which outweigh the solar mass is much lower than the temperature of the cosmic microwave background radiation. It would be also expected that tiny black holes created in the early universe will evaporate at some point [10].

After Hawking's original derivation, Hawking radiation has been discussed by many researchers and various derivations of Hawking radiation have been suggested over the past few decades. All of them reproduce the same result that the black hole entropy is described by the surface area of the black hole and the temperature of the black hole is described by a surface gravity of the black hole. Hawking radiation is thus one of the most striking effects that has been widely accepted now. And, moreover, presents us with not only a new realization about properties of black holes but also very interesting physics.

Recently, through the progress in theory of relativity, various solutions of higher dimensional black holes have been found[1]. As already stated, there exists the black hole uniqueness theorem in 4-dimensional spacetime. However, a variety of black hole solutions is permitted in higher dimensional spacetimes because the restriction associated with topologies of the event horizon is relaxed. Note that the results of Hawking radiation are basically described by quantum f eld theory in curved spacetime with a black hole. Therefore, as complicated as the metric of spacetime is, the action of a matter f eld is also involved.

In this chapter, we will present an approach to simplify this discussion for a complicated black hole spacetime. We consider an effective theory for a matter f eld near the horizon of a black hole. It is then shown that the effective theory near the horizon becomes a 2-dimensional theory in a spherically symmetric background spacetime. We refer to the approach as the technique of dimensional reduction near the horizon. We discuss derivations of Hawking radiation and Unruh radiation by using the technique of dimensional reduction near the horizon. And then we conf rm that the expected results which have been well known in previous works are reproduced by the use of the reduced metric instead of the full metric. As a consequence, this approach is very useful to understand thermal properties of black holes because the calculation is quite simplif ed by using the effective theory near the horizon. It however seems that little attention has been given to the value of the technique. Therefore, we would like to summarize the dimensional reduction approach. The approach presented here would be able to evaluate its calculation for discussions which have not ever been calculated concretely because of the complexity of the spacetime metric.

This chapter is organized as follows. In Section 2, we review the effective theory of a matter f eld near the horizon. We show that the effective theory near the horizon becomes the 2-dimensional theory in the spherically symmetric black hole spacetime by using the technique of dimensional reduction near the horizon. In Section 3, we present examples

[1]See, for example, [11].

of the application of the effective theory and argue its usability. Section 4 is devoted to discussion and conclusion.

2. Effective Theory of Matter Field near the Horizon

In this section, we discuss the effective theory of a matter f eld near the horizon. We show that a complicated metric of a black hole effectively is reduced to a 2-dimensional spherically symmetric metric by using the technique of dimensional reduction near the horizon. The essential idea of the technique is as follows: We consider the action for a scalar f eld. We can then ignore the mass, potential, and interaction terms in the action because the kinetic term dominates in the high-energy theory near the horizon. By expanding the scalar f eld in terms of the spherical harmonics and using these properties at the horizon, we f nd that the integrand in the action does not depend on angular variables. Thus, it is shown that the 4-dimensional action with the Kerr-Newman metric effectively becomes a 2-dimensional action with a spherically symmetric metric.

In contents of this section are as follows. In Subsection 2.1, we present the technique of dimensional reduction near the horizon. Actually, we illustrate the technique in the Kerr-Newman black hole spacetime and we show that the action for a scalar f eld is reduced to a 2-dimensional action in the reduced spherically symmetric metric. In Subsection 2.2, we collect results of the reduction for black hole solutions known until now.

2.1. Illustration of Dimensional Reduction near the Horizon

We show the technique of dimensional reduction near the horizon. As an example, we here consider a Kerr-Newman black hole background. The Kerr-Newman solution is known as the most general black hole solution which satisf es the black hole uniqueness theorem in four dimensional spacetime and is characterized by three parameters; namely, the mass M, charge Q, and angular momentum L, of the black hole. The space-time outside the Kerr-Newman black hole is represented by the Kerr-Newman metric;

$$\begin{aligned} ds^2 =& \frac{\Delta - a^2 \sin^2\theta}{\Sigma} dt^2 + \frac{2a\sin^2\theta}{\Sigma}(r^2+a^2-\Delta)dtd\varphi \\ &+ \frac{a^2\Delta\sin^2\theta - (r^2+a^2)^2}{\Sigma}\sin^2\theta d\varphi^2 - \frac{\Sigma}{\Delta}dr^2 - \Sigma d\theta^2, \end{aligned} \tag{2.1}$$

where a is def ned in order to adjust the dimensions by

$$a \equiv \frac{L}{M}, \tag{2.2}$$

and for simplicity, the symbols are respectively def ned by

$$\Sigma \equiv r^2 + a^2\cos^2\theta, \tag{2.3}$$

$$\Delta \equiv r^2 - 2Mr + a^2 + Q^2. \tag{2.4}$$

The event horizon of the Kerr-Newman black hole appears at the point $\Delta = 0$, i.e.,

$$r_\pm = M \pm \sqrt{M^2 - a^2 - Q^2}, \tag{2.5}$$

where r_+ and r_- are respectively called the outer event horizon and the inner event horizon. The inner event horizon r_- exists inside the outer event horizon. We do not care the existence of the inner event horizon since we cannot know information inside the outer horizon. In what follows, we simply describe the outer event horizon as the horizon. The Kerr-Newman geometry becomes spherically asymmetric because of its own rotation. We will show that the 4-dimensional Kerr-Newman metric effectively becomes a 2-dimensional spherically symmetric metric by using the technique of the dimensional reduction near the horizon.

For simplicity, we consider the 4-dimensional action for a complex scalar f eld

$$S = \int d^4x\sqrt{-g}g^{\mu\nu}(\nabla_\mu + ieV_\mu)\phi^*(\nabla_\nu - ieV_\nu)\phi + S_{\text{int}}, \tag{2.6}$$

where the f rst term is the kinetic term and the second term S_{int} represents the mass, potential, and interaction terms. The notation ∇_μ stands for the covariant derivative. The gauge f eld V_μ associated with the Coulomb potential of the black hole, is given by

$$(V_\mu) = \left(-\frac{Qr}{r^2+a^2}, 0, 0, 0\right). \tag{2.7}$$

The covariant derivative for a scalar f eld ϕ is consistent with a usual derivative,

$$\nabla_\mu\phi = \partial_\mu\phi. \tag{2.8}$$

The action (2.6) therefore becomes

$$S = \int d^4x\sqrt{-g}g^{\mu\nu}(\partial_\mu + ieV_\mu)\phi^*(\partial_\nu - ieV_\nu)\phi + S_{\text{int}}. \tag{2.9}$$

By substituting both the Kerr-Newman metric (2.1) and the gauge f eld (2.7) to the action (2.9), we obtain

$$\begin{aligned} S = \int dtdrd\theta d\varphi \sin\theta\phi^* \Bigg[&\left(\frac{(r^2+a^2)^2}{\Delta} - a^2\sin^2\theta\right)\left(\partial_t + \frac{ieQr}{r^2+a^2}\right)^2 \\ &+ 2a\left(\frac{r^2+a^2}{\Delta} - 1\right)\left(\partial_t + \frac{ieQr}{r^2+a^2}\right)\partial_\varphi \\ &- \partial_r\Delta\partial_r - \frac{1}{\sin\theta}\partial_\theta\sin\theta\partial_\theta - \frac{1}{\sin^2\theta}\partial_\varphi^2 + \frac{a^2}{\Delta}\partial_\varphi^2\Bigg]\phi + S_{\text{int}}, \end{aligned} \tag{2.10}$$

where we used

$$\sqrt{-g} \equiv \sqrt{-\det(g_{\mu\nu})} = \Sigma^2\sin\theta, \tag{2.11}$$

$$(g^{\mu\nu}) = \begin{pmatrix} -\frac{a^2\Delta\sin^2\theta-(r^2+a^2)^2}{\Delta\Sigma} & 0 & 0 & \frac{a(r^2+a^2-\Delta)}{\Delta\Sigma} \\ 0 & -\frac{\Delta}{\Sigma} & 0 & 0 \\ 0 & 0 & -\frac{1}{\Sigma} & 0 \\ \frac{a(r^2+a^2-\Delta)}{\Delta\Sigma} & 0 & 0 & -\frac{\Delta-a^2\sin^2\theta}{\Delta\Sigma\sin^2\theta} \end{pmatrix}. \tag{2.12}$$

The action (2.10) is also written as

$$S = \int dtdrd\theta d\varphi \sin\theta\phi^* \left[\left(\frac{(r^2+a^2)^2}{\Delta} - a^2\sin^2\theta \right) \left(\partial_t + \frac{ieQr}{r^2+a^2} \right)^2 \right.$$
$$\left. + 2ia\left(\frac{r^2+a^2}{\Delta} - 1 \right) \left(\partial_t + \frac{ieQr}{r^2+a^2} \right) \hat{L}_z - \partial_r \Delta \partial_r + \hat{\boldsymbol{L}}^2 - \frac{a^2}{\Delta}\hat{L}_z^2 \right] \phi + S_{\text{int}}, \quad (2.13)$$

where we used the angular momentum operators

$$\hat{\boldsymbol{L}}^2 = -\frac{1}{\sin\theta}\partial_\theta \sin\theta\partial_\theta - \frac{1}{\sin^2\theta}\partial_\varphi^2, \quad (2.14)$$
$$\hat{L}_z = -i\partial_\varphi. \quad (2.15)$$

By performing the partial wave decomposition of ϕ in terms of the spherical harmonics[2]

$$\phi = \sum_{l,m} \phi_{lm}(t,r)Y_{lm}(\theta,\varphi), \quad (2.16)$$

we obtain

$$S = \int dtdrd\theta d\varphi \sin\theta \sum_{l',m'} \phi^*_{l'm'}Y^*_{l'm'} \left[\frac{(r^2+a^2)^2}{\Delta} \left(\partial_t + \frac{ieQr}{r^2+a^2} \right)^2 \right.$$
$$- a^2\sin^2\theta \left(\partial_t + \frac{ieQr}{r^2+a^2} \right)^2 + 2ima\frac{r^2+a^2}{\Delta} \left(\partial_t + \frac{ieQr}{r^2+a^2} \right)$$
$$\left. - 2ima \left(\partial_t + \frac{ieQr}{r^2+a^2} \right) - \partial_r\Delta\partial_r + l(l+1) - \frac{m^2a^2}{\Delta} \right]$$
$$\times \sum_{l,m} \phi_{lm}Y_{lm} + S_{\text{int}}, \quad (2.17)$$

where we used eigenvalue equations for $\hat{\boldsymbol{L}}^2$ and $\hat{L}_z$

$$\hat{\boldsymbol{L}}^2 Y_{lm} = l(l+1)Y_{lm}, \quad (2.18)$$
$$\hat{L}_z Y_{lm} = mY_{lm}. \quad (2.19)$$

Here l is the azimuthal quantum number and m is the magnetic quantum number. Now, we transform the radial coordinate r into the tortoise coordinate r_* def ned by

$$\frac{dr_*}{dr} = \frac{r^2+a^2}{\Delta} \equiv \frac{1}{f(r)}. \quad (2.20)$$

[2]It is not clear whether a matter f eld is separable for appropriate variables in any higher dimensional black hole spacetime. It is however known that the f eld can be separated when the spacetime possesses "hidden symmetries" associated with the Killing-Yano tensors. As for details, see [12].

After this transformation, the action (2.17) is written by

$$
\begin{aligned}
S = \int dtdr_* d\theta d\varphi \sin\theta \sum_{l',m'} \phi^*_{l'm'} Y^*_{l'm'} \Bigg[(r^2+a^2) \left(\partial_t + \frac{ieQr}{r^2+a^2} \right)^2 \\
- f(r) a^2 \sin^2\theta \left(\partial_t + \frac{ieQr}{r^2+a^2} \right)^2 + 2ima \left(\partial_t + \frac{ieQr}{r^2+a^2} \right) \\
- f(r) 2ima \left(\partial_t + \frac{ieQr}{r^2+a^2} \right) - \partial_{r_*} (r^2+a^2) \partial_{r_*} \\
+ f(r) l(l+1) - \frac{m^2 a^2}{r^2+a^2} \Bigg] \sum_{l,m} \phi_{lm} Y_{lm} + S_{\text{int}}. \qquad (2.21)
\end{aligned}
$$

Here, we consider this action in the region near the horizon. Since $f(r_+) = 0$ at $r \to r_+$, we only retain dominant terms in (2.21). We thus obtain the effective action near the horizon $S_{(\text{H})}$

$$
\begin{aligned}
S_{(\text{H})} = \int dtdr_* d\theta d\varphi \sin\theta \sum_{l',m'} \phi^*_{l'm'} Y^*_{l'm'} \Bigg[(r^2+a^2) \left(\partial_t + \frac{ieQr}{r^2+a^2} \right)^2 \\
+ 2ima \left(\partial_t + \frac{ieQr}{r^2+a^2} \right) - \partial_{r_*} (r^2+a^2) \partial_{r_*} - \frac{m^2 a^2}{r^2+a^2} \Bigg] \sum_{l,m} \phi_{lm} Y_{lm}, \qquad (2.22)
\end{aligned}
$$

where we ignored S_{int} by using $f(r_+) = 0$ at $r \to r_+$. Because the theory becomes the high-energy theory near the horizon and the kinetic term dominates, we can ignore all the terms in S_{int}. For example, we consider the case of a mass term. In this case, a mass term is usually given by

$$
\begin{aligned}
\int dx^4 \left(\mu^2 \phi^* \phi \right) &= \int dtdrd\theta d\varphi \sin\theta \left(\mu^2 \phi^* \phi \right) \qquad (2.23) \\
&= \int dtdr_* d\theta d\varphi \sin\theta \left(f(r) \mu^2 \phi^* \phi \right), \qquad (2.24)
\end{aligned}
$$

where μ is a mass of the scalar f eld and we used (2.20) in the last line. We f nd that the term vanishes by using $f(r_+) = 0$ at $r \to r_+$. The same is equally true of other interaction terms S_{int}. After this analysis, we return to the expression written in terms of r, and we obtain

$$S_{(\mathrm{H})} = \int dtdrd\theta d\varphi \sin\theta \sum_{l',m'} \phi^*_{l'm'} Y^*_{l'm'} \left[\frac{(r^2+a^2)^2}{\Delta} \left(\partial_t + \frac{ieQr}{r^2+a^2} \right)^2 \right.$$
$$\left. + 2iam\frac{r^2+a^2}{\Delta}\left(\partial_t + \frac{ieQr}{r^2+a^2} \right) - \partial_r \Delta \partial_r + \frac{i^2a^2m^2}{\Delta} \right] \sum_{l,m} \phi_{lm} Y_{lm}$$
$$= \int dtdrd\theta d\varphi \sin\theta \sum_{l',m'} \phi^*_{l'm'} Y^*_{l'm'} \left[\frac{(r^2+a^2)^2}{\Delta} \left(\partial_t + \frac{ieQr}{r^2+a^2} + \frac{ima}{r^2+a^2} \right)^2 \right.$$
$$\left. - \partial_r \Delta \partial_r \right] \sum_{l,m} \phi_{lm} Y_{lm},$$
$$= \sum_{l,m} \int dtdr \phi^*_{lm} \left[\frac{(r^2+a^2)^2}{\Delta} \left(\partial_t + \frac{ieQr}{r^2+a^2} + \frac{ima}{r^2+a^2} \right)^2 - \partial_r \Delta \partial_r \right] \phi_{lm}, \quad (2.25)$$

where in the last line we used the orthonormal condition for the spherical harmonics[3]

$$\int d\theta d\varphi \sin\theta Y^*_{l'm'} Y_{lm} = \delta_{l',l} \delta_{m',m}. \quad (2.26)$$

Here we consider

$$(r^2+a^2)\partial_r \frac{\Delta}{r^2+a^2} \partial_r = -2rf(r)\partial_r + \partial_r \Delta \partial_r. \quad (2.27)$$

Since the f rst term in the right-hand side is ignored near the horizon, namely, $f(r \to r_+) = 0$, the relation becomes

$$\partial_r \Delta \partial_r = (r^2+a^2)\partial_r \frac{\Delta}{r^2+a^2} \partial_r \quad \text{at } r \to r_+. \quad (2.28)$$

By using the relation (2.28) into the action (2.25), we obtain

$$S_{(\mathrm{H})} = \sum_{l,m} \int dtdr (r^2+a^2) \phi^*_{lm} \left[\frac{r^2+a^2}{\Delta} \left(\partial_t + \frac{ieQr}{r^2+a^2} + \frac{ima}{r^2+a^2} \right)^2 \right.$$
$$\left. + \partial_r \left(-\frac{\Delta}{r^2+a^2} \right) \partial_r \right] \phi_{lm}. \quad (2.29)$$

From (2.29), we f nd that ϕ_{lm} can be considered as a (1+1)-dimensional complex scalar f eld in the backgrounds of the dilaton Φ, metric $g_{\mu\nu}$ and two $U(1)$ gauge f elds V_μ, U_μ

$$\Phi = r^2 + a^2, \quad (2.30)$$

$$g_{tt} = f(r), \quad g_{rr} = -\frac{1}{f(r)}, \quad g_{rt} = 0, \quad (2.31)$$

$$V_t = -\frac{Qr}{r^2+a^2}, \quad V_r = 0, \quad (2.32)$$

$$U_t = -\frac{a}{r^2+a^2}, \quad U_r = 0. \quad (2.33)$$

[3]Note, that the terms in square brackets are independent of angular variables θ and φ.

There are two $U(1)$ gauge fields: One is the original gauge field as in (2.7) while the other is the induced gauge field associated with the isometry along the φ direction. The induced $U(1)$ charge of the 2-dimensional field ϕ_{lm} is given by m. Then the gauge potential A_t is a sum of these two fields,

$$A_t \equiv eV_t + mU_t = -\frac{eQr}{r^2+a^2} - \frac{ma}{r^2+a^2}, \quad A_r = 0. \tag{2.34}$$

By using these notations, the action (2.29) is rewritten as

$$S_{(\mathrm{H})} = \sum_{l,m} \int dtdr\Phi\phi^*_{lm} \left[g^{tt} \left(\partial_t - iA_t\right)^2 + \partial_r g^{rr} \partial_r \right] \phi_{lm}. \tag{2.35}$$

From (2.31), we find that the 4-dimensional spherically non-symmetric Kerr-Newman metric (2.1) effectively behaves as a 2-dimensional spherically symmetric metric in the region near the horizon only

$$ds^2 = f(r)dt^2 - \frac{1}{f(r)}dr^2. \tag{2.36}$$

For confirmation, we show how to derive the surface gravity on the horizon of the Kerr-Newman black hole from $f(r)$ defined by (2.20). Actually by calculating the surface gravity, we can obtain

$$\kappa_\pm = \frac{1}{2}\partial_r f(r)\bigg|_{r=r_\pm} = \frac{r_\pm - r_\mp}{2(r_\pm^2 + a^2)}. \tag{2.37}$$

This result agrees with the well-known surface gravity on the horizon of the original Kerr-Newman metric.

2.2. Results after Reduction

In this subsection, we provide the effective theory near the horizon for various black hole solutions. We use the following notations in this subsection:

M : Mass of black hole,

Q (Q_e) : Electric charge of black hole,

Q_m : Magnetic monopole of black hole,

L (a) : Angular momentum of black hole, $\quad a = \frac{L}{M}$,

Λ : Cosmological constant,

l : (Anti) de Sitter radius, $\Lambda = (-)\frac{n(n+1)}{2l^2}$,

n : NUT parameter of black hole,

j : Rotating parameter of Gödel background,

$d\Omega_D^2$: $D-$ dimensional unit sphere,

G_D : Newton's constant in the $D-$ dimensional spacetime.

Schwarzschild Black Hole

- The D-dimensional Schwarzschild metric:

$$ds^2 = \left(1 - \frac{2M}{r}\right) dt^2 - \frac{1}{1 - \frac{2M}{r}} dr^2 - r^2 d\Omega^2_{(D-2)}. \tag{2.38}$$

The horizon is located at

$$r_+ = 2M. \tag{2.39}$$

There is no gauge f eld.

- The reduced Schwarzschild metric:

$$ds^2 = \left(1 - \frac{2M}{r}\right) dt^2 - \frac{1}{1 - \frac{2M}{r}} dr^2. \tag{2.40}$$

There is also no gauge f eld in the effective theory.

Reissner-Nordoström Black Hole

- The D-dimensional Reissner-Nordoström metric:

$$ds^2 = \left(1 - \frac{2M}{r} + \frac{Q^2}{r^2}\right) dt^2 - \frac{1}{1 - \frac{2M}{r} + \frac{Q^2}{r^2}} dr^2 - r^2 d\Omega^2_{(D-2)}. \tag{2.41}$$

The horizons are located at

$$r_\pm = M \pm \sqrt{M^2 - Q^2}. \tag{2.42}$$

There is a gauge f eld associated with the Coulomb potential of the black hole:

$$(A_\mu) = \left(-\frac{Q}{r}, 0, \cdots, 0\right). \tag{2.43}$$

- The reduced Reissner-Nordoström metric [13]:

$$ds^2 = \left(1 - \frac{2M}{r} + \frac{Q^2}{r^2}\right) dt^2 - \frac{1}{1 - \frac{2M}{r} + \frac{Q^2}{r^2}} dr^2. \tag{2.44}$$

There is a gauge f eld:

$$A_t = -\frac{Q}{r}, \quad A_r = 0. \tag{2.45}$$

Schwarzschild de Sitter Black Hole

- The $D = n + 2$-dimensional Schwarzschild de Sitter metric [14]:

$$ds^2 = -\left(1 - \frac{\omega_n M}{r^{n-1}} - \frac{r^2}{l^2}\right) dt^2 + \left(1 - \frac{\omega_n M}{r^{n-1}} - \frac{r^2}{l^2}\right)^{-1} dr^2 + r^2 d\Omega_n^2. \quad (2.46)$$

Notations:

$$\omega_n = \frac{16\pi G_{n+2}}{nV_n},$$

where

$$V_n : \text{Volume of } d\Omega_n^2.$$

The horizons are located at the positive solutions of

$$1 - \frac{\omega_n M}{r^{n-1}} - \frac{r^2}{l^2} = 0. \quad (2.47)$$

There is no gauge f eld.

- The reduced Schwarzschild de Sitter metric:

$$ds^2 = -\left(1 - \frac{\omega_n M}{r^{n-1}} - \frac{r^2}{l^2}\right) dt^2 + \left(1 - \frac{\omega_n M}{r^{n-1}} - \frac{r^2}{l^2}\right)^{-1} dr^2. \quad (2.48)$$

There are no gauge and dilaton f elds.

Schwarzschild anti de Sitter Black Hole

- The $D = n + 2$-dimensional Schwarzschild anti de Sitter metric [15]:

$$ds^2 = -\left(1 - \frac{\omega_n M}{r^{n-1}} + \frac{r^2}{l^2}\right) dt^2 + \left(1 - \frac{\omega_n M}{r^{n-1}} + \frac{r^2}{l^2}\right)^{-1} dr^2 + d\Omega_n^2. \quad (2.49)$$

Notations:

$$\omega_n = \frac{16\pi G_{n+2}}{nA_n},$$

where

$$A_n : \text{Volume of a unit } n - \text{sphere}, \quad \left(A_n = \frac{2\pi^{\frac{n+1}{2}}}{\Gamma(\frac{n+1}{2})}\right).$$

The horizons are located at

$$1 - \frac{\omega_n M}{r^{n-1}} + \frac{r^2}{l^2} = 0. \quad (2.50)$$

There is no gauge f eld.

- The reduced Schwarzschild anti de Sitter black hole [16]:

$$ds^2 = -\left(1 - \frac{\omega_n M}{r^{n-1}} + \frac{r^2}{l^2}\right) dt^2 + \left(1 - \frac{M\omega_n}{r^{n-1}} + \frac{r^2}{l^2}\right)^{-1} dr^2. \quad (2.51)$$

There are no gauge and dilaton f elds.

Kerr Black Hole

- The 4-dimensional Kerr metric:

$$\begin{aligned} ds^2 =& \frac{\Delta - a^2 \sin^2\theta}{\Sigma} dt^2 + \frac{2a\sin^2\theta}{\Sigma}(r^2 + a^2 - \Delta) dt d\varphi \\ &+ \frac{a^2\Delta\sin^2\theta - (r^2+a^2)^2}{\Sigma}\sin^2\theta d\varphi^2 - \frac{\Sigma}{\Delta}dr^2 - \Sigma d\theta^2. \end{aligned} \quad (2.52)$$

Notations:

$$\Delta = r^2 - 2Mr + a^2, \quad \Sigma = r^2 + a^2\cos^2\theta. \quad (2.53)$$

The horizons are located at

$$r_\pm = M \pm \sqrt{M^2 - a^2}. \quad (2.54)$$

There is no gauge f eld.

- The reduced Kerr metric [17]:

$$ds^2 = \frac{\Delta}{r^2 + a^2}dt^2 - \frac{r^2 + a^2}{\Delta}dr^2. \quad (2.55)$$

The induced gauge f eld associated with the angular velocity of the black hole:

$$A_t = -\frac{a}{r^2 + a^2}, \quad A_r = 0. \quad (2.56)$$

The dilaton f eld:

$$\Phi = r^2 + a^2. \quad (2.57)$$

Myers-Perry Black Hole

- The $D = 2n + 1 + \epsilon$ ($\epsilon = 0$ or 1)-dimensional Myers-Perry metric [18]:

$$\begin{aligned} ds^2 =& dt^2 - \epsilon r^2 d\alpha^2 - \sum_{i=1}^{n}(r^2 + a_i^2)\left(d\mu_i^2 + \mu_i^2 d\varphi_i^2\right) \\ &- \frac{\mu r^{2-\epsilon}}{\Pi F}\left(dt - \sum_{i=1}^{n} a_i \mu_i^2 d\varphi_i\right)^2 - \frac{\Pi F}{\Pi - \mu r^{2-\epsilon}}dr^2. \end{aligned} \quad (2.58)$$

Notations:

$$F = 1 - \sum_{i=1}^{n}\frac{a_i^2\mu_i^2}{r^2 + a_i^2}, \quad \Pi = \prod_{i=1}^{n}(r^2 + a_i^2), \quad (2.59)$$

where

$$\begin{aligned} &M = \frac{(D-2)A_{(D-2)}}{16\pi G}\mu : \text{Mass of black hole}, \\ &\frac{2}{D-2}Ma_i : \text{Angular momenta in the } \varphi_i\text{−directions}, \\ &A_{(D-2)} : \text{Volume of } S^{D-2}. \end{aligned}$$

The constraint for μ_i and α:

$$\sum_{i=1}^{n} \mu_i^2 + \epsilon\alpha^2 = 1, \quad (0 \le \mu_i \le 1, \ -1 \le \alpha \le 1). \tag{2.60}$$

The horizons are located at

$$\Pi - \mu r^{2-\epsilon} = 0. \tag{2.61}$$

There is no gauge f eld.

- The reduced Myers-Perry metric [19]:

$$ds^2 = \frac{\Pi - \mu r^{2-\epsilon}}{\Pi} dt^2 - \frac{\Pi}{\Pi - \mu r^{2-\epsilon}} dr^2. \tag{2.62}$$

The gauge f eld associated with the angular velocity of the black hole:

$$A_t^{(\varphi_i)} = -\frac{a_i}{r^2 + a_i^2}, \quad A_r^{(\varphi_i)} = 0. \tag{2.63}$$

The dilaton f eld:

$$\Phi = \mu r. \tag{2.64}$$

Rotating Black Hole with Cosmological Constant

- The metric of a $D = 2n + 1 + \epsilon$ ($\epsilon = 0$ or 1)-dimensional rotating black hole with with cosmological constant [20]:

$$\begin{aligned} ds^2 &= -W(1 - \Lambda r^2)dt^2 + \frac{2M}{U}\left(dt - \sum_{i=1}^{n} \frac{a_i \mu_i^2}{1 + \Lambda a_i^2} d\varphi_i\right)^2 \\ &+ \sum_{i=1}^{n+\epsilon} \frac{r^2 + a_i^2}{1 + \Lambda a_i^2} d\mu_i^2 + \frac{U}{V - 2M} dr^2 + \sum_{i=1}^{n} \frac{r^2 + a_i^2}{1 + \Lambda a_i^2} \mu_i^2 (d\varphi_i - \Lambda a_i dt)^2 \\ &+ \frac{\Lambda}{W(1 - \Lambda r^2)} \left(\sum_{i=1}^{n+\epsilon} \frac{r^2 + a_i^2}{1 + \Lambda a_i^2} \mu_i d\mu_i\right)^2. \end{aligned} \tag{2.65}$$

Notations:

$$V = r^{\epsilon-2}(1 - \Lambda r^2) \prod_{i=1}^{n} (r^2 + a_i^2) = \frac{U}{F},$$

$$F = \frac{r^3}{1 - \Lambda r^2} \sum_{i=1}^{n+\epsilon} \frac{\mu_i^2}{r^2 + a_i^2}, \quad W = \sum_{i=1}^{n+\epsilon} \frac{\mu_i^2}{1 + \Lambda a_i^2},$$

where Λ is a cosmological constant. For different Λ,

$$\begin{aligned} \Lambda > 0, &\quad \text{Kerr de Sitter}, \\ \Lambda = 0, &\quad \text{Myers} - \text{Perry}, \\ \Lambda < 0, &\quad \text{Kerr anti de Sitter}. \end{aligned}$$

The constraint condition is given by

$$\sum_{i=1}^{n} \mu_i^2 = 1. \tag{2.66}$$

The horizons are located at

$$V - 2M = 0. \tag{2.67}$$

There is no gauge f eld.

- The reduced metric of a rotating black hole with cosmological constant [21]:

$$ds^2 = -\frac{(1-\Lambda r^2)(V-2M)}{2M}dt^2 + \frac{2M}{(1-\Lambda r^2)(V-2M)}dr^2. \tag{2.68}$$

There is an induced gauge f eld:

$$A_t = -\frac{a(1-\Lambda r^2)}{r^2+a^2}\sum_{i=1}^{n} L_i, \quad A_r = 0. \tag{2.69}$$

The dilaton f eld:

$$\Phi = B\frac{2M}{U(1-\Lambda r^2)}\frac{r(r^2+a^2)^{n-1}}{(1+\Lambda a^2)^n}, \tag{2.70}$$

where B is the integral value.

Black String

- The metric of a $D = 2n + 1$-dimensional black string in the anti de Sitter spacetime [22]:

$$\begin{aligned} ds^2 = &-\left[\Xi^2 f(r) + \left(1-\Xi^2\right)\frac{r^2}{l^2}\right]dt^2 - 2\sum_{i=1}^{n}\left[\frac{r^2}{l^2} - f(r)\right]\Xi a_i dt d\varphi_i \\ &+\sum_{i=1}^{n}\left\{\left[\frac{r^2}{l^2} - f(r)\right]a_i^2 + r^2\right\}d\varphi_i^2 + \frac{1}{f(r)}dr^2 + \sum_{i=1}^{D-2-n} r^2 dz_i^2. \end{aligned} \tag{2.71}$$

Notations:

$$f(r) = \frac{r^2}{l^2} - \frac{2M}{r^{D-3}} + \frac{Q^2}{r^{2D-6}}, \quad \Xi = \left(1 + \sum_{i=1}^{n}\frac{a_i^2}{l^2}\right)^{\frac{1}{2}}.$$

The horizons are located at

$$f(r) = \frac{r^2}{l^2} - \frac{2M}{r^{D-3}} + \frac{Q^2}{r^{2D-6}} = 0. \tag{2.72}$$

There is a gauge f eld:

$$A = \frac{Q}{r^{D-3}}\sqrt{\frac{D-2}{2D-6}}\left(\Xi - \sum_{i=1}^{n} a_i d\varphi_i\right). \tag{2.73}$$

- The reduced metric of a black string in the anti de Sitter spacetime [23]:

$$ds^2 = -\left(\frac{rf(r)}{\sqrt{\Xi^2 r^2 + l^2(1-\Xi^2)f(r)}}\right) dt^2 + \left(\frac{rf(r)}{\sqrt{\Xi^2 r^2 + l^2(1-\Xi^2)f(r)}}\right)^{-1} dr^2. \tag{2.74}$$

There are two gauge f elds:

$$A_t = -e\Psi - \sum_{i=1}^{n} m_i \Omega_i, \quad A_r = 0, \tag{2.75}$$

where

$$\Psi = \frac{Q\Xi r^{5-D}}{\Xi^2 r^2 + l^2(1-\Xi^2)f(r)}\sqrt{\frac{D-2}{2D-6}}, \quad \Omega_i = \frac{a_i \Xi\left[\frac{r^2}{l^2} - f(r)\right]}{\Xi^2 r^2 + l^2(1-\Xi^2)f(r)}. \tag{2.76}$$

The dilaton f eld:

$$\Phi = r^{D-3}\sqrt{\Xi^2 r^2 + l^2(1-\Xi^2)f(r)}. \tag{2.77}$$

Black Ring

- The 5-dimensional dipole black ring metric [24]:

$$ds^2 = -\frac{F(y)}{F(x)}\left[\frac{H(x)}{H(y)}\right]^{\frac{N}{3}}\left[dt - CR\frac{1+y}{F(y)}d\psi\right]^2 + \frac{R^2}{(x-y)^2}F(x)\left[H(x)H^2(y)\right]^{\frac{N}{3}} \times\left[-\frac{G(y)}{F(y)H^N(y)}d\psi^2 - \frac{dy^2}{G(y)} + \frac{dx^2}{G(x)} + \frac{G(x)}{F(x)H^N(x)}d\varphi^2\right]. \tag{2.78}$$

Notations:

$$F(s) = 1 + \lambda s, \quad G(s) = (1-s^2)(1+\nu s),$$
$$H(s) = 1 - \mu s, \quad C = \sqrt{\lambda(\lambda-\nu)\frac{1+\lambda}{1-\lambda}},$$

where

$\nu,\ \lambda,\ \mu$: Dimensionless constants which lie in the range,

$$0 < \nu \leq \lambda < 1, \quad 0 \leq \mu,$$

R : Constant with the dimension of length which corresponds roughly to the radius of the ring circle,

N : Dimensionless constant which is related to the dilaton coupling by,

$$\alpha^2 = \frac{4}{N} - \frac{4}{3}, \quad 0 < N \leq 3.$$

The coordinates ψ and φ are two cycles of the black ring and the ranges of the coordinates x and y are

$$-1 \leq x \leq 1, \quad -\infty \leq y \leq -1.$$

The horizons are located at

$$G(y_+) = 0. \tag{2.79}$$

There is no gauge f eld.

- The reduced black ring metric [25]:

$$ds^2 = -\frac{\sqrt{-F(r_+)}G(r)}{CR(1+r_+)H^{\frac{N}{2}}(r_+)}dt^2 + \frac{CR(1+r_+)H^{\frac{N}{2}}(r_+)}{\sqrt{-F(r_+)}G(r)}dr^2, \tag{2.80}$$

where we rewrote y and y_+ in the original metric (2.78) as r and r_+ respectively. There is an induced gauge f eld:

$$A_t = -\frac{2\pi F(r)}{CR(1+r)\Delta\psi}, \quad A_r = 0. \tag{2.81}$$

The dilaton f eld:

$$\Phi = \frac{CR^3(1+r_+)H^{\frac{N}{2}}(r_+)\Delta\psi\Delta\varphi}{2\sqrt{-F(r_+)}}, \tag{2.82}$$

where

$$\Delta\psi = \Delta\varphi = 2\pi\frac{(1+\mu)^{\frac{N}{2}}\sqrt{1-\lambda}}{1-\nu}. \tag{2.83}$$

Squashed Kaluza-Klein Black Hole

- The 5-dimensional squashed Kaluza-Klein metric [26]:

$$\begin{aligned} ds^2 &= F(\rho)dt^2 - \frac{K^2(\rho)}{F(\rho)}d\rho^2 - \rho^2 K^2(\rho)d\Omega^2_{S2} \\ &\quad - \frac{r^2_\infty}{4K^2(\rho)}\left(d\psi + \cos\theta d\phi\right)^2. \end{aligned} \tag{2.84}$$

Notations:

$$F(\rho) = \frac{\rho^2 - 2M\rho + Q^2}{\rho^2} = \frac{(\rho-\rho_+)(\rho-\rho_-)}{\rho^2}, \quad K^2(\rho) = \frac{\rho+\rho_0}{\rho}, \tag{2.85}$$

where the parameters r_∞, $\rho_\pm$, and ρ_0 are related as

$$r^2_\infty = 4\left(\rho_+ + \rho_0\right)\left(\rho_- + \rho_0\right). \tag{2.86}$$

The horizons are located at

$$\rho_{\pm} = M \pm \sqrt{M^2 - Q^2}. \tag{2.87}$$

The gauge f eld:

$$(A_\mu) = \left(\pm\frac{\sqrt{3\rho_+\rho_-}}{2\rho}, 0, , 0, 0, 0\right). \tag{2.88}$$

- The reduced squashed Kaluza-Klein metric [27]:

$$ds^2 = \frac{F}{K}dt^2 - \frac{K}{F}d\rho^2. \tag{2.89}$$

The gauge potential:

$$A_t = \pm\frac{\sqrt{3\rho_+\rho_-}}{2\rho}, \quad A_\rho = 0. \tag{2.90}$$

The dilaton f eld:

$$\Psi = \rho^2 K. \tag{2.91}$$

Einstein-Maxwell-Chern-Simons-Gödel Black Hole

- The 5-dimensional Einstein-Maxwell-Chern-Simons-Gödel metric [28]:

$$\begin{aligned} ds^2 =& f(r)\left[dt + \frac{h(r)}{f(r)}(d\varphi + \cos\theta d\psi)\right]^2 - \frac{1}{4}r^2(d\theta^2 + \sin^2\theta d\psi^2) \\ &- \frac{dr^2}{V(r)} - \frac{r^2V(r)}{4f(r)}(d\varphi + \cos\theta d\psi)^2. \end{aligned} \tag{2.92}$$

Notations:

$$f(r) = 1 - \frac{2M}{r^2} + \frac{Q^2}{r^4}, \tag{2.93}$$

$$h(r) = jr^2 + 3jQ + \frac{(2M - Q)a}{2r^2} - \frac{Q^2a}{2r^4}, \tag{2.94}$$

$$\begin{aligned} V(r) =& 1 - \frac{2M}{r^2} + \frac{8j(M+Q)\left[a + 2j(M+2Q)\right]}{r^2} + \frac{2(M-Q)a^2}{r^4} \\ &+ \frac{Q^2\left[1 - 16ja - 8j^2(M+3Q)\right]}{r^4}. \end{aligned} \tag{2.95}$$

The horizon is located at positive solutions of

$$V(r) = 0. \tag{2.96}$$

The gauge f eld:

$$(A_\mu) = (B(r), 0, 0, C(r), C(r)\cos\theta). \tag{2.97}$$

Notations:

$$B(r) = \frac{\sqrt{3}Q}{2r^2}, \quad C(r) = \frac{\sqrt{3}}{2}\left(jr^2 + 2jQ - \frac{Qa}{2r^2}\right). \tag{2.98}$$

- The reduced Einstein-Maxwell-Chern-Simons-Gödel metric [29]:

$$ds^2 = \frac{rV(r)}{2\sqrt{U(r)}}dt^2 - \frac{2\sqrt{U(r)}}{rV(r)}dr^2, \tag{2.99}$$

where

$$U(r) = \frac{r^2V(r) - 4h^2(r)}{4f(r)}. \tag{2.100}$$

The gauge f eld:

$$A_t = e\big[\hat{B}(r) + \Omega(r)C(r)\big] + m\Omega(r), \quad A_r = 0, \tag{2.101}$$

where

$$\hat{B}(r) = B(r) + \frac{\sqrt{3}}{2}, \quad \Omega(r) = \frac{h(r)}{U(r)}. \tag{2.102}$$

The dilaton f eld:

$$\Phi = r^2\sqrt{U(r)}. \tag{2.103}$$

NUT-Kerr-Newman-Kusuya Black Hole

- The 4-dimensional NUT(Newman-Unti-Tamburino)-Kerr-Newman-Kusuya metric [30]:

$$ds^2 = \frac{\Delta}{\Sigma}(dt - Bd\varphi)^2 - \frac{\Sigma}{\Delta}dr^2 - \Sigma d\theta^2 - \frac{\sin^2\theta}{\Sigma}(adt - \eta d\varphi)^2. \tag{2.104}$$

Notations:

$$\Sigma = r^2 + (n + a\cos\theta)^2, \quad B = a\sin^2\theta - 2n\cos\theta, \quad \eta = r^2 + a^2 + n^2,$$
$$\Delta = r^2 + a^2 + n^2 - 2Mr - 2n^2 + Q_{\mathrm{e}}^2 + Q_{\mathrm{m}}^2. \tag{2.105}$$

The gauge f eld:

$$(A_\mu) = \left(-\frac{Q_{\mathrm{e}}r}{\Sigma} + \frac{Q_{\mathrm{m}}\cos\theta}{\Sigma}a, 0, 0, \frac{Q_{\mathrm{e}}r}{\Sigma}B - \frac{Q_{\mathrm{m}}\cos\theta}{\Sigma}\eta\right). \tag{2.106}$$

- The reduced NUT-Kerr-Newman-Kusuya metric [31]:

$$\begin{aligned} ds^2 =& \frac{r^2 + a^2 + n^2 - 2Mr - 2n^2 + Q_{\mathrm{e}}^2 + Q_{\mathrm{m}}^2}{r^2 + (r^2 + a^2 + n^2)}dt^2 \\ &- \frac{r^2 + (r^2 + a^2 + n^2)}{r^2 + a^2 + n^2 - 2Mr - 2n^2 + Q_{\mathrm{e}}^2 + Q_{\mathrm{m}}^2}dr^2. \end{aligned} \tag{2.107}$$

The gauge f eld:

$$A_t(r) = \frac{ma + e\sqrt{Q_{\mathrm{e}}^2 + Q_{\mathrm{m}}^2}r}{r^2 + a^2 + n^2}. \tag{2.108}$$

Nut-Kerr-Newman de Sitter Black Hole

- The 4-dimensional Nut-Kerr-Newman de Sitter black hole [32]:

$$\begin{aligned} ds^2 = &-\frac{\Delta_r - \Delta_\theta a^2 \sin^2\theta}{\rho^2\Xi^2}dt^2 + \frac{\rho^2}{\Delta_r}dr^2 + \frac{\rho^2}{\Delta_\theta}d\theta^2 \\ &- 2\frac{\Delta_\theta a(r^2+a^2)\sin^2\theta - \Delta_r\left(a - \frac{(n-a\cos\theta)^2}{a}\right)}{\rho^2\Xi^2}dtd\varphi \\ &+ \frac{\Delta_\theta (r^2-a^2)^2\sin^2\theta - \Delta_r\left(a - \frac{(n-a\cos\theta)^2}{a}\right)^2}{\rho^2\Xi^2}d\varphi^2. \end{aligned} \tag{2.109}$$

Notations:

$$\begin{aligned} &\rho^2 = r^2 + (n - a^2\cos^2\theta)^2, \quad \Xi = 1 + \frac{1}{3}\Lambda(a^2 - n^2) \\ &\Delta_r = (r^2 + a^2 - n^2)\left(1 - \frac{1}{3}\Lambda r^2\right) - 2Mr + Q^2, \\ &\Delta_\theta = 1 + \frac{1}{3}\Lambda(n - a\cos\theta)^2. \end{aligned}$$

The horizons are located at

$$\Delta_r = 0. \tag{2.110}$$

There is a gauge f eld associated with the electromagnetic potential:

$$(A_\mu) = \left(-\frac{Qr}{\rho^2\Xi}, 0, 0, -\frac{Qra\sin^2\theta}{\rho^2\Xi}\right). \tag{2.111}$$

- The reduced Nut-Kerr-Newman de Sitter black hole [32]:

$$ds^2 = \frac{\Delta_r^2}{(r^2+a^2)\Xi}dt^2 - \frac{(r^2+a^2)\Xi}{\Delta_r^2}dr^2. \tag{2.112}$$

There are two gauge f elds associated with the original gauge f eld and the induced gauge f eld connected to the isometry along the φ direction:

$$A_t = -e\Psi - m\Omega, \quad A_r = 0, \tag{2.113}$$

where

$$\Psi = \frac{Qr}{(r^2+a^2)\Xi}, \quad \Omega = \frac{a}{r^2+a^2}. \tag{2.114}$$

The dilaton f eld:

$$\Phi = \frac{r^2+a^2}{2\Xi}. \tag{2.115}$$

MTZ Black Hole

- The 4-dimensional MTZ (Martinez-Trincoso-Zanelli) metric [33]:

$$ds^2 = \frac{r(r+2G_4\mu)}{(r+G_4\mu)^2}\left[-\left(\frac{r^2}{l^2}-\left(1+\frac{G_4\mu}{r}\right)^2\right)dt^2 + \left(\frac{r^2}{l^2}-\left(1+\frac{G_4\mu}{r}\right)^2\right)^{-1} dr^2 + r^2 d\sigma^2\right]. \tag{2.116}$$

Notations:

$$d\sigma^2 = d\theta^2 + \sinh^2\theta d\varphi^2,$$

where

σ : Area of Σ,
Σ : 2 − dimensional manifold ($\Sigma = H^2/\Gamma$),
H^2 : Hyperbolic manifold,
Γ : Freely acting discrete subgroup of isometries ($\Gamma \in O(2,1)$),
μ : Constant which is bounded from below ($\mu > -l/4G_4$).

The horizon is located at

$$r_+ = \frac{l}{2}\left(1+\sqrt{1+\frac{4G_4\mu}{l}}\right). \tag{2.117}$$

There is no gauge f eld.

- The reduced MTZ (Martinez-Trincoso-Zanelli) metric [34]:

$$ds^2 = -\left[\frac{r^2}{l^2}-\left(1+\frac{G_4\mu}{r}\right)^2\right]dt^2 + \left[\frac{r^2}{l^2}-\left(1+\frac{G_4\mu}{r}\right)^2\right]^{-1} dr^2. \tag{2.118}$$

There is no gauge and dilaton f eld.

Acoustic Black Hole

- The 3-dimensional rotating acoustic metric [35]:

$$ds^2 = \frac{\tilde{\beta}_+}{\tilde{\beta}_-}\left[\frac{\tilde{\beta}_-}{\tilde{\beta}_+^2}\left(1-\frac{\tilde{\beta}_-(A^2+B^2)}{c^2r^2}\right)dt^2 - \left(1-\frac{\tilde{\beta}_-A^2}{c^2r^2}\right)^{-1} dr^2 + \frac{2\tilde{\beta}_-B}{\tilde{\beta}_+c}d\varphi dt - r^2 d\varphi\right]. \tag{2.119}$$

Notations:

$$\tilde{\beta}_{\pm} = 1 \pm \beta, \quad c = \sqrt{\frac{dh}{d\rho}}, \tag{2.120}$$

where

$$A,\ B,\ \beta : \text{Real constants,}$$
$$c : \text{Velocity of sound.}$$

The horizon is located at

$$r_{+} = \frac{\sqrt{\tilde{\beta}_{-}}|A|}{c}. \tag{2.121}$$

There is no gauge field.

- The reduced rotating acoustic metric [36]:

$$ds^2 = \frac{\sqrt{\tilde{\beta}_{-}F(r)}}{\tilde{\beta}_{+}}dt^2 - \frac{\tilde{\beta}_{+}}{\sqrt{\tilde{\beta}_{-}F(r)}}dr^2, \tag{2.122}$$

where

$$F(r) = 1 - \frac{r_{+}^2}{r^2}. \tag{2.123}$$

The induced gauge field associated with the angular velocity of the black hole:

$$A_t = -\frac{\tilde{\beta}_{-}B}{cr^2\tilde{\beta}_{+}}, \quad A_r = 0. \tag{2.124}$$

The dilaton field:

$$\Phi = r\left(\frac{\tilde{\beta}_{+}}{\tilde{\beta}_{-}}\right)^{\frac{3}{2}}. \tag{2.125}$$

3. Applications of Effective Theory

In the previous section, we showed that the action for a matter field in the region near the horizon is effectively reduced to the 2-dimensional action in the spherical symmetric black hole spacetime. The effective theory near the horizon is very useful when we discuss thermal properties of black holes. In this section, as applications of the effective theory, we actually derive the Hawking and Unruh temperature. Even in the use of the reduced metric, we can obtain the correct result and we would find that its calculation is considerably simplified.

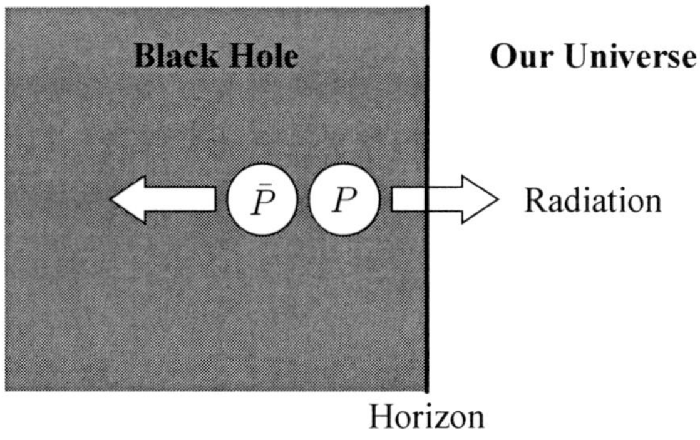

Figure 1. Intuitive picture of the tunneling mechanism.

This section is organized as follows. In Subsection 3.1, we present the method deriving Hawking radiation on the basis of the tunneling mechanism. The approach using the dimensional reduction technique in the tunneling mechanism was f rst put forwarded by the author [37]. We show to obtain the accurate Hawking temperature and black body spectrum in the case of using the reduced Kerr-Newman metric obtained in 2.1. In Subsection 3.2, we discuss the global embedding Minkowski spacetime approach which is the unif ed description between the Hawking effect and Unruh effect. We also evaluate the Unruh temperature from the reduced Kerr-Newman metric and we show that the Unruh temperature becomes connected with the Hawking temperature.

3.1. Hawking Radiation as Tunneling

Tunneling mechanism is a method of deriving Hawking radiation based on quantum tunneling. The mechanism was proposed by Srinivasan and Padmanabhan [38]. This derivation using the tunneling mechanism is intuitive and it is also capable of wide application.

The essential idea of the tunneling mechanism is as follows. We consider that a particle-antiparticle pair is formed close to the horizon inside a black hole. We can divide the f eld associated with a particle into ingoing modes moving toward the center of the black hole and outgoing modes trying to move outside the black hole. The ingoing modes are trapped inside the horizon. On the other hand, although the horizon plays a role of a barrier when the outgoing modes try to move outside the black hole, a part of the outgoing modes move outside the black hole by the quantum tunneling effect. If the particle which comes out to our universe has positive energy, such a particle (and also an antiparticle) can stably exist and we can regard the particle outside the horizon as the radiation from the black hole (Fig. 1).

This phenomenon is inspired by quantum tunneling with an inf nite barrier in quantum mechanics. This inf nite barrier may be written as a potential of the delta-function type. Particles cannot classically pass through the potential. According to quantum theory, it is well known that a part of particles can pass through the potential by the quantum tunneling effect. We might consider that for the outgoing particles inside a black hole, the horizon plays a role as the potential.

In this subsection, we derive the result of Hawking radiation from the tunneling mechanism by using the effective theory near the horizon. To begin with, we consider the Klein-Gordon equation obtained from the action (2.35)

$$\left[\frac{1}{f(r)}(\partial_t - iA_t)^2 - f(r)\partial_r^2 - \{\partial_r f(r)\}\partial_r\right]\phi = 0, \tag{3.1}$$

where $f(r)$ is defined in (2.20) and A_t is defined in (2.34). Note that the mass, potential and interaction terms of a matter field can be ignored because the kinetic term dominates in the high-energy theory near the horizon. Taking the standard WKB approximation

$$\phi(r,t) = e^{\frac{i}{\hbar}S(r,t)}, \tag{3.2}$$

and substituting the expansion for $S(r,t)$ in terms of the Planck constant $\hbar$

$$S(r,t) = S_0(r,t) + \sum_{i=1}^{\infty}\hbar^i S_i(r,t), \tag{3.3}$$

in (3.1) where we write the Planck constant explicitly, we obtain, in the semiclassical limit (i.e., keeping only S_0),

$$\partial_t S_0(r,t) = \pm f(r)\partial_r S_0(r,t). \tag{3.4}$$

To solve the classical action $S_0(r,t)$, we consider the classical Hamilton-Jacobi equation

$$\partial_t S_0 + H = 0, \tag{3.5}$$

where H is the Hamiltonian. The Hamiltonian asymptotically is given by

$$H = \omega - e\Phi_{\rm H} - m\Omega_{\rm H} \equiv \omega', \tag{3.6}$$

where ω is the characteristic frequency, e is the charge and m is the magnetic quantum number of the scalar particle. The other notations $\Phi_{\rm H}$ and $\Omega_{\rm H}$ are respectively the electric potential and the angular frequency on the horizon of the Kerr-Newman black hole

$$\Phi_{\rm H} = \frac{r_+ Q}{r_+^2 + a^2}, \quad \Omega_{\rm H} = \frac{a}{r_+^2 + a^2}. \tag{3.7}$$

Note that these potentials are consistent with both the potentials obtained from the original Kerr-Newman metric and the induced gauge potentials on the horizon as in (2.32) and (2.33). This is identified as the effective energy experienced by the particle at asymptotic infinity. By the Hamilton-Jacobi equation (3.5), we obtain

$$S_0 = -\omega' t + \tilde{S}_0(r), \tag{3.8}$$

where $\tilde{S}_0(r)$ is a time-independent arbitrary function. By substituting (3.8) into (3.4), we obtain

$$\omega' = \pm f(r)\partial_r \tilde{S}_0(r). \tag{3.9}$$

By using the relation of the tortoise coordinate defined by (2.20),

$$\partial_{r_*} = f(r)\partial_r, \tag{3.10}$$

the Hamilton-Jacobi equation (3.9) becomes

$$\partial_{r_*}\tilde{S}_0(r) = \pm\omega'. \tag{3.11}$$

By integrating (3.11) over r_*, we obtained

$$\tilde{S}_0(r) = \pm\omega' r_* + C, \tag{3.12}$$

where C is an integration constant and we ignore it since it is included in a normalization constant of the wave function. By substituting this into (3.8), the classical action becomes

$$S_0(r,t) = -\omega'(t \mp r_*). \tag{3.13}$$

Thus we can obtain the semiclassical solution for the scalar field

$$\phi(r,t) = \exp\left[-\frac{i}{\hbar}\omega'(t \mp r_*)\right]. \tag{3.14}$$

Here, we introduce both the retarded time u and the advanced time v defined by

$$u \equiv t - r_*, \quad v \equiv t + r_*, \tag{3.15}$$

where we can regard u (v) as the outgoing (ingoing) modes of particles. We can then separate the scalar field (3.14) into the ingoing (left handed) modes and outgoing (right handed) modes. Since the tunneling effect is the quantum effect arising within the Planck length in the near horizon region, we have to consider both the inside and outside regions which are very close to the horizon. In the regions $r_+ - \varepsilon < r < r_+$, and $r_+ \leq r < r_+ - \varepsilon$, respectively, we express the field ϕ as

$$\left.\begin{aligned} \phi_{\text{in}}^R &= \exp\left(-\frac{i}{\hbar}\omega' u_{\text{in}}\right), \\ \phi_{\text{in}}^L &= \exp\left(-\frac{i}{\hbar}\omega' v_{\text{in}}\right), \end{aligned}\right\} \quad \text{when} \quad r_+ - \varepsilon < r < r_+, \tag{3.16}$$

$$\left.\begin{aligned} \phi_{\text{out}}^R &= \exp\left(-\frac{i}{\hbar}\omega' u_{\text{out}}\right), \\ \phi_{\text{out}}^L &= \exp\left(-\frac{i}{\hbar}\omega' v_{\text{out}}\right), \end{aligned}\right\} \quad \text{when} \quad r_+ < r < r_+ + \varepsilon, \tag{3.17}$$

where ε is an arbitrarily small constant, "R (L)" stands for the right (left) modes and "in (out)" stands for the inside (outside) of the black hole, respectively (Fig. 2). Here we note that these fields are defined both in the inside and outside regions which are very close to the horizon.

Now, we consider that the outgoing particles inside the black hole quantum mechanically tunnel through the horizon. However, it is well known that both the Schwarzschild

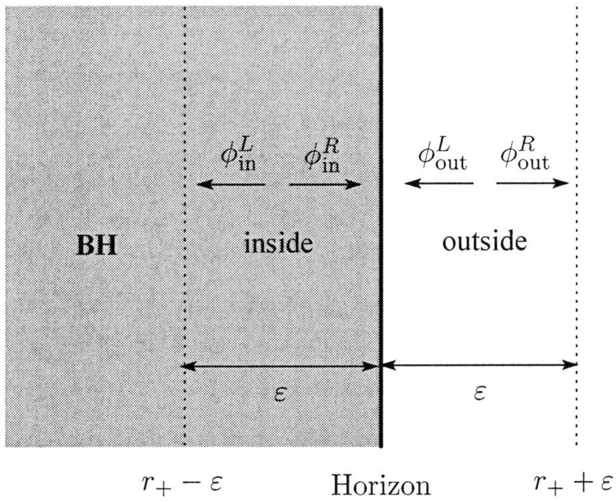

Figure 2. Intuitive picture of the scalar f eld near the horizon.

coordinates (t, r) and the Eddington-Finkelstein coordinates (u, v) are singular at the horizon. To describe horizon crossing phenomena, we introduce the Kruskal-like coordinates def ned by

$$\left.\begin{aligned} T &\equiv \left(\frac{r-r_+}{r_+}\right)^{\frac{1}{2}} \left(\frac{r-r_-}{r_-}\right)^{\frac{\kappa_+}{2\kappa_-}} e^{\kappa_+ r} \sinh(\kappa_+ t), \\ R &\equiv \left(\frac{r-r_+}{r_+}\right)^{\frac{1}{2}} \left(\frac{r-r_-}{r_-}\right)^{\frac{\kappa_+}{2\kappa_-}} e^{\kappa_+ r} \cosh(\kappa_+ t), \end{aligned}\right\} \quad \text{when} \quad r > r_+, \tag{3.18}$$

$$\left.\begin{aligned} T &\equiv \left(\frac{r_+ - r}{r_+}\right)^{\frac{1}{2}} \left(\frac{r-r_-}{r_-}\right)^{\frac{\kappa_+}{2\kappa_-}} e^{\kappa_+ r} \cosh(\kappa_+ t), \\ R &\equiv \left(\frac{r_+ - r}{r_+}\right)^{\frac{1}{2}} \left(\frac{r-r_-}{r_-}\right)^{\frac{\kappa_+}{2\kappa_-}} e^{\kappa_+ r} \sinh(\kappa_+ t), \end{aligned}\right\} \quad \text{when} \quad r_+ > r > r_-. \tag{3.19}$$

These coordinates are Kruskal coordinates for the reduced metric and have no singularity at the horizon. For details of these coordinates, see Appendix A. By using the tortoise coordinates def ned by (2.20), i.e.,

$$r_* = r + \frac{1}{2\kappa_+} \ln \frac{r-r_+}{r_+} + \frac{1}{2\kappa_-} \ln \frac{r-r_-}{r_-}, \quad \text{when} \quad r > r_+, \tag{3.20}$$

$$r_* = r + \frac{1}{2\kappa_+} \ln \frac{r_+ - r}{r_+} + \frac{1}{2\kappa_-} \ln \frac{r-r_-}{r_-}, \quad \text{when} \quad r_+ > r > r_-, \tag{3.21}$$

in both (A.15) and (A.20), we obtain

$$\left.\begin{aligned} T_{\rm out} &= \exp\left[\kappa_+ (r_{*\rm out})\right] \sinh\left(\kappa_+ t_{\rm out}\right), \\ R_{\rm out} &= \exp\left[\kappa_+ (r_{*\rm out})\right] \cosh\left(\kappa_+ t_{\rm out}\right), \end{aligned}\right\} \quad \text{when} \quad r > r_+, \tag{3.22}$$

$$\left.\begin{aligned} T_{\rm in} &= \exp\left[\kappa_+ (r_{*\rm in})\right] \cosh\left(\kappa_+ t_{\rm in}\right), \\ R_{\rm in} &= \exp\left[\kappa_+ (r_{*\rm in})\right] \sinh\left(\kappa_+ t_{\rm in}\right), \end{aligned}\right\} \quad \text{when} \quad r_+ > r > r_-, \tag{3.23}$$

where we wrote the suff xes “out” and “in” to emphasize that they respectively express outer and inner coordinates of the black hole.

In general, a metric of a black hole describes the spacetime outside the black hole. We have no real way of knowing the metric of the spacetime inside the black hole. Consequently, the readers may wonder if these metrics def ned by the Kruskal-like coordinates can really describe the spacetime inside the black hole. In our case, however, we study the tunneling effect across the horizon and thus we study the behavior in the very small regions near the horizon. In such an analysis, due to the reasons of continuity, it may not be unnatural to assume that the Kruskal-like coordinates (A.15) and (A.20) can be used in both of outside and inside regions of near the horizon.

A set of coordinates (3.23) is connected with the other coordinates (3.22) by the relations,

$$t_{\rm in} \rightarrow t_{\rm out} - i\frac{\pi}{2\kappa_+}, \quad (r_*)_{\rm in} \rightarrow (r_*)_{\rm out} + i\frac{\pi}{2\kappa_+}, \tag{3.24}$$

so that, with this mapping, $T_{\rm in} \rightarrow T_{\rm out}$ and $R_{\rm in} \rightarrow R_{\rm out}$ smoothly. Following the def nition (3.15), we obtain the relations connecting the null coordinates def ned inside and outside the horizon,

$$u_{\rm in} \equiv t_{\rm in} - (r_*)_{\rm in} \rightarrow u_{\rm out} - i\frac{\pi}{\kappa_+}, \tag{3.25}$$

$$v_{\rm in} \equiv t_{\rm in} + (r_*)_{\rm in} \rightarrow v_{\rm out}. \tag{3.26}$$

This mapping is not def ned if these coordinates are restricted to be real numbers. However, we can def ne it by extending these coordinates to complex numbers as in (3.24) and (3.25). We regard the appearance of the complex coordinates as a manifestation of quantum tunneling. In analogy to the quantum tunneling in ordinary quantum mechanics, our picture is that the inside solutions (3.16) and the outside solutions (3.17) of the inf nitely high but very thin barrier, which is located on top of the horizon, are connected via the complex coordinates (i.e., tunneling); the precise connection of two coordinates is determined by asking the smooth connection of the Kruskal-like coordinates at the horizon. Under these transformations the inside and outside modes are connected by,

$$\phi^R_{\rm in} \equiv \exp\left(-\frac{i}{\hbar}\omega' u_{\rm in}\right) \rightarrow \exp\left(-\frac{\pi\omega'}{\hbar\kappa_+}\right)\phi^R_{\rm out}, \tag{3.27}$$

$$\phi^L_{\rm in} \equiv \exp\left(-\frac{i}{\hbar}\omega' v_{\rm in}\right) \rightarrow \phi^L_{\rm out}, \tag{3.28}$$

where $\exp\left(-\frac{\pi\omega'}{\hbar\kappa_+}\right)$ of (3.27) stands for the effect of the tunneling mechanism. We would like to note that these scalar f elds are still identif ed with Schrödinger amplitude for a single particle state. Therefore, the squared absolute value of the wave function represents the probability. The probability of the tunneling effect is given by

$$\Gamma = \left|\exp\left(-\frac{\pi\omega'}{\hbar\kappa_+}\right)\phi^R_{\rm out}\right|^2 = e^{\left(-\frac{2\pi\omega'}{\hbar\kappa_+}\right)}, \tag{3.29}$$

where we used $\left|\phi_{\rm out}^{R}\right|^{2}=1$. It is natural to impose the normalization for the state outside the black hole, because the observer exists outside the black hole. In comparison with the Boltzmann factor in a thermal equilibrium state at a temperature T,

$$\Gamma_{\rm B}=e^{-\frac{\omega}{T}}, \tag{3.30}$$

we f nd that this result in the system of units $c=k_{\rm B}=G=1$ agrees with Hawking temperature (1.1), i.e.,

$$T_{\rm BH}=\frac{\hbar\kappa_{+}}{2\pi}. \tag{3.31}$$

Here we recall the essential idea of the tunneling mechanism. Among particles formed by pair creations inside a black hole, the ingoing mode is trapped inside the horizon while the outgoing mode can quantum mechanically tunnel through the horizon. The outgoing mode is then observed in the our universe as Hawking radiation. We f nd that the effect of the ingoing mode inside the horizon do not appear outside the horizon as in (3.28) since $v_{\rm in}$ changes to $v_{\rm out}$ without an extra term under the transformation connecting the null coordinates def ned inside and outside the horizon as in (3.26). On the other hand, we f nd that the effect of the outgoing mode inside the horizon appear with a non-negligible probability by tunneling through the horizon quantum mechanically as in (3.27). These results match up well with the concept of tunneling mechanism.

Furthermore, we will derive the black body spectrum from a black hole. A method deriving the black body spectrum from the consideration of the tunneling mechanism was proposed by Banerjee and Majhi [39]. We now consider n number of non-interacting virtual pairs that are created inside the black hole. According to quantum f eld theory, a wave function of one-particle system is related to the second-quantized f eld operator. By following the approach of Banerjee and Majhi [39], each of these pairs is represented by the modes def ned in the f rst set of (3.16) and (3.17). Then they def ned the physical state of the system, observed from outside, as

$$|\Psi\rangle=N\sum_{n}|n_{\rm in}^{L}\rangle\otimes|n_{\rm in}^{R}\rangle, \tag{3.32}$$

where $|n_{\rm in}^{L(R)}\rangle$ is the number state of ingoing (outgoing) modes inside the black hole and N is a normalization constant. From the transformations of both (3.27) and (3.28), we obtain

$$|\Psi\rangle=N\sum_{n}e^{-\frac{\pi n\omega'}{\hbar\kappa_{+}}}|n_{\rm out}^{L}\rangle\otimes|n_{\rm out}^{R}\rangle. \tag{3.33}$$

Here N can be determined by using the normalization condition $\langle\Psi|\Psi\rangle=1$. It is natural to determine the normalization constant N for the state outside the black hole, because the observer exists outside the black hole. Thus, we obtain

$$N=\frac{1}{\left(\sum_{n}e^{-\frac{2\pi n\omega'}{\hbar\kappa_{+}}}\right)^{\frac{1}{2}}}. \tag{3.34}$$

For bosons[4] ($n = 0, 1, 2, \cdots$), N is calculated as

$$N = \frac{1}{\left(\sum_{n=0}^{\infty} e^{-\frac{2\pi n\omega'}{\hbar\kappa_+}}\right)^{\frac{1}{2}}} = \left(1 - e^{-\frac{2\pi\omega'}{\hbar\kappa_+}}\right)^{\frac{1}{2}}. \tag{3.35}$$

By substituting (3.35) into (3.33), we obtain the normalized physical states of a system of bosons

$$|\Psi\rangle = \left(1 - e^{-\frac{2\pi\omega}{\hbar\kappa}}\right)^{\frac{1}{2}} \sum_n e^{-\frac{\pi n\omega}{\hbar\kappa}} |n^L_{\text{out}}\rangle \otimes |n^R_{\text{out}}\rangle. \tag{3.36}$$

The density matrix operator of the system is defined by

$$\begin{aligned} \hat{\rho} \equiv& |\Psi\rangle\langle\Psi| \\ =& \left(1 - e^{-\frac{2\pi\omega'}{\hbar\kappa_+}}\right) \sum_{n,m} e^{-\frac{\pi\omega'}{\hbar\kappa_+}(n+m)} |n^L_{\text{out}}\rangle \otimes |n^R_{\text{out}}\rangle\langle m^R_{\text{out}}| \otimes \langle m^L_{\text{out}}|. \end{aligned} \tag{3.37}$$

By tracing out the ingoing modes, we obtain the reduced density matrix for the right going modes,

$$\hat{\rho}^{(R)} = \text{Tr}\,(\hat{\rho}) \tag{3.38}$$

$$= \sum_n \langle n^L_{\text{out}}|\hat{\rho}|n^L_{\text{out}}\rangle \tag{3.39}$$

$$= \left(1 - e^{-\frac{2\pi\omega'}{\hbar\kappa_+}}\right) \sum_n e^{-\frac{2\pi n\omega'}{\hbar\kappa_+}} |n^R_{\text{out}}\rangle\langle n^R_{\text{out}}|. \tag{3.40}$$

Then, the expectation value of the number operator $\hat{n}$ is given by

$$\langle n\rangle_{\text{boson}} = \text{Tr}\left(\hat{n}\hat{\rho}^{(R)}\right) \tag{3.41}$$

$$= \sum_n \langle n^R_{\text{out}}|\hat{n}\hat{\rho}|n^R_{\text{out}}\rangle \tag{3.42}$$

$$= \left(1 - e^{-\frac{2\pi\omega'}{\hbar\kappa_+}}\right) \sum_n n e^{-\frac{2\pi n\omega'}{\hbar\kappa+}} \tag{3.43}$$

$$= \left(1 - e^{-\frac{2\pi\omega'}{\hbar\kappa_+}}\right) \left(-\frac{\hbar\kappa_+}{2\pi}\right) \frac{\partial}{\partial\omega'} \left(\sum_{n=0}^{\infty} e^{-\frac{2\pi n\omega'}{\hbar\kappa_+}}\right) \tag{3.44}$$

[4]For fermions ($n = 0, 1$), $N_{(\text{fermion})}$ is similarly calculated as

$$N_{(\text{fermion})} = \left(1 + e^{-\frac{2\pi\omega}{\hbar\kappa}}\right)^{-\frac{1}{2}}.$$

One might be surprised by the sudden appearance of fermions. However, we can explicitly present an answer to the question. As shown by the technique of dimensional reduction near the horizon, we know that the effective theory near the horizon becomes a 2-dimensional theory. According to 2-dimensional quantum field theory, it is known that there exists the boson-fermion duality [40]. Namely, the 2-dimensional boson theory can be treated as the 2-dimensional fermion theory by the fermionization. We can also discuss the tunneling mechanism for fermions in a manner similar to bosons.

$$= \left(1 - e^{-\frac{2\pi\omega'}{\hbar\kappa_+}}\right)\left(-\frac{\hbar\kappa_+}{2\pi}\right)\frac{\partial}{\partial\omega'}\left(\frac{1}{1 - e^{-\frac{2\pi\omega'}{\hbar\kappa_+}}}\right) \tag{3.45}$$

$$= \frac{1}{e^{\frac{2\pi\omega'}{\hbar\kappa_+}} - 1} \tag{3.46}$$

$$= \frac{1}{e^{\frac{1}{T_{\mathrm{BH}}}(\omega - e\Phi_{\mathrm{H}} - m\Omega_{\mathrm{H}})} - 1}, \tag{3.47}$$

where in the last line we used the def nition (3.6) and we identify the Hawking temperature by (3.31). This result corresponds to the black body spectrum with the Hawking temperature and agrees with the result of Hawking's original derivation in the Kerr-Newman black hole background [9].

Before closing this subsection, we would like to discuss the black hole entropy. Since a particle emitted by the black hole has the Hawking temperature, it is natural to consider that the black hole itself has the same temperature from the corresponding relationships between thermodynamics and black hole physics. According to black hole thermodynamics, the f rst law of black hole thermodynamics[5] is given by [6],

$$dM = \frac{\kappa_+}{8\pi}dA_{\mathrm{BH}} + \Phi_{\mathrm{H}}dQ + \Omega_{\mathrm{H}}dL, \tag{3.48}$$

where Φ_{H} and Ω_{H} are consistent with (3.7). The black hole entropy can be obtained by f xing both Q and L,

$$dS_{\mathrm{BH}} = \left(\frac{dM}{T_{\mathrm{BH}}}\right)_{Q,L}, \tag{3.49}$$

which is an analogue of the thermodynamics relation

$$dS = \left(\frac{dU}{T}\right)_{V,N}. \tag{3.50}$$

By integrating Eq. (3.49), the entropy agrees with the Bekenstein-Hawking entropy

$$S_{\mathrm{BH}} = \frac{A_{\mathrm{BH}}}{4\hbar}. \tag{3.51}$$

Hawking radiation decreases its black hole area and the classical Hawking's black hole area theorem [41] is violated. Thus one needs to generalize the second law as

$$\Delta S_{\mathrm{BH}} + \Delta S_{\mathrm{C}} = \Delta(S_{\mathrm{BH}} + S_{\mathrm{C}}) \geq 0, \tag{3.52}$$

where S_{C} is the entropy of the matter outside the black hole. The relation was given by Bekenstein and is called the generalized second law [42]. The generalized second law always holds in any physical process.

---

[5]The result (3.48) is obtained by taking the total differentiation in the equation of the black hole area, i.e., $A_{\mathrm{BH}} = 4\pi(2Mr_+ - Q^2)$.

3.2. Global Embedding Minkowski Spacetime Approach

As a similar phenomenon with Hawking radiation, it is known that there is Unruh radiation [43]. Unruh effect is that the Minkowski vacuum seems as thermally excited for a uniformly accelerating observer. Unruh radiation is the radiation associated with the thermally excited effect.

A unif ed description of both Hawking radiation and Unruh radiation is given by the global embedding Minkowski spacetime (GEMS) approach [44]. The essential idea of this approach is as follows. We consider to embed a black hole spacetime into an appropriate higher dimensional Minkowski spacetime globally. Then the relevant detector in the black hole spacetime with its horizon map of the uniformly accelerated detector corresponds to the f at higher dimensional embedding spacetime and its horizon which is the apparent event horizon formed by the accelerating observer. By identifying the acceleration of the Unruh detector, the Unruh temperature can be evaluated. The Hawking temperature is accurately derived from the Unruh temperature by using the Tolman formula.

This unif ed approach has been discussed only in spherically symmetric black hole spacetimes and there is no discussion in cases of complicated metric such as the Kerr-Newman black hole which lacks spherical symmetry because it is diff cult to f nd the corresponding f at higher dimensional embedding spacetime.

The technique of the dimensional reduction near the horizon is useful in the GEMS approach [45]. We have well known how to embed a spherically symmetric black hole spacetime into a higher dimensional Minkowski spacetime [46]. As shown in Section 2, we recall that a complicated black hole spacetime can be effectively regarded as a 2-dimensional spherically symmetric black hole spacetime when we consider an effective theory of a matter f eld near a black hole. We brief y can discuss the GEMS approach by using the reduced 2-dimensional metric.

In this subsection, we would like to demonstrate the GEMS approach by using the reduced Kerr-Newman metric obtained in Subsection 2.1. The reduced Kerr-Newman metric was given by (2.36)

$$ds^2 = \frac{\Delta}{r^2+a^2}dt^2 - \frac{r^2+a^2}{\Delta}dr^2 \tag{3.53}$$

$$= \left(1 - \frac{2Mr}{r^2+a^2} + \frac{Q^2}{r^2+a^2}\right)dt^2 - \frac{1}{1-\frac{2Mr}{r^2+a^2}+\frac{Q^2}{r^2+a^2}}dr^2. \tag{3.54}$$

This 2-dimensional metric can be embedded in the 5-dimensional f at space as [45],

$$ds^2 = \left(dz^0\right)^2 - \left(dz^1\right)^2 - \left(dz^2\right)^2 + \left(dz^3\right)^2 + \left(dz^4\right)^2, \tag{3.55}$$

by the following relations among the f at and curved coordinates[6]:

$$\left.\begin{aligned} z^0_{\text{out}} &= \tfrac{1}{\kappa_+}\left(1-\tfrac{2Mr}{r^2+a^2}+\tfrac{Q^2}{r^2+a^2}\right)^{\frac{1}{2}}\sinh(\kappa_+ t),\\ z^1_{\text{out}} &= \tfrac{1}{\kappa_+}\left(1-\tfrac{2Mr}{r^2+a^2}+\tfrac{Q^2}{r^2+a^2}\right)^{\frac{1}{2}}\cosh(\kappa_+ t), \end{aligned}\right\} \quad \text{when} \quad r > r_+,$$

[6]For $Q = 0$ and $a = 0$, the transformations (3.56) agree with the Schwarzschild case while only for $a = 0$ these are consistent with the Reissner-Nordström case [46].

$$\left.\begin{aligned} z^0_{\rm in} &= \tfrac{1}{\kappa_+}\left(\tfrac{2Mr}{r^2+a^2} - \tfrac{Q^2}{r^2+a^2} - 1\right)^{\frac{1}{2}} \cosh(\kappa_+ t), \\ z^1_{\rm in} &= \tfrac{1}{\kappa_+}\left(\tfrac{2Mr}{r^2+a^2} - \tfrac{Q^2}{r^2+a^2} - 1\right)^{\frac{1}{2}} \sinh(\kappa_+ t), \end{aligned}\right\} \quad \text{when} \quad r_+ > r > r_-,$$

$$z^2 = \int dr \left[1 + \frac{(r^2+a^2)(r_+ + r_-) + r_+^2(r + r_+)}{(r^2+a^2)(r - r_-)}\right]^{\frac{1}{2}},$$

$$z^3 = \int dr \left[\frac{4r_+^5 r_-}{(r^2+a^2)^2(r_+ - r_-)^2}\right]^{\frac{1}{2}},$$

$$z^4 = \int dra \left[\frac{r_+ + r_-}{(a^2+r_-^2)(r_- - r)} + \frac{4(a^2+r_+^2)(a^2 - r_+ r_- + (r_+ + r_-)r)}{(r_+ - r_-)^2(a^2+r^2)^3}\right.$$
$$\left. + \frac{4r_+ r_-(a^2 + 2r_+^2)}{(r_+ - r_-)^2(a^2+r^2)^2} + \frac{rr_- - a^2 + r_+(r + r_-)}{(a^2+r_-^2)(a^2+r^2)}\right]^{\frac{1}{2}}, \tag{3.56}$$

where the suffxes "out" and "in" stand for outer and inner coordinates of the horizon respectively while coordinates without any suffx imply that these are valid on both sides of the horizon. If the Hawking detector moves according to a constant r surface outside the horizon in the curved space, the corresponding Unruh detector then moves on the constant (z^2, z^3, z^4) surface and it will follow the hyperbolic trajectory

$$\left(z^1_{\rm out}\right)^2 - \left(z^0_{\rm out}\right)^2 = \frac{1}{\kappa_+^2}\left(1 - \frac{2Mr}{r^2+a^2} + \frac{Q^2}{r^2+a^2}\right) = \frac{1}{\tilde{a}^2}, \tag{3.57}$$

where $\tilde{a}$ is the uniform acceleration defned by

$$\tilde{a} \equiv \frac{1}{\kappa_+}\sqrt{1 - \frac{2Mr}{r^2+a^2} + \frac{Q^2}{r^2+a^2}}. \tag{3.58}$$

The relation (3.57) means that the Unruh detector is moving with the uniform acceleration $\tilde{a}$ in the $(z^0_{\rm out}, z^1_{\rm out})$ fat plane. According to Unruh [43], the accelerated detector in the Minkowski vacuum sees a thermal spectrum with the Unruh temperature given by

$$T = \frac{\hbar\tilde{a}}{2\pi} = \frac{\hbar\kappa_+}{2\pi\sqrt{1 - \frac{2Mr}{r^2+a^2} + \frac{Q^2}{r^2+a^2}}}. \tag{3.59}$$

By using Tolman's formula which ensure constancy between the product of temperatures and corresponding Tolman factor $\sqrt{g_{tt}}$ measured at two different points in spacetime [47], i.e.,

$$\sqrt{g_{tt}}T = \sqrt{g'_{tt}}T' = \text{const.}, \tag{3.60}$$

the Hawking temperature measured by any observer away from the horizon can be obtained

$$T_{\rm BH} = \sqrt{g_{tt}}T = \frac{\hbar\kappa_+}{2\pi}, \tag{3.61}$$

where we used the fact that both the original Kerr-Newman metric $g_{tt}^{(\mathrm{o})}(r,\theta) = \frac{\Delta - a^2 \sin^2\theta}{\Sigma}$ and the reduced Kerr-Newman metric $g_{tt}^{(\mathrm{r})}(r) = \frac{\Delta}{r^2+a^2}$ become asymptotically flat namely,

$$g_{tt}^{(\mathrm{o})}(r \to \infty) = g_{tt}^{(\mathrm{r})}(r \to \infty) = 1. \tag{3.62}$$

Of course, we should use the original metric $g_{tt}^{(o)}$ at infinit because the effective theory does not hold in the region away from the horizon. The result (3.61) is consistent with the previous one (3.31). Therefore, we fin that the desired result can be derived even in the case of the reduced metric and the effective theory near the horizon is very useful in the GEMS approach.

Conclusion

We have reviewed the effective theory of a matter fiel near the event horizon of a black hole and its applications. In the case of a complicated black hole solution obtained by considering a higher dimensional spacetime, we showed that the effective theory near the horizon becomes a 2-dimensional theory in the spherically symmetric background spacetime by using the technique of dimensional reduction near the horizon. Therefore, we can briefl discuss the behavior of a matter fiel near the horizon without major differences in comparison with discussions in the Schwarzschild case.

In particular, the derivation of Hawking radiation and Unruh radiation were quite simplifie by using the effective theory. Since the horizon plays an important role in both cases of Hawking effect and Unruh effect, the theory near the horizon is significant Actually, we demonstrated the derivation of Hawking radiation on the basis of the tunneling mechanism and the global embedding Minkowski spacetime approach by using properties of the effective theory near the horizon. As a consequence, in the use of the reduced metric we can produce the expected results which would be obtained by the full metric. In this chapter, although we mainly argued over the case of a Kerr-Newman black hole, we can similarly calculate in cases of more general black holes.

Finally, whether the technique of dimensional reduction is possible for all black hole solutions or not remains as a matter to be discussed further. If new black hole solutions are discovered in higher dimensional spacetime, it is necessary to show that the effective theory of a matter fiel in the region near the horizon can indeed reduce for the respective solutions.

Acknowledgment

I would like to thank K. Fujikawa and I.S. Sogami for helpful comments and encouragement.

Appendix

A. Kruskal-Like Coordinates for the Reduced Metric

In this appendix, we show how to obtain the Kruskal-like coordinates (3.22) and (3.23) from the $(t-r)$-coordinates which appears in the 2-dimensional spherically-symmetric metric (2.36) by performing several coordinate transformations. As a preliminary analysis, we would like to summarize the Schwarzschild metric and Kruskal-Szekeres coordinates brief y [48, 49]. The well-known Schwarzschild metric is given by

$$ds^2 = -\left(1 - \frac{2M}{r}\right) dt^2 + \frac{1}{1-\frac{2M}{r}} dr^2 + r^2(d\theta^2 + \sin^2\theta d\varphi^2). \tag{A.1}$$

It follows from the expression (A.1) that there are two singularities $r = 0$ and $r = 2M$ in the Schwarzschild metric. A singularity at $r = 0$ is the curvature singularity which cannot be removed while the other at $r = 2M$ is a f ctitious singularity arising merely from an improper choice of coordinates. Therefore the singularity at $r = 2M$ can be removed by using appropriate coordinates. Kruskal and Szekeres use a dimensionless time coordinate T and a dimensionless radial coordinate R related to the Schwarzschild t and r by

$$\left.\begin{aligned} T &= \left(\tfrac{r}{2M} - 1\right)^{\frac{1}{2}} e^{\frac{r}{4M}} \sinh\left(\tfrac{t}{4M}\right), \\ R &= \left(\tfrac{r}{2M} - 1\right)^{\frac{1}{2}} e^{\frac{r}{4M}} \cosh\left(\tfrac{t}{4M}\right), \end{aligned}\right\} \quad \text{when} \quad r > 2M, \tag{A.2}$$

$$\left.\begin{aligned} T &= \left(1 - \tfrac{r}{2M}\right)^{\frac{1}{2}} e^{\frac{r}{4M}} \cosh\left(\tfrac{t}{4M}\right), \\ R &= \left(1 - \tfrac{r}{2M}\right)^{\frac{1}{2}} e^{\frac{r}{4M}} \sinh\left(\tfrac{t}{4M}\right), \end{aligned}\right\} \quad \text{when} \quad r < 2M. \tag{A.3}$$

By making this change of coordinates in the Schwarzschild metric (A.1), we obtain

$$ds^2 = \frac{32M^3}{r} e^{-\frac{r}{2M}} (-dT^2 + dR^2) + r^2(d\theta^2 + \sin^2\theta d\varphi^2). \tag{A.4}$$

It follows from the expression (A.4) that there is only the curvature singularity at $r = 0$, i.e., the singularity at $r = 2M$ does not appear. The coordinates def ned as in (A.2) and (A.3), are called Kruskal coordinates. Similarly, we can also perform related discussions for the spherically-symmetric metric such as the Reissner-Nordström metric, and the def ned coordinates are then called "the Kruskal-like coordinates" in common parlance.

In the body of the paper, we showed that the 4-dimensional Kerr-Newman metric becomes the 2-dimensional spherically-symmetric metric (2.36) by using a technique of the dimensional reduction near the horizon. In what follows, we derive the Kruskal-like coordinates for this reduced metric following the derivation of the Kruskal coordinates in a standard textbook[7]. In this derivation, we apply a series of coordinate transformations. Note that all the variables which appear in our transformations are def ned to be real numbers.

Now the metric is given by

$$ds^2 = -f(r)dt^2 + \frac{1}{f(r)} dr^2, \tag{A.5}$$

[7] See, for example, [50]

where $f(r)$ is defined by (2.20)

$$f(r) = \frac{\Delta}{r^2 + a^2} = \frac{(r - r_+)(r - r_-)}{r^2 + a^2}, \tag{A.6}$$

and $r_{+(-)}$ is the outer (inner) horizon given by (2.5).

As a first step of coordinate transformations, we use the tortoise coordinate defined by (2.20)

$$dr_* \equiv \frac{1}{f(r)} dr. \tag{A.7}$$

The metric (A.5) is then written by

$$ds^2 = -f(r)(dt - dr_*)(dt + dr_*). \tag{A.8}$$

By integrating (A.7) over r from 0 to r, we obtain

$$r_* = r + \frac{1}{2\kappa_+} \ln \frac{|r - r_+|}{r_+} + \frac{1}{2\kappa_-} \ln \frac{|r - r_-|}{r_-} + C, \tag{A.9}$$

where $\kappa_{+(-)}$ is the surface gravity on the outer (inner) horizon and C is generally a pure imaginary integral constant which appears in the analytic continuation. However, as already stated, we would like to treat the case where all the variables (or parameters) are defined in the range of real numbers. Thus, we need to consider the three cases $r > r_+$, $r_+ < r < r_-$ and $r_- < r < 0$ with respect to the range of r. Actually, we have only to consider the two cases $r > r_+$ and $r_+ < r < r_-$ because of the consideration near the outer horizon. When $r > r_+$, the relation (A.9) becomes

$$r_* = r + \frac{1}{2\kappa_+} \ln \frac{r - r_+}{r_+} + \frac{1}{2\kappa_-} \ln \frac{r - r_-}{r_-}. \tag{A.10}$$

As the second step, we use the retarded time u and the advanced time v defined by

$$\left.\begin{aligned} u &\equiv t - r_* = t - r + \tfrac{1}{2\kappa_+} \ln \tfrac{r-r_+}{r_+} + \tfrac{1}{2\kappa_-} \ln \tfrac{r-r_-}{r_-}, \\ v &\equiv t + r_* = t + r + \tfrac{1}{2\kappa_+} \ln \tfrac{r-r_+}{r_+} + \tfrac{1}{2\kappa_-} \ln \tfrac{r-r_-}{r_-}, \end{aligned}\right\} \quad \text{when} \quad r > r_+. \tag{A.11}$$

The metric (A.8) is then written as

$$ds^2 = -f(r)dudv. \tag{A.12}$$

As the third step we use the following coordinate transformations U and V defined by

$$\left.\begin{aligned} U &\equiv -e^{-\kappa_+ u} = -\left(\tfrac{r-r_+}{r_+}\right)^{\frac{1}{2}} \left(\tfrac{r-r_-}{r_-}\right)^{\frac{\kappa_+}{2\kappa_-}} e^{\kappa_+ r} e^{-\kappa_+ t}, \\ V &\equiv e^{\kappa_+ v} = \left(\tfrac{r-r_+}{r_+}\right)^{\frac{1}{2}} \left(\tfrac{r-r_-}{r_-}\right)^{\frac{\kappa_+}{2\kappa_-}} e^{\kappa_+ r} e^{\kappa_+ t}, \end{aligned}\right\} \quad \text{when} \quad r > r_+. \tag{A.13}$$

The metric (A.12) is then written as

$$ds^2 = -\frac{r_+ r_-}{\kappa_+^2}\frac{e^{-2\kappa_+ r}}{r^2+a^2}\left(\frac{r_-}{r-r_-}\right)^{\frac{\kappa_+}{\kappa_-}-1} dUdV. \tag{A.14}$$

As the f nal step we use the following coordinate transformations T, R def ned by

$$\left.\begin{aligned} T &\equiv \tfrac{1}{2}(V+U) = \left(\tfrac{r-r_+}{r_+}\right)^{\frac{1}{2}}\left(\tfrac{r-r_-}{r_-}\right)^{\frac{\kappa_+}{2\kappa_-}} e^{\kappa_+ r}\sinh(\kappa_+ t),\\ R &\equiv \tfrac{1}{2}(V-U) = \left(\tfrac{r-r_+}{r_+}\right)^{\frac{1}{2}}\left(\tfrac{r-r_-}{r_-}\right)^{\frac{\kappa_+}{2\kappa_-}} e^{\kappa_+ r}\cosh(\kappa_+ t),\end{aligned}\right\}\quad \text{when} \quad r > r_+. \tag{A.15}$$

The metric (A.14) is then written as

$$ds^2 = \frac{r_+ r_-}{\kappa_+^2}\frac{e^{-2\kappa_+ r}}{r^2+a^2}\left(\frac{r_-}{r-r_-}\right)^{\frac{\kappa_+}{\kappa_-}-1} (-dT^2+dR^2). \tag{A.16}$$

Similarly, we consider the case of $r_+ > r > r_-$. When $r_+ > r > r_-$, the relation (A.9) becomes

$$r_* = r + \frac{1}{2\kappa_+}\ln\frac{r_+ - r}{r_+} + \frac{1}{2\kappa_-}\ln\frac{r-r_-}{r_-}. \tag{A.17}$$

As for the remaining coordinate transformations, we use the following ones

$$\left.\begin{aligned} u &\equiv t - r_* = t - r + \tfrac{1}{2\kappa_+}\ln\tfrac{r_+-r}{r_+} + \tfrac{1}{2\kappa_-}\ln\tfrac{r-r_-}{r_-},\\ v &\equiv t + r_* = t + r + \tfrac{1}{2\kappa_+}\ln\tfrac{r_+-r}{r_+} + \tfrac{1}{2\kappa_-}\ln\tfrac{r-r_-}{r_-},\end{aligned}\right\}\quad \text{when } r_+ > r > r_-, \tag{A.18}$$

$$\left.\begin{aligned} U &\equiv e^{-\kappa_+ u} = \left(\tfrac{r_+-r}{r_+}\right)^{\frac{1}{2}}\left(\tfrac{r-r_-}{r_-}\right)^{\frac{\kappa_+}{2\kappa_-}} e^{\kappa_+ r}e^{-\kappa_+ t},\\ V &\equiv e^{\kappa_+ v} = \left(\tfrac{r_+-r}{r_+}\right)^{\frac{1}{2}}\left(\tfrac{r-r_-}{r_-}\right)^{\frac{\kappa_+}{2\kappa_-}} e^{\kappa_+ r}e^{\kappa_+ t},\end{aligned}\right\}\quad \text{when } r_+ > r > r_-, \tag{A.19}$$

$$\left.\begin{aligned} T &\equiv \tfrac{1}{2}(V+U) = \left(\tfrac{r_+-r}{r_+}\right)^{\frac{1}{2}}\left(\tfrac{r-r_-}{r_-}\right)^{\frac{\kappa_+}{2\kappa_-}} e^{\kappa_+ r}\cosh(\kappa_+ t),\\ R &\equiv \tfrac{1}{2}(V-U) = \left(\tfrac{r_+-r}{r_+}\right)^{\frac{1}{2}}\left(\tfrac{r-r_-}{r_-}\right)^{\frac{\kappa_+}{2\kappa_-}} e^{\kappa_+ r}\sinh(\kappa_+ t),\end{aligned}\right\}\quad \text{when } r_+ > r > r_-. \tag{A.20}$$

Of course, by performing these coordinate transformations, the corresponding metrics (A.12), (A.14), and (A.16) do not change. The two sets of def ned coordinates, both (A.15) and (A.20), are in fact the Kruskal-like coordinates. In the Schwarzschild case ($a = Q = 0$), (A.15) and (A.20) respectively agree with (A.2) and (A.3). Finally, we obtain (3.22) and (3.23) by rewriting the expressions written in terms of r to the ones in terms of r_* in the formulas (A.15) and (A.20).

References

[1] A. Einstein, "The foundation of the general theory of relativity," *Annalen Phys.* **49**, 769 (1916) [Annalen Phys. **14**, 517 (2005)].

[2] J. R. Oppenheimer and H. Snyder, "On continued gravitational contraction," *Phys. Rev.* **56**, 455 (1939).

[3] W. Israel, "Event horizons in static vacuum space-times," *Phys. Rev.* **164**, 1776 (1967).

[4] B. Carter, "Axisymmetric black hole has only two degrees of freedom," *Phys. Rev. Lett.* **26**, 331 (1971).

[5] R. Ruff ni, J. A. Wheeler, "Introducing the black hole," Phys. Today, **24**, 1, 30 (1971). [Phys. Today, **62**, 4, 47-53 (2009)].

[6] J. M. Bardeen, B. Carter, and S. W. Hawking, "The four laws of black hole mechanics," *Commun. Math. Phys.* **31**, 161 (1973).

[7] J. D. Bekenstein, "Black holes and entropy," *Phys. Rev. D* **7**, 2333 (1973).

[8] L. Brillouin, "Science and Information Theory," Acad. Press, Inc., New York, (1956) §12.1.

[9] S. W. Hawking, "Particle creation by black holes," *Commun. Math. Phys.* **43**, 199 (1975) [Erratum-ibid. **46**, 206 (1976)].

[10] S. W. Hawking, "Black hole explosions," *Nature* **248**, 30 (1974).

[11] R. Emparan and H. S. Reall, "Black holes in higher dimensions," *Living Rev. Rel.* **11**, 6 (2008) [arXiv:0801.3471 [hep-th]].

[12] Y. Yasui and T. Houri, "Hidden symmetry and exact solutions in Einstein gravity," Prog. Theor. Phys. Suppl. **189**, 126-164 (2011). [arXiv:1104.0852 [hep-th]].

[13] S. Iso, H. Umetsu, and F. Wilczek, "Hawking radiation from charged black holes via gauge and gravitational anomalies," *Phys. Rev. Lett.* **96**, 151302 (2006) [hep-th/0602146].

[14] A. J. M. Medved, "Radiation via tunneling from a de Sitter cosmological horizon," *Phys. Rev. D* **66**, 124009 (2002) [hep-th/0207247].

[15] G. T. Horowitz and V. E. Hubeny, "Quasinormal modes of AdS black holes and the approach to thermal equilibrium," *Phys. Rev. D* **62**, 024027 (2000) [hep-th/9909056].

[16] X. -X. Zeng, K. Lin, and S. -Z. Yang, "Hawking radiation from the higher-dimensional Schwarzschild de Sitter and anti-de Sitter black holes via covariant anomaly," *Int. J. Theor. Phys.* **47**, 2533 (2008).

[17] K. Umetsu, "Ward identities in the derivation of Hawking radiation from anomalies," *Prog. Theor. Phys.* **119**, 849 (2008) [arXiv:0804.0963 [hep-th]].

[18] R. C. Myers and M. J. Perry, "Black holes in higher dimensional space-times," *Annals Phys.* **172**, 304 (1986).

[19] S. Iso, T. Morita, and H. Umetsu, "Quantum anomalies at horizon and Hawking radiations in Myers-Perry black holes," *JHEP* **0704**, 068 (2007) [hep-th/0612286].

[20] G. W. Gibbons, H. Lu, D. N. Page, and C. N. Pope, "Rotating black holes in higher dimensions with a cosmological constant," *Phys. Rev. Lett.* **93**, 171102 (2004) [hep-th/0409155].

[21] L. -C. Zhang, H. -F. Li, and R. Zhao, "Tunneling mechanism in higher-dimensional rotating black hole with a cosmological constant in the approach of dimensional reduction," *Astrophys. Space Sci.* **333**, 457 (2011).

[22] A. M. Awad, "Higher dimensional charged rotating solutions in (A)dS space-times," *Class. Quant. Grav.* **20**, 2827 (2003) [hep-th/0209238].

[23] J. -J. Peng and S. -Q. Wu, "Covariant anomalies and Hawking radiation from charged rotating black strings in anti-de Sitter spacetimes," *Phys. Lett. B* **661**, 300 (2008) [arXiv:0801.0185 [hep-th]].

[24] R. Emparan, "Rotating circular strings, and inf nite nonuniqueness of black rings," *JHEP* **0403**, 064 (2004) [hep-th/0402149].

[25] U. Miyamoto and K. Murata, "On Hawking radiation from black rings," *Phys. Rev. D* **77**, 024020 (2008) [arXiv:0705.3150 [hep-th]].

[26] H. Ishihara and K. Matsuno, "Kaluza-Klein black holes with squashed horizons," *Prog. Theor. Phys.* **116**, 417 (2006) [hep-th/0510094].

[27] K. Matsuno and K. Umetsu, "Hawking radiation as tunneling from squashed Kaluza-Klein black hole," *Phys. Rev. D* **83**, 064016 (2011) [arXiv:1101.2091 [hep-th]].

[28] S. -Q. Wu, "General non-extremal rotating charged Godel black holes in minimal f ve-dimensional gauged supergravity," *Phys. Rev. Lett.* **100**, 121301 (2008) [arXiv:0709.1749 [hep-th]].

[29] S. -Q. Wu and J. -J. Peng, "Thermodynamics and Hawking radiation of f ve-dimensional rotating charged Godel black holes," *Phys. Rev. D* **83**, 044028 (2011) [arXiv:1101.5474 [hep-th]].

[30] M. H. Ali, "Hawking radiation via tunneling from hot NUT-Kerr-Newman-Kasuya spacetime," *Class. Quant. Grav.* **24**, 5849 (2007) [arXiv:0706.3890 [gr-qc]].

[31] K. Lin and S. Yang, "Hawking radiation from NUT Kerr Newman Kusuya black hole via effective action and covariant anomalies," *Int. J. Theor. Phys.* **49**, 927 (2010).

[32] K. Lin, S. W. Chen, and S. Z. Yang, "Anomalies and Hawking radiation of NUT-Kerr-Newman de Sitter black hole," *Int. J. Theor. Phys.* **47**, 2453 (2008).

[33] C. Martinez, R. Troncoso, and J. Zanelli, "Exact black hole solution with a minimally coupled scalar f eld," *Phys. Rev. D* **70**, 084035 (2004) [hep-th/0406111].

[34] E. Papantonopoulos and P. Skamagoulis, "Hawking Radiation via gravitational anomalies in non-spherical topologies," *Phys. Rev. D* **79**, 084022 (2009) [arXiv:0812.1759 [hep-th]].

[35] M. A. Anacleto, F. A. Brito, and E. Passos, "Acoustic black holes from Abelian Higgs model with Lorentz symmetry breaking," *Phys. Lett. B* **694**, 149 (2010) [arXiv:1004.5360 [hep-th]].

[36] L. -C. Zhang, H. -F. Li, and R. Zhao, "Hawking radiation from a rotating acoustic black hole," *Phys. Lett. B* **698**, 438 (2011).

[37] K. Umetsu, "Hawking radiation from Kerr-Newman black hole and tunneling mechanism," *Int. J. Mod. Phys. A* **25**, 4123 (2010) [arXiv:0907.1420 [hep-th]].

[38] K. Srinivasan and T. Padmanabhan, "Particle production and complex path analysis," *Phys. Rev. D* **60**, 024007 (1999) [gr-qc/9812028].

[39] R. Banerjee and B. R. Majhi, "Hawking black body spectrum from tunneling mechanism," *Phys. Lett. B* **675**, 243 (2009) [arXiv:0903.0250 [hep-th]].

[40] S. R. Coleman, "The quantum Sine-Gordon equation as the massive Thirring model," *Phys. Rev. D* **11**, 2088 (1975).

[41] S. W. Hawking, "Black holes in general relativity," *Commun. Math. Phys.* **25**, 152 (1972).

[42] J. D. Bekenstein, "Generalized second law of thermodynamics in black hole physics," *Phys. Rev. D* **9**, 3292 (1974).

[43] W. G. Unruh, "Notes on black hole evaporation," *Phys. Rev. D* **14**, 870 (1976).

[44] S. Deser and O. Levin, "Equivalence of Hawking and Unruh temperatures through f at space embeddings," *Class. Quant. Grav.* **15**, L85 (1998) [hep-th/9806223].

[45] R. Banerjee and B. R. Majhi, "A new global embedding approach to study Hawking and Unruh effects," *Phys. Lett. B* **690**, 83 (2010) [arXiv:1002.0985 [gr-qc]].

[46] S. Deser and O. Levin, "Mapping Hawking into Unruh thermal properties," *Phys. Rev. D* **59**, 064004 (1999) [hep-th/9809159] and references therein.

[47] R. C. Tolman, "On the weight of heat and thermal equilibrium in general relativity," *Phys. Rev.* **35**, 904 (1930).

[48] M. D. Kruskal, "Maximal extension of Schwarzschild metric," *Phys. Rev.* **119**, 1743 (1960).

[49] G. Szekeres, "On the singularities of a Riemannian manifold," *Publ. Math. Debrecen* **7**, 285 (1960).

[50] C. W. Misner, K. S. Thorne, and J. A. Wheeler, "Gravitation," San Francisco 1973, 1279p, §31.

In: Black Holes: Evolution, Theory and Thermodynamics ISBN: 978-1-61942-929-1
Editors: A. J. Bauer and D. G. Eiffel, pp. 41-79

Chapter 2

BLACK HOLES AND STELLAR STRUCTURES IN $f(R)$-GRAVITY

***Mariafelicia De Laurentis*[1]* *and Salvatore Capozziello*[2]†**
[1]Dipartimento di Scienze Fisiche, Universitá di Napoli "Federico II"
[2]INFN Sezione di Napoli, Compl. Univ. di Monte S. Angelo,
Edificio G, Napoli, Italy

Abstract

In the last ten years, increasing attention has been devoted to Extended Theories of Gravity with the aim to understand several cosmological and astrophysical issues such as the today observed accelerated expansion of the universe and the presence of Dark Matter in self-gravitating structures. Some of these models assume modifications of General Relativity by adding higher order terms of curvature invariants like the Ricci scalar R, the Ricci tensor $R_{\mu\nu}$ and the Riemann tensors $R_{\mu\nu\lambda\sigma}$, or the presence of suitable scalar fields like the former Brans-Dicke theory. It is therefore natural to ask for black hole solutions in this context since, on the one hand, black holes signatures may be the test-bed to compare new models to the Einstein gravity; on the other hand, they may lead to rule out models which disagree with observations. Although black holes are one of the most striking predictions of General Relativity, they remain one of its least tested concepts. Electromagnetic observations allow indirectly to infer their existence, but direct evidences remains elusive. In the next decade, data coming from very long-baseline interferometry and gravitational wave detectors should allow to image and study black holes in detail. Such observations will test General Relativity in the non-linear and strong-field regimes where data are currently lacking. Testing strong-field features of General Relativity is of utmost importance to physics and astrophysics as a whole. This is because the black holes solutions, such as the Schwarzschild and Kerr metrics, enter several calculations, including accretion disk structure, gravitational lensing, cosmology and gravitational waves theory. These black hole solutions could indicate strong-field departures from General Relativity with deep implications for the still unknown fundamental theory of gravity. Beside the physical interest, black hole solutions represent a very active area for mathematical physics investigations. Here we review the problem of black holes in a particular

*E-mail address: felicia@na.infn.it
†E-mail address: capozziello@na.infn.it

class of Extended Theories of Gravity, the so called $f(R)$-gravity, discussing some resolution techniques, obtaining exact solutions and comparing results with standard General Relativity. Furthermore, we discuss the problems of hydrostatic equilibrium and stellar structure in the context of $f(R)$-gravity showing that new features could emerge. The observation of such features could both explain the physics of exotic self-gravitating objects and constitute a signature for Extended Theories of Gravity.

PACS 05.45-a, 52.35.Mw, 96.50.Fm

Keywords: Extended Theories of Gravity; Black Hole; Exact Solutions; Stellar Structure

1. Introduction

The issue to extend General Relativity (GR) has recently become dramatically urgent due to the missing matter problem at all astrophysical scales and the accelerating behavior of cosmic fluid, detected by Super Novae Ia used as standard candles. Up to now, no final answer on new particles has been given at fundamental level so Dark Energy and Dark Matter constitute a puzzle to be solved in order to achieve a self-consistent picture of the observed Universe. $f(R)$-gravity, where $f(R)$ is a generic function of the Ricci scalar R, comes into the game as a straightforward extension of GR where further geometrical degrees of freedom are considered instead of searching for new material ingredients [1]. From an epistemological point of view, the action of gravity is not selected *a priori*, but it could be "reconstructed", in principle, by matching consistently the observations [2–4]. This approach can be adopted considering any function of the curvature invariants as $R_{\mu\nu}R^{\mu\nu}$, $R\Box R$ and so on. Several of these extended models reproduce Solar System tests so they are not in conflict with GR experimental results but actually extend them enclosing new features that could be, in principle, observed [5–7].

From a genuine mathematical point of view, Extended Theories of Gravity pose the problem to recover the well-established results of GR as the initial value problem [8], the stability of solutions and, in particular, the issue of finding out new solutions. As it is well known, beside cosmological solutions, spherically and axially symmetric solutions play a fundamental role in several astrophysical problems ranging from black holes to active galactic nuclei. Extended gravities, to be consistent with results of GR, should comprise solutions like Schwarzschild and Kerr ones but could present, in general, new solutions of physical interest. Due to this reason, methods to find out exact and approximate solutions are particularly relevant in order to check if observations can be framed in Extended Theories of Gravity [9].

Recently, the interest in spherically symmetric solutions of $f(R)$-gravity is growing up. In [10], solutions in vacuum have been found considering relations among functions that define the spherical metric or imposing a costant Ricci curvatue scalar. The authors have reconstructed the form of some $f(R)$-models, discussing their physical relevance. In [11], the same authors have discussed static spherically symmetric solutions, in presence of perfect fluid matter, adopting the metric formalism. They have shown that a given matter distribution is not capable of globally determining the functional form of $f(R)$. Others authors have discussed in details the spherical symmetry of $f(R)$-gravity considering also the relations with the weak field limit. Exact solutions are obtained for constant Ricci curvature scalar and for Ricci scalar depending on the radial coordinate. In particular, it can be considered how to obtain results consistent with GR assuming the well-known post-Newtonian and post-Minkowskian limits as consistency checks [12].

As we will discuss below, a general method to find out rotating black hole solutions can be achieved by performing a complex coordinate transformation on the spherical black hole metrics. Since the discovery of the Kerr solution [13], many attempts have been made to find a physically reasonable interior matter distribution that may be considered as its source. For a review on these approaches see [14,15]. Though much progress has been made, results have been generally disappointing. As far as we know, nobody has obtained a physically satisfactory interior solution. This seems surprising given the success of matching internal spherically symmetric solutions to the Schwarzschild metric. The problem is that the loss of a degree of symmetry makes the derivation of analytic results much more difficult. Severe restrictions are placed on the interior metric by maintaining that it must be joined smoothly to the external axially symmetric metric. Further restrictions are placed on the interior solutions to ensure that they correspond to physical objects.

Furthermore since the axially symmetric metric has no radiation field associated with it, its source should be also non-radiating. This places even further constraints on the structure of the interior solution [16]. Given the strenuous nature of these limiting conditions, it is not surprising to learn that no satisfactory solution to the problem of finding sources for the Kerr metric has been obtained. In general, the failure is due to internal structures whose physical properties are unknown. This shortcoming makes hard to find consistent boundary conditions.

Newman and Janis showed that it is possible to obtain rotating solutions (like the Kerr metric) by making an elementary complex transformation on the Schwarzschild solution [17]. This same method has been used to obtain a new stationary and axially symmetric solution known as the Kerr-Newman metric [18]. The Kerr-Newman space-time is associated to the exterior geometry of a rotating massive and charged black-hole. For a review on the Newman-Janis method to obtain both the Kerr and Kerr-Newman metrics see [19].

By means of very elegant mathematical arguments, Schiffer et al. [20] have given a rigorous proof to show how the Kerr metric can be derived starting from a complex transformation on the Schwarzschild solution. We will not go into details of this demonstration, but point out that the proof relies on two main assumptions. The first is that metric belongs to the same algebraic class of the Kerr-Newman solution, namely the Kerr-Schild class [21]. The second assumption is that metric corresponds to an empty solution of the Einstein field equations [20]. It is clear, by the generation of the Kerr-Newman metric, that all the components of the stress-energy tensor need to be non-zero for the Newman-Janis method to be successful. Such a transformation can be extended to $f(R)$-gravity as discussed in [22].

On the other hand, the strong gravity regime is another way to check the viability of these theories [23]. In general the formation and the evolution of stars can be considered suitable test-beds for Extended Theories of Gravity. Considering the case of $f(R)$-gravity, divergences stemming from the functional form of $f(R)$ may prevent the existence of relativistic stars in these theories [24], but thanks to the chameleon mechanism, introduced by Khoury and Weltman [25], the possible problems jeopardizing the existence of these objects may be avoided [26]. Furthermore, there are also numerical solutions corresponding to static star configurations with strong gravitational fields [27] where the choice of the equation of state is crucial for the existence of solutions.

It is also important to stress that $f(R)$-gravity has interesting applications also in stellar astrophysics and could contribute to solve several puzzles related to observed peculiar objects (e.g. stars in the instability strips, protostars, etc. [28, 29]), structure and star formation [30, 31]. Furthermore some observed stellar systems are incompatible with the standard models of stellar structure. We refer to anomalous neutron stars, the so called "magnetars" [32] with masses larger than their expected Volkoff mass. It seems that, on particular length scales, the gravitational force is larger or smaller than the corresponding GR value. For example, a modification of the Hilbert-Einstein Lagrangian, consisting of R^2 terms, enables a major attraction while a $R_{\alpha\beta}R^{\alpha\beta}$ term gives a repulsive contribution [33]. Understanding on which scales the modifications to GR are working or what is the weight of corrections to gravitational potential is a crucial point that could confirm or rule out these extended approaches to gravitational interaction.

This Chapter is organized as follow. In the Sec.2, we introduce the $f(R)$-gravity action, the field equations and give some general remarks on spherical symmetry. In Sec 3, a summary is given on the Noether Simmetry Approach [9]. This technique is extremely useful to find exact solutions. In particular, we find spherically symmetric black hole solutions for $f(R)$-gravity. In Sec 4, we review the Newman-Janis method to obtain rotating solutions starting from spherically symmetric ones. The resulting metric is written in terms of two arbitrary functions. A further suitable coordinate transformation allows to write the metric in the so called Boyer-Lindquist coordinates. Such a transformation makes the physical interpretation much clearer and reduces the amount of algebra required to calculate the metric properties. In Sec.5, the Newman-Janis method is applied to a spherically symmetric exact solution, previously derived by the Noether Symmetry, and an axially symmetric exact solution is obtained. This result shows that the Newman-Janis method works also in $f(R)$-gravity. Physical applications of the result are also discussed.

In Sec.6, we review the classical hydrostatic problem for stellar structures. In Sec. 7 we derive the Newtonian limit of $f(R)$-gravity obtaining the modified Poisson equation. The modified Lané-Emden equation is obtained in Sec. 8 and its structure is compared with respect to the standard one. In Sec.9, we show how analytical solutions of standard Lané-Emden equation can be compared with those perturbatively obtained from $f(R)$-gravity. In order to apply the above results, in Section 10 the classical theory of gravitational collapse for dust-dominated systems is summarized. In Section 11, we discuss the weak field limit of $f(R)$-gravity obtaining corrections to the standard Newtonian potential that can be figured out as two Newtonian potentials contributing to the dynamics. In Section 12 we recover the dispersion relation and Jeans mass limit. Some self-gravitating dust system are discussed in this approach. The difference between GR and $f(R)$-gravity are put in evidence, in particular the Jeans mass profiles with respect to the temperature. We report a catalogue of observed molecular clouds in order to compare the classical Jeans mass to the $f(R)$-one. Finally, in Section 13, we discuss the results and draw conclusions.

2. Spherical Symmetry in $f(R)$-Gravity

Let us start by discussing exact solutions in $f(R)$-gravity with spherical symmetry. As we will see, a crucial role is played by the relation between the metric potentials and the Ricci scalar that can be regarded as a constraint assuming the form of a Bernoulli equation.

Let us consider an analytic function $f(R)$ of the Ricci scalar R in four dimensions. The variational principle for this action is:

$$\delta \int d^4x\sqrt{-g}\left[f(R)+\mathcal{X}\mathcal{L}_m\right] = 0 \tag{1}$$

where $\mathcal{X} = \frac{8\pi G}{c^4}$, $\mathcal{L}_m$ is the standard matter Lagrangian and g is the determinant of the metric[1].

By varying with respect to the metric, we obtain the field equations [2]

$$\begin{cases} H_{\mu\nu} = f'(R)R_{\mu\nu} - \frac{1}{2}f(R)g_{\mu\nu} - f'(R)_{;\mu\nu} + g_{\mu\nu}\Box f'(R) = \mathcal{X}T_{\mu\nu} \\ H = g^{\rho\sigma}H_{\rho\sigma} = 3\Box f'(R) + f'(R)R - 2f(R) = \mathcal{X}T \end{cases} \tag{2}$$

where $T_{\mu\nu} = \frac{-2}{\sqrt{-g}}\frac{\delta(\sqrt{-g}\mathcal{L}_m)}{\delta g^{\mu\nu}}$ is the energy-momentum tensor of standard fluid matter and the second equation is the trace. The most general spherically symmetric solution can be written as follows:

$$ds^2 = m_1(t',r')dt'^2 + m_2(t',r')dr'^2 + m_3(t',r')dt'dr' + m_4(t',r')d\Omega, \tag{3}$$

where m_i are functions of the radius r' and of the time t'. $d\Omega$ is the solid angle. We can consider a coordinate transformation that maps the metric (3) in a new one where the off-diagonal term vanishes and $m_4(t',r') = -r^2$, that is[3]:

$$ds^2 = g_{tt}(t,r)dt^2 - g_{rr}(t,r)dr^2 - r^2 d\Omega. \tag{4}$$

This expression can be considered, without loss of generality, as the most general definition of a spherically symmetric metric compatible with a pseudo-Riemannian manifold without torsion. Actually, by inserting this metric into the field Eqs. (76), one obtains:

$$\begin{cases} f'(R)R_{\mu\nu} - \frac{1}{2}f(R)g_{\mu\nu} + \mathcal{H}_{\mu\nu} = \mathcal{X}T_{\mu\nu} \\ f'(R)R - 2f(R) + \mathcal{H} = \mathcal{X}T \end{cases} \tag{5}$$

[1]We are adopting the convention $c = 1$. The convention $R_{\mu\nu} = R^{\rho}{}_{\mu\rho\nu}$ for the Ricci tensor and $R^{\alpha}{}_{\beta\mu\nu} = \Gamma^{\alpha}_{\beta\nu,\mu} - ...$, for the Riemann tensor. Connections are Levi-Civita :

$$\Gamma^{\mu}_{\alpha\beta} = \frac{1}{2}g^{\mu\rho}(g_{\alpha\rho,\beta} + g_{\beta\rho,\alpha} - g_{\alpha\beta,\rho}).$$

The signature is $(+ - - -)$.

[2]It is possible to take into account also the Palatini approach in which the metric g and the connection Γ are considered independent variables (see for example [34]). Here we will consider the Levi-Civita connection and will use the metric approach. See [3, 34, 35] for a detailed comparison between the two pictures.

[3]This condition allows to obtain the standard definition of the circumference with radius r.

where the two quantities $\mathcal{H}_{\mu\nu}$ and $\mathcal{H}$ read :

$$\begin{aligned}\mathcal{H}_{\mu\nu} &= -f''(R)\Big\{R_{,\mu\nu}-\Gamma^{t}_{\mu\nu}R_{,t}-\Gamma^{r}_{\mu\nu}R_{,r}-g_{\mu\nu}\Big[\Big({g^{tt}}_{,t}+\\ &+g^{tt}\left(\ln\sqrt{-g}\right)_{,t}\Big)R_{,t}+\Big({g^{rr}}_{,r}+g^{rr}\left(\ln\sqrt{-g}\right)_{,r}\Big)R_{,r}+\\ &+g^{tt}R_{,tt}+g^{rr}R_{,rr}\Big]\Big\}-f'''(R)\Big[R_{,\mu}R_{,\nu}-g_{\mu\nu}\Big(g^{tt}{R_{,t}}^{2}+g^{rr}{R_{,r}}^{2}\Big)\Big]\end{aligned} \tag{6}$$

$$\begin{aligned}\mathcal{H} &= g^{\sigma\tau}\mathcal{H}_{\sigma\tau}=3f''(R)\Big[\Big({g^{tt}}_{,t}+g^{tt}\left(\ln\sqrt{-g}\right)_{,t}\Big)R_{,t}+\\ &+\Big({g^{rr}}_{,r}+g^{rr}\left(\ln\sqrt{-g}\right)_{,r}\Big)R_{,r}+g^{tt}R_{,tt}+g^{rr}R_{,rr}\Big]+3f'''(R)\Big[g^{tt}{R_{,t}}^{2}+g^{rr}{R_{,r}}^{2}\Big].\end{aligned} \tag{7}$$

Our task is now to find out exact spherically symmetric solutions.

In the case of time-independent metric, i.e., $g_{tt}=a(r)$ and $g_{rr}=b(r)$, the Ricci scalar can be recast as a Bernoulli equation of index two with respect to the metric potential $b(r)$ (see [12] for details):

$$\begin{aligned}&b'(r)+\left\{\frac{r^2a'(r)^2-4a(r)^2-2ra(r)[2a(r)'+ra(r)'']}{ra(r)[4a(r)+ra'(r)]}\right\}b(r)+\\ &+\left\{\frac{2a(r)}{r}\left[\frac{2+r^2R(r)}{4a(r)+ra'(r)}\right]\right\}b(r)^2=0.\end{aligned} \tag{8}$$

where $R=R(r)$ is the Ricci scalar. A general solution of (8) is:

$$b(r)=\frac{\exp[-\int dr\,h(r)]}{K+\int dr\,l(r)\exp[-\int dr\,h(r)]}, \tag{9}$$

where K is an integration constant while $h(r)$ and $l(r)$ are two functions that, according to Eq.(8), define the coefficients of the quadratic and the linear term with respect to $b(r)$ [36]. We can fix $l(r)=0$; this choice allows to find out solutions with a Ricci scalar scaling as $-\frac{2}{r^2}$ in term of the radial coordinate. On the other hand, it is not possible to have $h(r)=0$ since, otherwise, we get imaginary solutions. A particular consideration deserves the limit $r\to\infty$. In order to achieve a gravitational potential $b(r)$ with the correct Minkowski limit, both $h(r)$ and $l(r)$ have to go to zero at infinity, provided that the quantity $r^2R(r)$ turns out to be constant: this result implies $b'(r)=0$, and, finally, also the metric potential $b(r)$ has a correct Minkowski limit.

In general, if we ask for the asymptotic flatness of the metric as a feature of the theory, the Ricci scalar has to evolve to infinity as r^{-n} with $n\geqslant 2$. Formally, it has to be:

$$\lim_{r\to\infty} r^2R(r)=r^{-n}, \tag{10}$$

with $n \in \mathbb{N}$. Any other behavior of the Ricci scalar could affect the requirement to achieve a correct asymptotic flatness.

The case of constant curvature is equivalent to GR with a cosmological constant and the solution is time independent. This result is well known (see, for example, [37]) but we report, for the sake of completeness, some considerations related with it in order to deal with more general cases where a radial dependence for the Ricci scalar is supposed. If the scalar curvature is constant ($R = R_0$), field Eqs.(5), being $\mathcal{H}_{\mu\nu} = 0$, reduce to:

$$\begin{cases} f'_0 R_{\mu\nu} - \frac{1}{2} f_0 g_{\mu\nu} = X T_{\mu\nu} \\ \\ f'_0 R_0 - 2 f_0 = X T \end{cases} \tag{11}$$

where $f(R_0) = f_0$, $f'(R_0) = f'_0$. A general solution, when one considers a stress-energy tensor of perfect-fluid $T_{\mu\nu} = (\rho + p) u_\mu u_\nu - p g_{\mu\nu}$, is

$$ds^2 = \left(1 + \frac{k_1}{r} + \frac{qX\rho - \lambda}{3} r^2\right) dt^2 - \frac{dr^2}{1 + \frac{k_1}{r} + \frac{qX\rho - \lambda}{3} r^2} - r^2 d\Omega . \tag{12}$$

when $p = -\rho$, $\lambda = -\frac{f_0}{2f'_0}$ and $q^{-1} = f'_0$. This result means that any $f(R)$-model, in the case of constant curvature, exhibits solutions with de Sitter-like behavior. This is one of the reasons why the dark energy issue can be addressed using these theories [1].

If $f(R)$ is analytic, it is possible to write the series:

$$f(R) = \Lambda + \Psi_0 R + \Psi(R) , \tag{13}$$

where Ψ_0 is a coupling constant, Λ plays the role of the cosmological constant and $\Psi(R)$ is a generic analytic function of R satisfying the condition

$$\lim_{R \to 0} R^{-2} \Psi(R) = \Psi_1 , \tag{14}$$

where Ψ_1 is a constant. If we neglect the cosmological constant Λ and Ψ_0 is set to zero, we obtain a new class of theories which, in the limit $R \longrightarrow 0$, does not reproduce GR (from Eq.(14), we have $\lim_{R \to 0} f(R) \sim R^2$). In such a case, analyzing the whole set of Eqs.(11), one can observe that both zero and constant $\neq 0$ curvature solutions are possible. In particular, if $R = R_0 = 0$ field equations are solved for any form of gravitational potential entering the spherically symmetric background, provided that the Bernoulli Eq. (8), relating these functions, is fulfilled for the particular case $R(r) = 0$. The solutions are thus defined by the relation

$$b(r) = \frac{\exp[-\int dr h(r)]}{K + 4 \int \frac{dr a(r) \exp[-\int dr h(r)]}{r[a(r) + r a'(r)]}} , \tag{15}$$

being $g_{tt}(t,r) = b(r)$ from Eq.(4). In [12], some examples of $f(R)$-models admitting solutions with constant$\neq 0$ or null scalar curvature are discussed.

3. The Noether Symmetry Approach

Besides spherically symmetric solutions with constant curvature scalar, also solutions with the Ricci scalar depending on radial coordinate r are possible in $f(R)$-gravity [12]. Furthermore, spherically symmetric solutions can be achieved starting from a point-like $f(R)$-Lagrangian [9]. Such a Lagrangian can be obtained by imposing the spherical symmetry directly in the action (1). As a consequence, the infinite number of degrees of freedom of the original field theory will be reduced to a finite number. The technique is based on the choice of a suitable Lagrange multiplier defined by assuming the Ricci scalar, argument of the function $f(R)$ in spherical symmetry.

Starting from the above considerations, a static spherically symmetric metric can be expressed as

$$ds^2 = A(r)dt^2 - B(r)dr^2 - M(r)d\Omega\,, \tag{16}$$

and then the point-like $f(R)$ Lagrangian[4] is

$$\begin{aligned}\mathcal{L} &= -\frac{A^{1/2}f'}{2MB^{1/2}}M'^2 - \frac{f'}{A^{1/2}B^{1/2}}A'M' - \frac{Mf''}{A^{1/2}B^{1/2}}A'R' + \\ &\quad -\frac{2A^{1/2}f''}{B^{1/2}}R'M' - A^{1/2}B^{1/2}[(2+MR)f' - Mf]\,,\end{aligned} \tag{17}$$

which is canonical since only the configuration variables and their first order derivatives with respect to the radial coordinate r are present. Details of calculations are in [9]. Eq. (17) can be recast in a more compact form introducing the matrix representation:

$$\mathcal{L} = \underline{q}'^t\hat{T}\underline{q}' + V \tag{18}$$

where $\underline{q} = (A,B,M,R)$ and $\underline{q}' = (A',B',M',R')$ are the generalized positions and velocities associated to $\mathcal{L}$. It is easy to check the complete analogy between the field equation approach and point-like Lagrangian approach [9].

In order to find out solutions for the Lagrangian (17), we can search for symmetries related to cyclic variables and then reduce dynamics. This approach allows, in principle, to select $f(R)$-gravity models compatible with spherical symmetry. As a general remark, the Noether Theorem states that conserved quantities are related to the existence of cyclic variables into dynamics [38–40].

It is worth noticing that the Hessian determinant of Eq. (17), $\left\|\frac{\partial^2\mathcal{L}}{\partial q'_i\partial q'_j}\right\|$, is zero. This result clearly depends on the absence of the generalized velocity B' into the point-like Lagrangian. As matter of fact, using a point-like Lagrangian approach implies that the metric variable B does not contributes to dynamics, but the equation of motion for B has to be considered as a further constraint equation. Then the Lagrangian (17) has three degrees of freedom and not four, as one should expect *a priori*.

[4]Obviously, the above choices are recovered for $A(r) = a(r)$, $B(r) = b(r)$, and $M(r) = r^2$. Here we deal with A,B,M as a set of coordinates in a configuration space.

Now, since the equation of motion describing the evolution of the metric potential B does not depend on its derivative, it can be explicitly solved in term of B as a function of the other coordinates :

$$B = \frac{2M^2 f'' A' R' + 2M f' A' M' + 4AM f'' M' R' + A f' M'^2}{2AM[(2+MR)f' - Mf]} . \tag{19}$$

By inserting Eq.(19) into the Lagrangian (17), we obtain a non-vanishing Hessian matrix removing the singular dynamics. The new Lagrangian reads[5]

$$\mathcal{L}^* = \mathbf{L}^{1/2} \tag{20}$$

with

$$\mathbf{L} = \underline{q}'^t \hat{\mathbf{L}} \underline{q}' = \frac{[(2+MR)f' - fM]}{M} [2M^2 f'' A' R' + 2MM'(f'A' + 2Af''R') + Af'M'^2] .$$

If one assumes the spherical symmetry, the role of the *affine parameter* is played by the coordinate radius r. In this case, the configuration space is given by $\mathcal{Q} = \{A, M, R\}$ and the tangent space by $\mathcal{TQ} = \{A, A', M, M', R, R'\}$. On the other hand, according to the Noether Theorem, the existence of a symmetry for dynamics described by Lagrangian (17) implies a constant of motion. Let us apply the Lie derivative to the (17), we have[6] :

$$L_{\mathbf{X}}\mathbf{L} = \underline{\alpha} \cdot \nabla_q \mathbf{L} + \underline{\alpha}' \cdot \nabla_{q'} \mathbf{L} = \underline{q}'^t \left[\underline{\alpha} \cdot \nabla_q \hat{\mathbf{L}} + 2 \left(\nabla_q \alpha \right)^t \hat{\mathbf{L}} \right] \underline{q}' , \tag{21}$$

that vanishes if the functions $\underline{\alpha}$ satisfy the following system

$$\underline{\alpha} \cdot \nabla_q \hat{\mathbf{L}} + 2(\nabla_q \underline{\alpha})^t \hat{\mathbf{L}} = 0 \longrightarrow \quad \alpha_i \frac{\partial \hat{\mathbf{L}}_{km}}{\partial q_i} + 2 \frac{\partial \alpha_i}{\partial q_k} \hat{\mathbf{L}}_{im} = 0 . \tag{22}$$

Solving the system (22) means to find out the functions α_i which assign the Noether vector [38, 39]. However the system (22) implicitly depends on the form of $f(R)$ and then, by solving it, we get also $f(R)$-models compatible with spherical symmetry. On the other hand, by choosing the $f(R)$-form, we can explicitly solve (22). As an example, one finds that the system (22) is satisfied if we choose

$$f(R) = f_0 R^s , \quad \underline{\alpha} = (\alpha_1, \alpha_2, \alpha_3) = \Big((3-2s)kA, \ -kM, \ kR \Big) , \tag{23}$$

with s a real number, k an integration constant and f_0 a dimensional coupling constant[7]. This means that, for any $f(R) = R^s$, exists, at least, a Noether symmetry and a related constant of motion Σ_0 :

$$\Sigma_0 = \underline{\alpha} \cdot \nabla_{q'} \mathbf{L} = 2skMR^{2s-3}[2s + (s-1)MR][(s-2)RA' - (2s^2 - 3s + 1)AR'] .$$

[5]Lowering the dimension of configuration space through the substitution (19) does not affect the dynamics since B is a non-evolving quantity. In fact, inserting Eq. (19) into the dynamical equations given by (17), they coincide with those derived by (17).

[6]From now on, $\underline{q}$ indicates the vector $\{A, M, R\}$.

[7]The dimensions are given by R^{1-s} in terms of the Ricci scalar. For the sake of simplicity, we will put $f_0 = 1$ in the forthcoming discussion.

A physical interpretation of Σ_0 is possible if one gives an interpretation of this quantity in GR, that means for $f(R) = R$ and $s = 1$. In other words, the above procedure has to be applied to the Lagrangian of GR. We obtain the solution

$$\underline{\alpha}_{GR} = (-kA,\, kM)\,. \tag{24}$$

The functions A and M give the Schwarzschild solution, and then the constant of motion acquires the standard form

$$\Sigma_0 = \frac{2GM}{c^2}\,. \tag{25}$$

In other words, in the case of Einstein gravity, the Noether symmetry gives, as a conserved quantity, the Schwarzschild radius or the mass of the gravitating system. This result can be assumed as a consistency check.

In the general case, $f(R) = R^s$, the Lagrangian (17) becomes

$$\begin{aligned}\mathbf{L} \;=\;& \frac{sR^{2s-3}[2s+(s-1)MR]}{M}[2(s-1)M^2A'R' + 2MRM'A' + \\ & 4(s-1)AMM'R' + ARM'^2]\,,\end{aligned} \tag{26}$$

and the expression (19) for B is

$$B = \frac{s[2(s-1)M^2A'R' + 2MRM'A' + 4(s-1)AMM'R' + ARM'^2]}{2AMR[2s+(s-1)MR]} \tag{27}$$

As it can be easily checked, GR is recovered for $s = 1$.

Using the constant of motion (24), we solve in term of A and obtain

$$A = R^{\frac{2s^2-3s+1}{s-2}}\left\{k_1 + \Sigma_0 \int \frac{R^{\frac{4s^2-9s+5}{2-s}}dr}{2ks(s-2)M[2s+(s-1)MR]}\right\} \tag{28}$$

for $s \neq 2$ and k_1 an integration constant. For $s = 2$, one finds

$$A = -\frac{\Sigma_0}{12kr^2(4+r^2R)RR'}\,. \tag{29}$$

These relations allow to find out general black hole solutions for the field equations giving the dependence of the Ricci scalar on the radial coordinate r. For example, a solution is found for

$$s = 5/4\,, \quad M = r^2\,, \quad R = 5r^{-2}\,, \tag{30}$$

obtaining the spherically symmetric space-time

$$ds^2 = (\alpha+\beta r)dt^2 - \frac{1}{2}\frac{\beta r}{\alpha+\beta r}dr^2 - r^2 d\Omega\,, \tag{31}$$

where α is a combination of Σ_0 and k and $\beta = k_1$. In principle, the same procedure can be worked out any time Noether symmetries are identified. Our task is now to show how, from a spherically symmetric solution, one can generate an axially symmetric solution adopting the Newman-Janis procedure that works in GR. In general, the approach is not immediately straightforward since, as soon as $f(R) \neq R$, we are dealing with fourth-order field equations which have, in principle, different existence theorems and boundary conditions. However, the existence of the Noether symmetry guarantees the consistency of the chosen $f(R)$-model with the field equations.

4. Axial Symmetry Derived from Spherical Symmetry

We want to show now how it is possible to obtain an rotating solution starting from a spherically symmetric one adopting the method developed by Newman and Janis in GR. Such an algorithm can be applied to a static spherically symmetric metric considered as a"seed" metric. Let us recast the spherically symmetric metric (4) in the form

$$ds^2 = e^{2\phi(r)}dt^2 - e^{2\lambda(r)}dr^2 - r^2 d\Omega, \tag{32}$$

with $g_{tt}(t,r) = e^{2\phi(r)}$ and $g_{rr}(t,r) = e^{2\lambda(r)}$. Such a form is suitable for the considerations below. Following Newman and Janis, Eq. (32) can be written in the so called Eddington–Finkelstein coordinates (u,r,θ,ϕ), i.e. the g_{rr} component is eliminated by a change of coordinates and a cross term is introduced [41]. Specifically this is achieved by defining the time coordinate as $dt = du + F(r)dr$ and setting $F(r) = \pm e^{\lambda(r)-\phi(r)}$. Once such a transformation is performed, the metric (32) becomes

$$ds^2 = e^{2\phi(r)}du^2 \pm 2e^{\lambda(r)+\phi(r)}dudr - r^2 d\Omega. \tag{33}$$

The surface $u =$ costant is a light cone starting from the origin $r = 0$. The metric tensor for the line element (33) in null-coordinates is

$$g^{\mu\nu} = \begin{pmatrix} 0 & \pm e^{-\lambda(r)-\phi(r)} & 0 & 0 \\ \pm e^{-\lambda(r)-\phi(r)} & -e^{-2\lambda(r)} & 0 & 0 \\ 0 & 0 & -1/r^2 & 0 \\ 0 & 0 & 0 & -1/(r^2\sin^2\theta) \end{pmatrix}. \tag{34}$$

The matrix (34) can be written in terms of a null tetrad as

$$g^{\mu\nu} = l^\mu n^\nu + l^\nu n^\mu - m^\mu \bar{m}^\nu - m^\nu \bar{m}^\mu, \tag{35}$$

where l^μ, n^μ, m^μ and $\bar{m}^\mu$ are the vectors satisfying the conditions

$$l_\mu l^\mu = m_\mu m^\mu = n_\mu n^\mu = 0, \quad l_\mu n^\mu = -m_\mu \bar{m}^\mu = 1, \quad l_\mu m^\mu = n_\mu m^\mu = 0. \tag{36}$$

The bar indicates the complex conjugation. At any point in space, the tetrad can be chosen in the following manner: l^μ is the outward null vector tangent to the cone, n^μ is the inward null vector pointing toward the origin, and m^μ and $\bar{m}^\mu$ are the vectors tangent to the two-dimensional sphere defined by constant r and u. For the spacetime (34), the tetrad null vectors can be

$$\begin{cases} l^\mu = \delta^\mu_1 \\ n^\mu = -\frac{1}{2}e^{-2\lambda(r)}\delta^\mu_1 + e^{-\lambda(r)-\phi(r)}\delta^\mu_0 \\ m^\mu = \frac{1}{\sqrt{2}r}\left(\delta^\mu_2 + \frac{i}{\sin\theta}\delta^\mu_3\right) \\ \bar{m}^\mu = \frac{1}{\sqrt{2}r}\left(\delta^\mu_2 - \frac{i}{\sin\theta}\delta^\mu_3\right) \end{cases} \tag{37}$$

Now we need to extend the set of coordinates $x^\mu = (u, r, \theta, \phi)$ replacing the real radial coordinate by a complex variable. Then the tetrad null vectors become [8]

$$\begin{cases} l^\mu = \delta^\mu_1 \\ n^\mu = -\frac{1}{2}e^{-2\lambda(r,\bar{r})}\delta^\mu_1 + e^{-\lambda(r,\bar{r})-\phi(r,\bar{r})}\delta^\mu_0 \\ m^\mu = \frac{1}{\sqrt{2}\bar{r}}(\delta^\mu_2 + \frac{i}{\sin\theta}\delta^\mu_3) \\ \bar{m}^\mu = \frac{1}{\sqrt{2}r}(\delta^\mu_2 - \frac{i}{\sin\theta}\delta^\mu_3) \end{cases} \tag{38}$$

A new metric is obtained by making a complex coordinates transformation

$$x^\mu \to \tilde{x}^\mu = x^\mu + iy^\mu(x^\sigma), \tag{39}$$

where $y^\mu(x^\sigma)$ are analityc functions of the real coordinates x^σ, and simultaneously let the null tetrad vectors $Z^\mu_a = (l^\mu, n^\mu, m^\mu, \bar{m}^\mu)$, with $a = 1, 2, 3, 4$, undergo the transformation

$$Z^\mu_a \to \tilde{Z}^\mu_a(\tilde{x}^\sigma, \bar{\tilde{x}}^\sigma) = Z^\rho_a \frac{\partial \tilde{x}^\mu}{\partial x^\rho}. \tag{40}$$

Obviously, one has to recover the old tetrads and metric as soon as $\tilde{x}^\sigma = \bar{\tilde{x}}^\sigma$. In summary, the effect of the "*tilde transformation*" (39) is to generate a new metric whose components are (real) functions of complex variables, that is

$$g_{\mu\nu} \to \tilde{g}_{\mu\nu} : \tilde{\mathbf{x}} \times \tilde{\mathbf{x}} \mapsto \mathbb{R} \tag{41}$$

with

$$\tilde{Z}^\mu_a(\tilde{x}^\sigma, \bar{\tilde{x}}^\sigma)|_{\mathbf{x}=\tilde{\mathbf{x}}} = Z^\mu_a(x^\sigma). \tag{42}$$

For our aims, we can make the choice

$$\tilde{x}^\mu = x^\mu + ia(\delta^\mu_1 - \delta^\mu_0)\cos\theta \to \begin{cases} \tilde{u} = u + ia\cos\theta \\ \tilde{r} = r - ia\cos\theta \\ \tilde{\theta} = \theta \\ \tilde{\phi} = \phi \end{cases} \tag{43}$$

where a is constant and the tetrad null vectors (38), if we choose $\tilde{r} = \bar{\tilde{r}}$, become

[8]It is worth noticing that a certain arbitrariness is present in the complexification process of the functions λ and ϕ. Obviously, we have to obtain the metric (34) as soon as $r = \bar{r}$.

$$\begin{cases} \tilde{l}^{\mu} = \delta^{\mu}_{1} \\ \tilde{n}^{\mu} = -\frac{1}{2}e^{-2\lambda(\tilde{r},\theta)}\delta^{\mu}_{1} + e^{-\lambda(\tilde{r},\theta)-\phi(\tilde{r},\theta)}\delta^{\mu}_{0} \\ \tilde{m}^{\mu} = \frac{1}{\sqrt{2}(\tilde{r}-ia\cos\theta)}\left[ia(\delta^{\mu}_{0}-\delta^{\mu}_{1})\sin\theta + \delta^{\mu}_{2} + \frac{i}{\sin\theta}\delta^{\mu}_{3}\right] \\ \bar{\tilde{m}}^{\mu} = \frac{1}{\sqrt{2}(\tilde{r}+ia\cos\theta)}\left[-ia(\delta^{\mu}_{0}-\delta^{\mu}_{1})\sin\theta + \delta^{\mu}_{2} - \frac{i}{\sin\theta}\delta^{\mu}_{3}\right] \end{cases} \tag{44}$$

From the transformed null tetrad vectors, a new metric is recovered using (35). For the null tetrad vectors given by (44) and the transformation given by (43), the new metric, with coordinates $\tilde{x}^{\mu} = (\tilde{u},\tilde{r},\theta,\phi)$, is

$$\tilde{g}^{\mu\nu} = \begin{pmatrix} -\frac{a^2\sin^2\theta}{\Sigma^2} & e^{-\lambda(\tilde{r},\theta)-\phi(\tilde{r},\theta)} + \frac{a^2\sin^2\theta}{\Sigma^2} & 0 & -\frac{a}{\Sigma^2} \\ \cdot & -e^{-2\lambda(\tilde{r},\theta)} - \frac{a^2\sin^2\theta}{\Sigma^2} & 0 & \frac{a}{\Sigma^2} \\ \cdot & \cdot & -\frac{1}{\Sigma^2} & 0 \\ \cdot & \cdot & \cdot & -\frac{1}{\Sigma^2\sin^2\theta} \end{pmatrix} \tag{45}$$

where $\Sigma = \sqrt{\tilde{r}^2 + a^2\cos^2\theta}$. In the covariant form, the metric (45) is

$$\begin{aligned} ds^2 \;=\; & e^{2\phi(\tilde{r},\theta)}d\tilde{u}^2 + 2e^{\lambda(\tilde{r},\theta)}d\tilde{u}d\tilde{r} + 2ae^{\phi(\tilde{r},\theta)}\left[e^{\lambda(\tilde{r},\theta)} - e^{\phi(\tilde{r},\theta)}\right]\sin^2\theta d\tilde{u}d\phi + \\ & -2ae^{\phi(\tilde{r},\theta)+\lambda(\tilde{r},\theta)}\sin^2\theta d\tilde{r}^2 d\phi - Sigma^2 d\theta^2 + \\ & -[\Sigma^2 + a^2\sin^2\theta e^{\phi(\tilde{r},\theta)}(2e^{\lambda(\tilde{r},\theta)} - e^{\phi(\tilde{r},\theta)})]\sin^2\theta d\phi^2 \end{aligned} \tag{46}$$

Since the metric is symmetric, the dots in the matrix are used to indicate $g^{\mu\nu} = g^{\nu\mu}$. The form of this metric gives the general result of the Newman-Janis algorithm starting from any spherically symmetric "seed" metric.

The metric given in Eq. (46) can be simplified by a further gauge transformation so that the only off-diagonal component is $g_{\phi t}$. This procedure makes it easier to compare with the standard Boyer-Lindquist form of the Kerr metric [41] and to interpret physical properties such as the frame dragging. The coordinates $\tilde{u}$ and ϕ can be redefined in such a way that the metric in the new coordinate system has the properties described above. More explicitly, if we define the coordinates in the following way

$$d\tilde{u} = dt + g(\tilde{r})d\tilde{r} \;\text{ and }\; d\phi = d\phi + h(\tilde{r})d\tilde{r} \tag{47}$$

where

$$\begin{cases} g(\tilde{r}) = -\frac{e^{\lambda(\tilde{r},\theta)}(\Sigma^2+a^2\sin^2\theta e^{\lambda(\tilde{r},\theta)+\phi(\tilde{r},\theta)})}{e^{\phi(\tilde{r},\theta)}(\Sigma^2+a^2\sin^2\theta e^{2\lambda(\tilde{r},\theta)})} \\ h(\tilde{r}) = -\frac{ae^{2\lambda(\tilde{r},\theta)}}{\Sigma^2+a^2\sin^2\theta e^{2\lambda(\tilde{r},\theta)}} \end{cases} \tag{48}$$

after some algebraic manipulations, one finds that, in $(t,\tilde{r},\theta,\phi)$ coordinates system, the metric (46) becomes

$$
\begin{aligned}
ds^2 &= e^{2\phi(\tilde{r},\theta)}dt^2 + ae^{\phi(\tilde{r},\theta)}[e^{\lambda(\tilde{r},\theta)} - e^{\phi(\tilde{r},\theta)}]\sin^2\theta dt d\phi - \\
& \frac{\Sigma^2}{(\Sigma^2 e^{-2\lambda(\tilde{r},\theta)} + a^2\sin^2\theta)} d\tilde{r}^2 - \Sigma^2 d\theta^2 + \\
& -[\Sigma^2 + a^2\sin^2\theta e^{\phi(\tilde{r},\theta)}(2e^{\lambda(\tilde{r},\theta)} - e^{\phi(\tilde{r},\theta)})]\sin^2\theta d\phi^2
\end{aligned} \tag{49}
$$

This metric represents the complete family of metrics that may be obtained by performing the Newman-Janis algorithm on any static spherically symmetric "seed" metric, written in Boyer-Lindquist type coordinates. The validity of these transformations requires the condition $\Sigma^2 + a^2\sin^2\theta e^{2\lambda(\tilde{r},\theta)} \neq 0$, where $e^{2\lambda(\tilde{r},\theta)} > 0$. Our task is now to show that such an approach can be used to derive axially symmetric solutions also in $f(R)$-gravity that, possibly, can be regarded as black hole solutions..

5. Axially Symmetric Solutions in $f(R)$-Gravity

Starting from the above spherically symmetric solution (31), the metric tensor, written in the Eddington–Finkelstein coordinates (u, r, θ, ϕ) of the form (34) is

$$
g^{\mu\nu} = \begin{pmatrix} 0 & \sqrt{\frac{2}{\beta r}} & 0 & 0 \\ . & -2 - \frac{2\alpha}{\beta r} & 0 & 0 \\ . & . & -1/r^2 & 0 \\ . & . & . & -1/(r^2\sin^2\theta) \end{pmatrix}. \tag{50}
$$

The complex tetrad null vectors (38) are now

$$
\begin{cases}
l^\mu = \delta_1^\mu \\
n^\mu = -\left[1 + \frac{\alpha}{\beta}\left(\frac{1}{\bar{r}} + \frac{1}{r}\right)\right]\delta_1^\mu + \sqrt{\frac{2}{\beta}}\frac{1}{\sqrt[4]{\bar{r}r}}\delta_0^\mu \\
m^\mu = \frac{1}{\sqrt{2}\bar{r}}(\delta_2^\mu + \frac{i}{\sin\theta}\delta_3^\mu)
\end{cases} . \tag{51}
$$

By computing the complex coordinates transformation (43), the tetrad null vectors become

$$
\begin{cases}
\tilde{l}^\mu = \delta_1^\mu \\
\tilde{n}^\mu = -\left[1 + \frac{\alpha}{\beta}\frac{\mathrm{Re}\{\tilde{r}\}}{\Sigma^2}\right]\delta_1^\mu + \sqrt{\frac{2}{\beta}}\frac{1}{\sqrt{\Sigma}}\delta_0^\mu \\
\tilde{m}^\mu = \frac{1}{\sqrt{2}(\tilde{r} + ia\cos\theta)}\left[ia(\delta_0^\mu - \delta_1^\mu)\sin\theta + \delta_2^\mu + \frac{i}{\sin\theta}\delta_3^\mu\right]
\end{cases} \tag{52}
$$

Now by performing the same procedure as in previous section, we derive an axially symmetric metric of the form (49) but starting from the spherically symmetric metric (31), that is

$$
\begin{aligned}
ds^2 &= \frac{r(\alpha+\beta r)+a^2\beta\cos^2\theta}{\Sigma}du^2+2\frac{a(-2\alpha r-2\beta\Sigma^2+\sqrt{2\beta}\Sigma^{3/2})\sin^2\theta}{2\Sigma}dud\phi+\\
&-\frac{\beta\Sigma^2}{2\alpha r+\beta(a^2+r^2+\Sigma^2)}dr^2-\Sigma^2d\theta^2+\\
&-\left[\Sigma^2-\frac{a^2(\alpha r+\beta\Sigma^2-\sqrt{2\beta}\Sigma^{3/2})\sin^2\theta}{\Sigma}\right]\sin^2\theta d\phi^2
\end{aligned}
\tag{53}
$$

It is worth noticing that the condition $a = 0$ immediately gives the metric (31). This is nothing else but an example: the method is general and can be extended to any spherically symmetric solution derived in $f(R)$-gravity.

5.1. Physical Applications: Geodesics and Orbits

Let us discuss now possible physical applications of the above results. We will take into account a freely falling particle moving in the space-time described by the metric (53). For our aims, we make explicit use of the Hamiltonian formalism. Given a metric $g_{\mu\nu}$, the motion along the geodesics is described by the Lagrangian

$$
\mathcal{L}(x^\mu,\dot{x}^\mu)=\frac{1}{2}g_{\mu\nu}\dot{x}^\mu\dot{x}^\nu\,, \tag{54}
$$

where the overdot stands for derivative with respect to an affine parameter λ used to parametrize the curve. The Hamiltonian description is achieved by considering the canonical momenta and the Hamiltonian function

$$
p_\mu=\frac{\partial\mathcal{L}}{\partial\dot{x}^\mu}=g^{\mu\nu}p_\mu p_\nu\,,\qquad \mathcal{H}=p_\mu\dot{x}^\mu-\mathcal{L}\,, \tag{55}
$$

that results $\mathcal{H}=\frac{1}{2}p_\mu p_\nu g^{\mu\nu}$. The advantage of the Hamiltonian formalism with respect to the Lagrangian one is that the resulting equations of motion do not contain any sign ambiguity coming from turning points in the orbits (see, for example, [42]) . The Hamiltonian results explicitly independent of time and it is $\mathcal{H}=-\frac{1}{2}m^2$, where the rest mass m is a constant ($m=0$ for photons). The geodesic equations are

$$
\frac{dx^\mu}{d\lambda}=\frac{\partial\mathcal{H}}{\partial p_\mu}=g^{\mu\nu}p_\nu=p^\mu, \tag{56}
$$

$$
\frac{dp_\mu}{d\lambda}=-\frac{\partial\mathcal{H}}{\partial x\mu}=-\frac{1}{2}\frac{\partial g^{\alpha\beta}}{\partial x^\mu}p_\alpha p_\beta=g^{\gamma\beta}\Gamma^\alpha_{\mu\gamma}p_\alpha p_\beta. \tag{57}
$$

In addition, since the Hamiltonian is independent of the affine parameter λ, one can directly use the coordinate time as integration parameter. The problem is so reduced to solve six equations of motion. Using the above definitions, it is easy to achieve the reduced Hamiltonian (now linear in the momenta)

$$H=-p_0=\left[\frac{p_i g^{0i}}{g^{00}}+\left[\left(\frac{p_i g^{0i}}{g^{00}}\right)^2-\frac{m^2+p_i p_j g^{ij}}{g^{00}}\right]^{1/2}\right] \tag{58}$$

with the equations of motion

$$\frac{dx^i}{dt}=\frac{\partial H}{\partial p_i}\,, \qquad \frac{dp_i}{dt}=-\frac{\partial H}{\partial x^i}\,, \tag{59}$$

that give the orbits. The method can be applied to the above solution (53) by which the elements of the inverse metric can be easily obtained:

$$\begin{aligned}
g^{tt} &= \left\{4\Sigma^2\left[\Sigma^2-\frac{a^2\sin^2\theta\left(r\alpha-\sqrt{2}\sqrt{\beta}\Sigma^{3/2}+\beta\Sigma^2\right)}{\Sigma}\right]\right\}\times\\
&\left\{a^2\sin^2\theta\left(2r\alpha-\sqrt{2}\sqrt{\beta}\Sigma^{3/2}+2\beta\Sigma^2\right)^2+4\Sigma\left[a^2\beta\cos^2\theta+r(r\beta+\alpha)\right]\times\right.\\
&\left.\left(\Sigma^2-\frac{a^2\sin^2\theta\left(r\alpha-\sqrt{2}\sqrt{\beta}\Sigma^{3/2}+\beta\Sigma^2\right)}{\Sigma}\right)\right\}^{-1},\\
g^{rr} &= -\frac{\beta\left(a^2+r^2+\Sigma^2\right)+2r\alpha}{\beta\Sigma^2},\\
g^{\theta\theta} &= -\frac{1}{\Sigma^2},\\
g^{t\phi} &= \left\{2a\Sigma\left(-2r\alpha+\sqrt{2}\sqrt{\beta}\Sigma^{3/2}-2\beta\Sigma^2\right)\right\}\times\\
&\left\{a^2\sin^2\theta\left(2r\alpha-\sqrt{2}\sqrt{\beta}\Sigma^{3/2}+2\beta\Sigma^2\right)^2+4\Sigma\left[a^2\beta\cos^2\theta+r(r\beta+\alpha)\right]\times\right.\\
&\left.\left(\Sigma^2-\frac{a^2\sin^2\theta\left(r\alpha-\sqrt{2}\sqrt{\beta}\Sigma^{3/2}+\beta\Sigma^2\right)}{\Sigma}\right)\right\}^{-1},\\
g^{\phi\phi} &= -\left\{4\Sigma\csc^2\theta\left[a^2\beta\cos^2\theta+r(r\beta+\alpha)\right]\right\}\times\\
&\left\{a^2\sin^2\theta\left(2r\alpha-\sqrt{2}\sqrt{\beta}\Sigma^{3/2}+2\beta\Sigma^2\right)^2+4\Sigma\left[a^2\beta\cos^2\theta+r(r\beta+\alpha)\right]\times\right.\\
&\left.\left(\Sigma^2-\frac{a^2\sin^2\theta\left(r\alpha-\sqrt{2}\sqrt{\beta}\Sigma^{3/2}+\beta\Sigma^2\right)}{\Sigma}\right)\right\}^{-1},
\end{aligned} \tag{60}$$

and the null ones

$$g^{tr}=g^{t\theta}=g^{r\theta}=g^{r\phi}=g^{\theta\phi}=0\,. \tag{61}$$

Let us consider the equatorial plane, *i.e.* $\theta=\frac{\pi}{2}$, $\dot{\theta}=0$, and assume $\alpha=1$ and $\beta=2$. The reduced Hamiltonian $H(r,\theta,\phi,p_r,p_\theta,p_\phi;t)=H$ can be written as

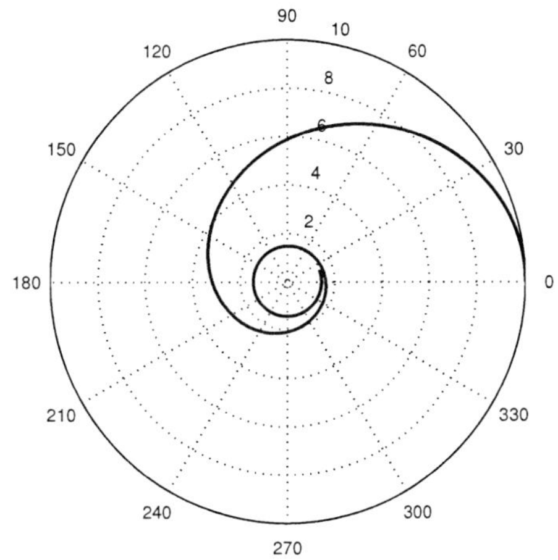

Figure 1. Relative motion of the test particle with $m = 1$.

$$H = \frac{2ap_\phi\left(-2r^3+r^2-1\right)}{a^2\left(-2(r-1)r^2-1\right)+r^5} + \Bigg\{\Bigg[\left(4a^2p_\phi^2\left(-2r^3+r^2-1\right)^2 + \right. \\ \left. -a^2\left(-2(r-1)r^2-1\right)-r^5\right)\left(a^2\left(r^2(r(2r-3)(2r+1)+6)-2\right)+ \right. \\ \left. +(2r+1)r^4\right) \times \\ \left(-\frac{p_\phi(2r+1)}{a^2\left(r^2(r(2r-3)(2r+1)+6)-2\right)+(2r+1)r^4} + \right. \\ \left. -\frac{p_r\left(a^2+r^2+r\right)+p_\theta}{r^4} - p_r + 1\right)\Bigg]\Bigg\}^{\frac{1}{2}}. \tag{62}$$

It is independent of ϕ (*i.e.* we are considering an azimuthally symmetric spacetime), and then the conjugate momentum p_ϕ is an integral of motion. From Eqs. (59), one can derive the coupled equations for $\{r, \theta, \phi, p_r, p_\theta\}$ and integrate them numerically (the expressions are very cumbersome and will not be reported here). To this goal, we have to specify the initial value of the position-momentum vector in the phase space. A Runge-Kutta method can be used to solve the differential equations. In Fig 1, the relative trajectories are sketched.

6. Hydrostatic Equilibrium of Stellar Structures

We have shown that it is possible to derive black hole spherically and axially symmetric solutions in $f(R)$-gravity. The physics of black holes contains a lot of unsolved problems. One of them is the question on the nature of the black holes as well as the nature of l stellar objects presenting anomalous behaviors (e.g. the magnetars) that do not obey the standard dynamics of (relativistic and non-relativistic) stellar structures. For this purpose, it is necessary to understand the hydrostatic equilibrium of such objects, that is one of fundamental properties (or one of the basic assumptions) of self-gravitating systems. Here we start by describing the standard hydrostatic equilibrium assuming Newtonian (i.e. the GR weak field limit) and then we compare these results with the equilibrium derived from the weak field limit of $f(R)$-gravity.

The condition of hydrostatic equilibrium for stellar structures in Newtonian dynamics is achieved by considering the equation

$$\frac{dp}{dr} = \frac{d\Phi}{dr}\rho, \tag{63}$$

where p is the pressure, $-\Phi$ is the gravitational potential, and ρ is the density [43]. Together with the above equation, the Poisson equation

$$\frac{1}{r^2}\frac{d}{dr}\left(r^2\frac{d\Phi}{dr}\right) = -4\pi G\rho, \tag{64}$$

gives the gravitational potential as solution for a given matter density ρ. Since we are taking into account only static and stationary situations, here we consider only time-independent solutions [9]. In general, the temperature τ appears in Eqs. (63) and (64) the density satisfies an equation of state of the form $\rho = \rho(p,\tau)$. In any case, we assume that there exists a polytropic relation between p and ρ of the form

$$p = K\rho^{\gamma}, \tag{65}$$

where K and γ are constant. Note that $\Phi > 0$ in the interior of the model since we define the gravitational potential as $-\Phi$. The polytropic constant K is fixed and can be obtained as a combination of fundamental constants. However there are several realistic cases where K is not fixed and another equation for its evolution is needed. The constant γ is the *polytropic exponent* . Inserting the polytropic equation of state into Eq. (63), we obtain

$$\frac{d\Phi}{dr} = \gamma K\rho^{\gamma-2}\frac{d\rho}{dr}. \tag{66}$$

For $\gamma \neq 1$, the above equation can be integrated giving

$$\frac{\gamma K}{\gamma-1}\rho^{\gamma-1} = \Phi \rightarrow \rho = \left[\frac{\gamma-1}{\gamma K}\right]^{\frac{1}{\gamma-1}}\Phi^{\frac{1}{\gamma-1}} \doteq A_n\Phi^n \tag{67}$$

where we have chosen the integration constant to give $\Phi = 0$ at surface ($\rho = 0$). The constant n is called the *polytropic index* and is defined as $n = \frac{1}{\gamma-1}$. Inserting the relation (67) into the Poisson equation, we obtain a differential equation for the gravitational potential

$$\frac{d^2\Phi}{dr^2} + \frac{2}{r}\frac{d\Phi}{dr} = -4\pi G A_n\Phi^n. \tag{68}$$

[9]The radius r is assumed as the spatial coordinate. It varies from $r = 0$ at the center to $r = \xi$ at the surface of the star

Let us define now the dimensionless variables

$$\begin{cases} z = |\mathbf{x}|\sqrt{\frac{XA_n\Phi_c^{n-1}}{2}} \\ w(z) = \frac{\Phi}{\Phi_c} = \left(\frac{\rho}{\rho_c}\right)^{\frac{1}{n}} \end{cases} \tag{69}$$

where the subscript c refers to the center of the star and the relation between ρ and Φ is given by Eq. (67). At the center $(r = 0)$, we have $z = 0$, $\Phi = \Phi_c$, $\rho = \rho_c$ and therefore $w = 1$. Then Eq. (68) can be written

$$\frac{d^2w}{dz^2} + \frac{2}{z}\frac{dw}{dz} + w^n = 0 \tag{70}$$

This is the standard *Lane-Embden equation* describing the hydrostatic equilibrium of stellar structures in the Newtonian theory [43]. Now we want to compare this standard result with the one coming from $f(R)$-gravity.

7. The Newtonian Limit of $f(R)$-Gravity

In order to achieve the Newtonian limit of the theory the metric tensor $g_{\mu\nu}$ have to be approximated as follows

$$g_{\mu\nu} \sim \begin{pmatrix} 1 - 2\Phi(t,\mathbf{x}) + O(4) & O(3) \\ O(3) & -\delta_{ij} + O(2) \end{pmatrix} \tag{71}$$

where $O(n)$ (with n = integer) denotes the order of the expansion (see [44] for details). The set of coordinates[10] adopted is $x^\mu = (t, x^1, x^2, x^3)$. The Ricci scalar formally becomes

$$R \sim R^{(2)}(t,\mathbf{x}) + O(4) \tag{72}$$

The n-th derivative of Ricci function can be developed as

$$f^n(R) \sim f^n(R^{(2)} + O(4)) \sim f^n(0) + f^{n+1}(0)R^{(2)} + O(4) \tag{73}$$

here $R^{(n)}$ denotes a quantity of order $O(n)$. From lowest order of field equations, we have $f(0) = 0$ which trivially follows from the above assumption (71) for the metric. This means that the space-time is asymptotically Minkowskian and we are discarding a cosmological constant term in this analysis[11]. Field eqautaions at $O(2)$-order, that is at Newtonian level, are

[10]The Greek index runs between 0 and 3; the Latin index between 1 and 3.

[11]This assumption is quite natural since the contribution of a cosmological constant term is irrelevant at stellar level.

$$\begin{cases} R_{tt}^{(2)} - \frac{R^{(2)}}{2} - f''(0)\triangle R^{(2)} = \mathcal{X}\, T_{tt}^{(0)} \\ \\ -3f''(0)\triangle R^{(2)} - R^{(2)} = \mathcal{X}\, T^{(0)} \end{cases} \tag{74}$$

where $\triangle$ is the Laplacian in the flat space, $R_{tt}^{(2)} = -\triangle\Phi(t,\mathbf{x})$ and, for the sake of simplicity, we set $f'(0) = 1$. We recall that the energy-momentum tensor for a perfect fluid is

$$T_{\mu\nu} = (\varepsilon + p)\, u_\mu u_\nu - p\, g_{\mu\nu} \tag{75}$$

where p is the pressure and ε is the energy density. Being the pressure contribution negligible in the field equations in Newtonian approximation, we have

$$\begin{cases} \triangle\Phi + \frac{R^{(2)}}{2} + f''(0)\triangle R^{(2)} = -\mathcal{X}\rho \\ \\ 3f''(0)\triangle R^{(2)} + R^{(2)} = -\mathcal{X}\rho \end{cases} \tag{76}$$

where ρ is now the mass density[12]. We note that for $f''(0) = 0$ we have the standard Poisson equation: $\triangle\Phi = -4\pi G\rho$. This means that as soon as the second derivative of $f(R)$ is different from zero, deviations from the Newtonian limit of GR emerge.

The gravitational potential $-\Phi$, solution of Eqs. (76), has in general a Yukawa-like behavior depending on a characteristic length on which it evolves [44]. Then as it is evident the Gauss theorem is not valid[13] since the force law is not $\propto |\mathbf{x}|^{-2}$. The equivalence between a spherically symmetric distribution and point-like distribution is not valid and how the matter is distributed in the space is very important [44–46].

Besides the Birkhoff theorem results modified at Newtonian level: the solution can be only factorized by a space-depending function and an arbitrary time-depending function [44]. Furthermore the correction to the gravitational potential is depending on the only first two derivatives of $f(R)$ in $R = 0$. This means that different analytical theories, from the third derivative perturbation terms on, admit the same Newtonian limit [44, 45].

Eqs. (76) can be considered the *modified Poisson equation* for $f(R)$-gravity. They do not depend on gauge condition choice [46]. We know that

$$R^{(2)} \simeq \frac{1}{2}\nabla^2 g_{00}^{(2)} - \frac{1}{2}\nabla^2 g_{ii}^{(2)}\,. \tag{77}$$

Inserting in the above result the $g_{\mu\nu}$ approximations (71) we obtain

$$R^{(2)} \simeq \nabla^2(\Phi - \Psi). \tag{78}$$

Finally, we obtain the field equations

$$\nabla^2\Phi + \nabla^2\Psi - 2f''(0)\nabla^4\Phi + 2f''(0)\nabla^4\Psi = 2\mathcal{X}\rho \tag{79}$$

$$\nabla^2\Phi - \nabla^2\Psi + 3f''(0)\nabla^4\Phi - 3f''(0)\nabla^4\Psi = -\mathcal{X}\rho\,. \tag{80}$$

[12]Generally it is $\varepsilon = \rho c^2$.

[13]It is worth noticing that also if the Gauss theorem does not hold, the Bianchi identities are always valid so the conservation laws are guaranteed.

By eliminating the higher-order terms, the standard Poisson equation is recovered. This last step, will be useful to calculate the Jeans instability.

8. Stellar Hydrostatic Equilibrium in $f(R)$-Gravity

From the Bianchi identity, satisfied by the field equations, we have

$$T^{\mu\nu}{}_{;\mu} = 0 \;\rightarrow\; \frac{\partial p}{\partial x^k} = -\frac{1}{2}(p+\varepsilon)\frac{\partial \ln g_{tt}}{\partial x^k} \tag{81}$$

If the dependence on the temperature τ is negligible, *i.e.* $\rho = \rho(p)$, this relation can be introduced into Eqs. (76), which become a system of three equations for p, Φ and $R^{(2)}$ and can be solved without the other structure equations.

Let us suppose that matter satisfies still a polytropic equation $p = K\rho^{\gamma}$. If we introduce Eq.(67) into Eqs. (76) we obtain an integro-differential equation for the gravitational potential $-\Phi$, that is

$$\triangle\Phi(\mathbf{x}) + \frac{2XA_n}{3}\Phi(\mathbf{x})^n = -\frac{m^2XA_n}{6}\int d^3\mathbf{x}'\,\mathcal{G}(\mathbf{x},\mathbf{x}')\Phi(\mathbf{x}')^n \tag{82}$$

where $\mathcal{G}(\mathbf{x},\mathbf{x}') = -\frac{1}{4\pi}\frac{e^{-m|\mathbf{x}-\mathbf{x}'|}}{|\mathbf{x}-\mathbf{x}'|}$ is the Green function of the field operator $\triangle_{\mathbf{x}} - m^2$ for systems with spherical symmetry and $m^2 = -\frac{1}{3f''(0)}$ (for details see [45,46]). The integro-differential nature of Eq.(82) is the proof of the non-viability of Gauss theorem for $f(R)$-gravity. Adopting again the dimensionless variables

$$\begin{cases} z = \frac{|\mathbf{x}|}{\xi_0} \\ \\ w(z) = \frac{\Phi}{\Phi_c} \end{cases} \tag{83}$$

where

$$\xi_0 \doteq \sqrt{\frac{3}{2XA_n\Phi_c^{n-1}}} \tag{84}$$

is a characteristic length linked to stellar radius ξ, Eq. (82) becomes

$$\frac{d^2w(z)}{dz^2} + \frac{2}{z}\frac{dw(z)}{dz} + w(z)^n = \frac{m\xi_0}{8}\frac{1}{z}\int_0^{\xi/\xi_0} dz'\,z'\left\{e^{-m\xi_0|z-z'|} - e^{-m\xi_0|z+z'|}\right\}w(z')^n \tag{85}$$

which is the *modified Lané-Emden equation* deduced from $f(R)$-gravity. Clearly the particular $f(R)$-model is specified by the parameters m and ξ_0. If $m \rightarrow \infty$ (*i.e.* $f(R) \rightarrow R$),

Eq. (85) becomes Eq. (70). We are only interested in solutions of Eq. (85) that are finite at the center, that is for $z = 0$. Since the center must be an equilibrium point, the gravitational acceleration $|\mathbf{g}| = -d\Phi/dr \propto dw/dz$ must vanish for $w'(0) = 0$. Let us assume we have solutions $w(z)$ of Eq.(85) that fulfill the boundary conditions $w(0) = 1$ and $w(\xi/\xi_0) = 0$; then according to the choice (83), the radial distribution of density is given by

$$\rho(|\mathbf{x}|) = \rho_c w^n, \quad \rho_c = A_n \Phi_c{}^n \tag{86}$$

and the pressure by

$$p(|\mathbf{x}|) = p_c w^{n+1}, \quad p_c = K \rho_c{}^\gamma \tag{87}$$

For $\gamma = 1$ (or $n = \infty$) the integro-differential Eq. (85) is not correct. This means that the theory does not contain the case of isothermal sphere of ideal gas. In this case, the polytropic relation is $p = K\rho$. Putting this relation into Eq.(81) we have

$$\frac{\Phi}{K} = \ln\rho - \ln\rho_c \quad \rightarrow \quad \rho = \rho_c e^{\Phi/K} \tag{88}$$

where the constant of integration is chosen in such a way that the gravitational potential is zero at the center. If we introduce Eq.(88) into Eqs. (76), we have

$$\triangle\Phi(\mathbf{x}) + \frac{2X\rho_c}{3} e^{\Phi(\mathbf{x})/K} = -\frac{m^2 X\rho_c}{6} \int d^3\mathbf{x}' \mathcal{G}(\mathbf{x},\mathbf{x}') e^{\Phi(\mathbf{x}')/K} \tag{89}$$

Assuming the dimensionless variables $z = \frac{|\mathbf{x}|}{\xi_1}$ and $w(z) = \frac{\Phi}{K}$ where $\xi_1 \doteq \sqrt{\frac{3K}{2X\rho_c}}$, Eq. (89) becomes

$$\frac{d^2w(z)}{dz^2} + \frac{2}{z}\frac{dw(z)}{dz} + e^{w(z)} = \frac{m\xi_1}{8}\frac{1}{z}\int_0^{\xi/\xi_1} dz' z' \left\{ e^{-m\xi_1|z-z'|} - e^{-m\xi_1|z+z'|} \right\} e^{w(z')} \tag{90}$$

which is the *modified "isothermal" Lané-Emden equation* derived $f(R)$-gravity.

9. Solutions of the Standard and Modified Lané-Emden Equations

The task is now to solve the modified Lané-Emden equation and compare its solutions to those of standard Newtonian theory. Only for three values of n, the solutions of Eq.(70) have analytical expressions [43]

$$\begin{aligned} n = 0 \quad &\rightarrow \quad w_{GR}^{(0)}(z) = 1 - \frac{z^2}{6} \\ n = 1 \quad &\rightarrow \quad w_{GR}^{(1)}(z) = \frac{\sin z}{z} \\ n = 5 \quad &\rightarrow \quad w_{GR}^{(5)}(z) = \frac{1}{\sqrt{1+\frac{z^2}{3}}} \end{aligned} \tag{91}$$

We label these solution with $_{GR}$ since they agree with the Newtonian limit of GR. The surface of the polytrope of index n is defined by the value $z = z^{(n)}$, where $\rho = 0$ and thus $w = 0$. For $n = 0$ and $n = 1$ the surface is reached for a finite value of $z^{(n)}$. The case $n = 5$ yields a model of infinite radius. It can be shown that for $n < 5$ the radius of polytropic models is finite; for $n > 5$ they have infinite radius. From Eqs.(91) one finds $z_{GR}^{(0)} = \sqrt{6}$, $z_{GR}^{(1)} = \pi$ and $z_{GR}^{(5)} = \infty$. A general property of the solutions is that $z^{(n)}$ grows monotonically with the polytropic index n. In Fig. 2 we show the behavior of solutions $w_{GR}^{(n)}$ for $n = 0, 1, 5$. Apart from the three cases where analytic solutions are known, the classical Lané-Emden Eq. (70) has to be be solved numerically, considering with the expression

$$w_{GR}^{(n)}(z) = \sum_{i=0}^{\infty} a_i^{(n)} z^i \tag{92}$$

for the neighborhood of the center. Inserting Eq.(92) into Eq. (70) and by comparing coefficients one finds, at lowest orders, a classification of solutions by the index n, that is

$$w_{GR}^{(n)}(z) = 1 - \frac{z^2}{6} + \frac{n}{120} z^4 + \ldots \tag{93}$$

The case $\gamma = 5/3$ and $n = 3/2$ is the non-relativistic limit while the case $\gamma = 4/3$ and $n = 3$ is the relativistic limit of a completely degenerate gas.

Also for modified Lané-Emden Eq. (85), we have an exact solution for $n = 0$. In fact, it is straightforward to find out

$$w_{f(R)}^{(0)}(z) = 1 - \frac{z^2}{8} + \frac{(1+m\xi)e^{-m\xi}}{4m^2{\xi_0}^2}\left[1 - \frac{\sinh m\xi_0 z}{m\xi_0 z}\right] \tag{94}$$

where the boundary conditions $w(0) = 1$ and $w'(0) = 0$ are satisfied. A comment on the GR limit (that is $f(R) \to R$) of solution (94) is necessary. In fact when we perform the limit $m \to \infty$, we do not recover exactly $w_{GR}^{(0)}(z)$. The difference is in the definition of quantity ξ_0. In $f(R)$-gravity we have the definition (84) while in GR it is $\xi_0 = \sqrt{\frac{2}{XA_n\Phi_c^{n-1}}}$, since in the first equation of (76), when we perform $f(R) \to R$, we have to eliminate the trace equation condition. In general, this means that the Newtonian limit and the Eddington parameterization of different relativistic theories of gravity cannot coincide with those of GR (see [47] for further details on this point).

The point $z_{f(R)}^{(0)}$ is calculated by imposing $w_{f(R)}^{(0)}(z_{f(R)}^{(0)}) = 0$ and by considering the Taylor expansion

$$\frac{\sinh m\xi_0 z}{m\xi_0 z} \sim 1 + \frac{1}{6}(m\xi_0 z)^2 + O(m\xi_0 z)^4 \tag{95}$$

we obtain $z_{f(R)}^{(0)} = \frac{2\sqrt{6}}{\sqrt{3+(1+m\xi)e^{-m\xi}}}$. Since the stellar radius ξ is given by definition $\xi = \xi_0 z_{f(R)}^{(0)}$, we obtain the constraint

$$\xi = \sqrt{\frac{3\Phi_c}{2\pi G}}\frac{1}{\sqrt{1+\frac{1+m\xi}{3}e^{-m\xi}}} \tag{96}$$

By solving numerically the constraint[14] Eq.(96), we find the modified expression of the radius ξ. If $m \to \infty$ we have the standard expression $\xi = \sqrt{\frac{3\Phi_c}{2\pi G}}$ valid for the Newtonian limit of GR. Besides, it is worth noticing that in the $f(R)$-gravity case, for $n = 0$, the radius is smaller than in GR. On the other hand, the gravitational potential $-\Phi$ gives rise to a deeper potential well than the corresponding Newtonian potential derived from GR [45].

In the case $n = 1$, Eq. (85) can be recast as follows

$$\frac{d^2\tilde{w}(z)}{dz^2}+\tilde{w}(z) = \frac{m\xi_0}{8}\int_0^{\xi/\xi_0} dz'\left\{e^{-m\xi_0|z-z'|}-e^{-m\xi_0|z+z'|}\right\}\tilde{w}(z') \tag{97}$$

where $\tilde{w} = zw$. If we consider the solution of (97) as a small perturbation to the one of GR, we have

$$\tilde{w}^{(1)}_{f(R)}(z) \sim \tilde{w}^{(1)}_{GR}(z)+e^{-m\xi}\Delta\tilde{w}^{(1)}_{f(R)}(z) \tag{98}$$

The coefficient $e^{-m\xi} < 1$ is the parameter with respect to which we perturb Eq. (97). Besides these position ensure us that when $m \to \infty$ the solution converge to something like $\tilde{w}^{(1)}_{GR}(z)$. Substituting Eq.(98) in Eq.(97), we have

$$\frac{d^2\Delta\tilde{w}^{(1)}_{f(R)}(z)}{dz^2}+\Delta\tilde{w}^{(1)}_{f(R)}(z) = \frac{m\xi_0\, e^{m\xi}}{8}\int_0^{\xi/\xi_0} dz'\left\{e^{-m\xi_0|z-z'|}-e^{-m\xi_0|z+z'|}\right\}\tilde{w}^{(1)}_{GR}(z') \tag{99}$$

and the solution is easily found

$$\begin{aligned} w^{(1)}_{f(R)}(z) \sim{}& \frac{\sin z}{z}\left\{1+\frac{m^2{\xi_0}^2}{8(1+m^2{\xi_0}^2)}\left[1+\frac{2\,e^{-m\xi}}{1+m^2{\xi_0}^2}(\cos\xi/\xi_0+m\xi_0\sin\xi/\xi_0)\right]\right\} \\ &-\frac{m^2{\xi_0}^2}{8(1+m^2{\xi_0}^2)}\left[\frac{2\,e^{-m\xi}}{1+m^2{\xi_0}^2}(\cos\xi/\xi_0+m\xi_0\sin\xi/\xi_0)\frac{\sinh m\xi_0 z}{m\xi_0 z}+\cos z\right] \end{aligned} \tag{100}$$

Also in this case, for $m \to \infty$, we do not recover exactly $w^{(1)}_{GR}(z)$. The reason is the same of previous $n = 0$ case [47]. Analytical solutions for other values of n are not available.

[14] In principle, there is a solution for any value of m.

To conclude this section, we report the gravitational potential profile generated by a spherically symmetric source of uniform mass with radius ξ. We can impose a mass density of the form

$$\rho = \frac{3M}{4\pi\xi^3}\Theta(\xi - |\mathbf{x}|) \tag{101}$$

where Θ is the Heaviside function and M is the mass [45,46]. By solving field Eqs. (76) *inside the star* and considering the boundary conditions $w(0) = 1$ and $w'(0) = 0$, we get

$$\begin{aligned} w_{f(R)}(z) &= \left[\frac{3}{2\xi} + \frac{1}{m^2\xi^3} - \frac{e^{-m\xi}(1+m\xi)}{m^2\xi^3}\right]^{-1} \times \\ &\times \left[\frac{3}{2\xi} + \frac{1}{m^2\xi^3} - \frac{\xi_0{}^2 z^2}{2\xi^3} - \frac{e^{-m\xi}(1+m\xi)}{m^2\xi^3}\frac{\sinh m\xi_0 z}{m\xi_0 z}\right] \end{aligned} \tag{102}$$

In the limit $m \to \infty$, we recover the GR case $w_{GR}(z) = 1 - \frac{\xi_0{}^2 z^2}{3\xi^2}$. In Fig. 2 we show the behaviors of $w^{(0)}_{f(R)}(z)$ and $w^{(1)}_{f(R)}(z)$ with respect to the corresponding GR cases. Furthermore, we plot the potential generated by a uniform spherically symmetric mass distribution in GR and $f(R)$-gravity and the case $w^{(5)}_{GR}(z)$.

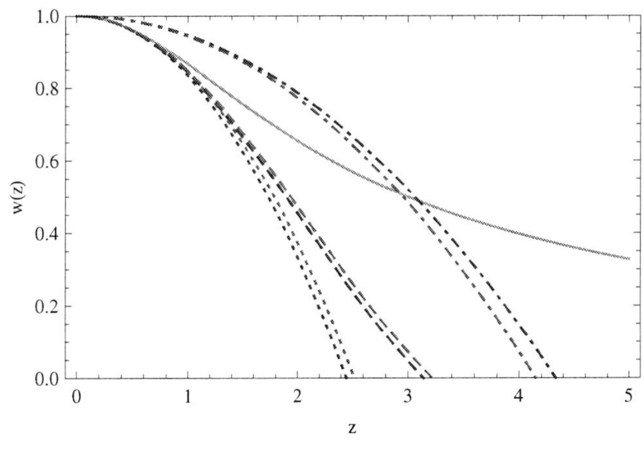

Figure 2. Plot of solutions (blue lines) of standard Lané-Emden Eq. (70): $w^{(0)}_{GR}(z)$ (dotted line) and $w^{(1)}_{GR}(z)$ (dashed line). The green line corresponds to $w^{(5)}_{GR}(z)$. The red lines are the solutions of modified Lané-Emden Eq. (85): $w^{(0)}_{f(R)}(z)$ (dotted line) and $w^{(1)}_{f(R)}(z)$ (dashed line). The blue dashed-dotted line is the potential derived from GR ($w_{GR}(z)$) and the red dashed-dotted line the potential derived from $f(R)$-gravity ($w_{f(R)}(z)$) for a uniform spherically symmetric mass distribution. The assumed values are $m\xi = 1$ and $m\xi_0 = 4$. From a rapid inspection of these plots, the differences between GR and $f(R)$ gravitational potentials are clear and the tendency is that at larger radius z they become more evident.

10. Examples: Dust-dominated Self-gravitating Systems

From the above equations of hydrostatic equilibrium in $f(R)$-gravity, one can study the formation and collapse of self gravitating objects like stars and black holes. It is well known that the scenario involving the formation of cosmic structures (*i.e.* from stars up to galaxies and clusters of galaxies) occurs for gravitational instability. In the standard picture, the first objects that form are the dark matter halos, which aggregate in a hierarchical way due to gravitational collapse. If such systems reach a virial equilibrium, the collapse can stop otherwise it can indefinitely continue. Subsequently, the baryons are affected by the gravity of the halo potential wells: gas, of collisional nature, converts the kinetic energy of the 'fall' into thermal energy and heat reaching the virial temperature. Subsequently, the radiation losses cause the cooling of the baryonic component, its condensation and subsequent formation of molecular clouds, finally of stars. The standard theory of gravitational collapse for dust dominated systems and can be compared to that for $f(R)$-gravity. The collapse of self-gravitational collisionless systems can be dealt with the introduction of coupled collisionless Boltzmann and Poisson equations (for details, see [48]):

$$\frac{\partial f(\vec{r},\vec{v},t)}{\partial t}+\left(\vec{v}\cdot\vec{\nabla}_r\right)f(\vec{r},\vec{v},t)-\left(\vec{\nabla}\Phi\cdot\vec{\nabla}_v\right)f(\vec{r},\vec{v},t)=0 \tag{103}$$

$$\vec{\nabla}^2\Phi(\vec{r},t)=4\pi G\int f(\vec{r},\vec{v},t)d\vec{v}. \tag{104}$$

A self-gravitating system at equilibrium is described by a time-independent distribution function $f_0(x,v)$ and a potential $\Phi_0(x)$ that are solutions of Eq.(103) and (104). Considering a small perturbation to this equilibrium:

$$f(\vec{r},\vec{v},t)=f_0(\vec{r},\vec{v})+\varepsilon f_1(\vec{r},\vec{v},t), \tag{105}$$

$$\Phi(\vec{r},t)=\Phi_0(\vec{r})+\varepsilon\Phi_1(\vec{r},t), \tag{106}$$

where $\varepsilon \ll 1$ and by substituting in Eq. (103) and (104) and by linearizing, one obtains:

$$\frac{\partial f_1(\vec{r},\vec{v},t)}{\partial t}+\vec{v}\cdot\frac{\partial f_1(\vec{r},\vec{v},t)}{\partial\vec{r}}-\vec{\nabla}\Phi_1(\vec{r},t)\cdot\frac{\partial f_0(\vec{r},\vec{v})}{\partial\vec{v}}-\vec{\nabla}\Phi_0(\vec{r})\cdot\frac{\partial f_1(\vec{r},\vec{v},t)}{\partial\vec{v}}=0, \tag{107}$$

$$\vec{\nabla}^2\Phi_1(\vec{r},t)=4\pi G\int f_1(\vec{r},\vec{v},t)d\vec{v}, \tag{108}$$

Since the equilibrium state is assumed to be homogeneous and time-independent, one can set $f_0(\vec{x},\vec{v},t)=f_0(\vec{v})$, and the so-called Jeans "swindle" to set $\Phi_0=0$. In Fourier components, Eqs.(107) and (108) become:

$$-i\omega f_1+\vec{v}\cdot\left(i\vec{k}f_1\right)-\left(i\vec{k}\Phi_1\right)\cdot\frac{\partial f_0}{\partial\vec{v}}=0, \tag{109}$$

$$-k^2\Phi_1=4\pi G\int f_1 d\vec{v}. \tag{110}$$

By combining these equations, the dispersion relation

$$1+\frac{4\pi G}{k^2}\int\frac{\vec{k}\cdot\dfrac{\partial f_0}{\partial\vec{v}}}{\vec{v}\cdot\vec{k}-\omega}d\vec{v}=0; \tag{111}$$

is obtained. In the case of stellar systems, by assuming a Maxwellian distribution function for f_0, we have

$$f_0 = \frac{\rho_0}{(2\pi\sigma^2)^{\frac{3}{2}}} e^{-\frac{v^2}{2\sigma^2}}, \tag{112}$$

imposing that $\vec{k} = (k, 0, 0)$ and substituting in Eq.(111), one gets:

$$1 - \frac{2\sqrt{2\pi}G\rho_0}{k\sigma^3} \int \frac{v_x e^{-\frac{v_x^2}{2\sigma^2}}}{kv_x - \omega} dv_x = 0. \tag{113}$$

By setting $\omega = 0$, the limit for instability is obtained:

$$k^2(\omega = 0) = \frac{4\pi G\rho_0}{\sigma^2} = k_J^2, \tag{114}$$

by which it is possible to define the Jeans mass (M_J) as the mass originally contained within a sphere of diameter λ_J:

$$M_J = \frac{4\pi}{3}\rho_0 \left(\frac{1}{2}\lambda_J\right)^3, \tag{115}$$

where

$$\lambda_J^2 = \frac{\pi\sigma^2}{G\rho_0} \tag{116}$$

is the Jeans length. Substituting Eq. (116) into Eq. (115), we recover

$$M_J = \frac{\pi}{6}\sqrt{\frac{1}{\rho_0}\left(\frac{\pi\sigma^2}{G}\right)^3}. \tag{117}$$

All perturbations with wavelengths $\lambda > \lambda_J$ are unstable in the stellar system. In order to evaluate the integral in the dispersion relation for real and nonzero values of ω, the dispersion relation has to be rewritten as

$$1 - \frac{k_J^2}{k^2} W\left(\frac{\omega}{k\sigma}\right) = 0, \tag{118}$$

defining

$$W\left(\frac{\omega}{k\sigma}\right) \equiv \frac{1}{\sqrt{2\pi}} \int \frac{x e^{-\frac{x^2}{2}}}{x - Z} dx, \tag{119}$$

and setting $\omega = i\omega_I$ and $Re(W\left(\frac{\omega}{k\sigma}\right)) = 0$. In order to study unstable modes (for details, see Appendix B in [48]) we replace the following identities

$$\begin{cases} \int\limits_0^\infty \frac{x^2 e^{-x^2}}{x^2 + \beta^2} dx = \frac{1}{2}\sqrt{\pi} - \frac{1}{2}\pi\beta e^{\beta^2}\left[1 - \text{erf}\beta\right], \\ \\ \text{erf}z = \frac{2}{\sqrt{\pi}} \int\limits_0^z e^{-t^2} dt. \end{cases}$$

into the dispersion relation obtaining:

$$k^2 = k_J^2 \left\{ 1 - \frac{\sqrt{\pi}\omega_I}{\sqrt{2}k\sigma} e^{\left(\frac{\omega_I}{\sqrt{2}k\sigma}\right)^2} \left[1 - \text{erf}\left(\frac{\omega_I}{\sqrt{2}k\sigma} \right) \right] \right\}. \tag{120}$$

This is the standard dispersion relation describing the criterion to collapse for infinite homogeneous fluid and stellar systems [48].

11. Jeans Criterion for Gravitational Instability in $f(R)$-Gravity

Our task is now to check how the Jeans instability occurs in $f(R)$-gravity. Let us approach the Jeans instability with the Poisson equations given by Eqs. (79) and (80) after assuming the collisionless Boltzmann equation:

$$\frac{\partial f(\vec{r},\vec{v},t)}{\partial t} + \left(\vec{v}\cdot\vec{\nabla}_r\right) f(\vec{r},\vec{v},t) - \left(\vec{\nabla}\Phi\cdot\vec{\nabla}_v\right) f(\vec{r},\vec{v},t) = 0\,. \tag{121}$$

Then we have

$$\nabla^2(\Phi+\Psi) - 2\alpha\nabla^4(\Phi-\Psi) = 16\pi G \int f(\vec{r},\vec{v},t) d\vec{v} \tag{122}$$

$$\nabla^2(\Phi-\Psi) + 3\alpha\nabla^4(\Phi-\Psi) = -8\pi G \int f(\vec{r},\vec{v},t) d\vec{v}\,. \tag{123}$$

In the previous equations, we have replaced $f''(0)$ with the greek letter α. As in standard case, we consider small perturbation to the equilibrium and linearize the equations. After we write equations in Fourier space so they became

$$-i\omega f_1 + \vec{v}\cdot\left(i\vec{k}f_1\right) - \left(i\vec{k}\Phi_1\right)\cdot\frac{\partial f_0}{\partial \vec{v}} = 0, \tag{124}$$

$$-k^2(\Phi_1+\Psi_1) - 2\alpha k^4(\Phi_1-\Psi_1) = 16\pi G \int f_1 d\vec{v}, \tag{125}$$

$$k^2(\Phi_1-\Psi_1) - 3\alpha k^4(\Phi_1-\Psi_1) = 8\pi G \int f_1 d\vec{v}. \tag{126}$$

Combining Eqs. (125) and(126), we obtain a relation between Φ_1 and Ψ_1,

$$\Psi_1 = \frac{3-4\alpha k^2}{1-4\alpha k^2}\Phi_1$$

inserting this relation in Eq. (125) and combining it with Eq. (124), we obtain the dispersion relation

$$1 - 4\pi G\frac{1-4\alpha k^2}{3\alpha k^4 - k^2} \int \left(\frac{\vec{k}\cdot\frac{\partial f_0}{\partial \vec{v}}}{\vec{v}\cdot\vec{k}-\omega} \right) d\vec{v} = 0. \tag{127}$$

If we assume, as in standard case, that f_0 is given by (112) and $\vec{k} = (k,0,0)$, one can write

$$1 + \frac{2\sqrt{2\pi}G\rho_0}{\sigma^3}\frac{1-4\alpha k^2}{3\alpha k^4 - k^2}\left[\int \frac{kv_x e^{-\frac{v_x^2}{2\sigma^2}}}{kv_x - \omega} dv_x \right] = 0. \tag{128}$$

By eliminating the higher-order terms (imposing $\alpha = 0$), we obtain again the standard dispersion Eq. (111). In order to compute the integral in the dispersion relation (128), we consider the same approach used in the classical case, and finally we obtain:

$$1 + \mathcal{G}\frac{1 - 4\alpha k^2}{3\alpha k^4 - k^2}\left[1 - \sqrt{\pi} x e^{x^2}\left(1 - \text{erf}[x]\right)\right] = 0, \tag{129}$$

where $x = \frac{\omega_i}{\sqrt{2}k\sigma}$ and $\mathcal{G} = \frac{4G\pi\rho_0}{\sigma^2}$. In order to evaluate Eq. (129) comparing it with the classical one, given by Eq. (111), it is very useful to normalize the equation to the classical Jeans length showed in Eq. (116), by fixing the parameter of $f(R)$-gravity, that is

$$\alpha = -\frac{1}{k_j^2} = -\frac{\sigma^2}{4\pi G\rho_0}. \tag{130}$$

This parameterization is correct because the dimension α (an inverse of squared length) allows us to parametrize as in the standard case. Finally we write

$$\frac{3k^4}{k_j^4} + \frac{k^2}{k_j^2} = \left(\frac{4k^2}{k_j^2} + 1\right)\left[1 - \sqrt{\pi} x e^{x^2}\left(1 - \text{erf}[x]\right)\right] = 0. \tag{131}$$

The function is plotted in Fig.3, where Eq. (129) and the standard dispersion [48] are confronted in order to see the difference between $f(R)$ and Newtonian gravity.

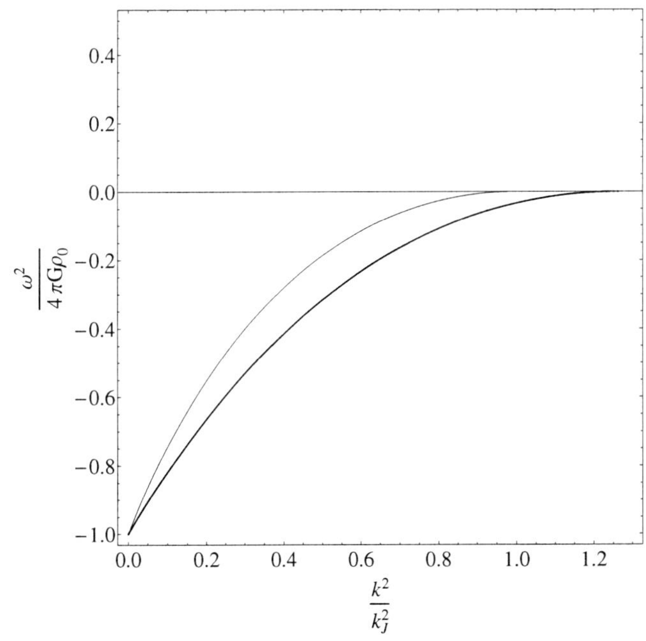

Figure 3. The bold line indicates the plot of the dispersion relation (129) in which we imposed the value for α given by (130). The thin line indicates the plot of the standard dispersion equation [48].

As shown in the Figure 3, the effects of a different theory of gravity changes the limit of instability. The limit is higher than the classical case and the curve has a greater slope. This fact is important because the mass limit value of interstellar clouds decreases changing the initial conditions to start the collapse. This feature could have dramatic effects for star and black hole formation.

12. The Jeans Mass Limit in $f(R)$-Gravity

A numerical estimation of the $f(R)$-instability length in terms of the standard Newtonian one can be achieved. By solving numerically Eq. (131) with the condition $\omega = 0$, we obtain that the collapse occurs for

$$k^2 = 1.2637 k_J^2. \tag{132}$$

However we can estimate also analytically the limit for the instability. In order to evaluate the Jeans mass limit in $f(R)$-gravity, we set $\omega = 0$ in Eq. (128) and then

$$3\sigma^2 \alpha k^4 - \left(16\pi G\rho_0\alpha + \sigma^2\right) k^2 + 4\pi G\rho_0 = 0. \tag{133}$$

It is worth stressing that the additional condition $\alpha < 0$ discriminates the class of viable $f(R)$ models: in such a case we obtain stable cosmological solution and positively defined massive states [49]. In other words, this condition selects the physically viable models allowing to solve Eq.(133) for real values of k. In particular, the above numerical solution can be recast as

$$k^2 = \frac{2}{3}\left(3+\sqrt{21}\right)\pi\frac{G\rho}{\sigma^2}. \tag{134}$$

The relation to the Newtonian value of the Jeans instability is

$$k^2 = \frac{1}{6}\left(3+\sqrt{21}\right)k_J^2. \tag{135}$$

Now, we can define the new Jeans mass as:

$$\tilde{M}_J = 6\sqrt{\frac{6}{\left(3+\sqrt{21}\right)^3}}M_J, \tag{136}$$

that is proportional to the standard Newtonian value. We will confront this specific solutions with some observed structures.

Before this comparison, some considerations are in order. Star formation is one of the best settled problems of modern astrophysics. However, some shortcomings emerge as soon as one faces dynamics of diffuse gas evolving into stars and star formation in galactic environment. One can deal with the star formation problem in two ways: *i*) we can take into account the formation of individual stars and *ii*) we can discuss the formation of the whole star system starting from interstellar clouds [50]. To answer these problems it is very important to study the interstellar medium (ISM) and its properties.The ISM physical conditions in the galaxies change in a very wide range, from hot X-ray emitting plasma to cold molecular gas, so it is very complicated to classify the ISM by its properties. However, we can distinguish, in the first approximation, between [51–54]:

- **Diffuse Hydrogen Clouds**. The most powerful tool to measure the properties of these clouds is the 21cm line emission of HI. They are cold clouds so the temperature is in the range $10 \div 50$ K, and their extension is up to $50 \div 100$ kpc from galactic center.

- **Diffuse Molecular Clouds** are generally self-gravitating, magnetized, turbulent fluids systems, observed in sub-mm. The most of the molecular gas is H_2, and the rest is CO. Here, the conditions are very similar to the HI clouds but in this case, the cloud can be more massive. They have, typically, masses in the range $3 \div 100 M_\odot$, temperature in $15 \div 50$K and particle density in $(5 \div 50) \times 10^8$ m^{-3}.

- **Giant Molecular Clouds** are very large complexes of particles (dust and gas), in which the range of the masses is typically $10^5 \div 10^6 M_\odot$ but they are very cold. The temperature is ~ 15K, and the number of particles is $(1 \div 3) \times 10^8$ m^{-3} [50,55–57]. However, there exist also small molecular clouds with masses $M < 10^4 M_\odot$ [57]. They are the best sites for star formation, despite the mechanism of formation does not recover the star formation rate that would be $250 M_\odot yr^{-1}$ [55].

- **HII regions**. They are ISM regions with temperatures in the range $10^3 \div 10^4$ K, emitting primarily in the radio and IR regions. At low frequencies, observations are associated to free-free electron transition (thermal Bremsstrahlung). Their densities range from over a million particles per cm^3 in the ultra-compact H II regions to only a few particles per cm^3 in the largest and most extended regions. This implies total masses between 10^2 and 10^5 M$_\odot$ [58].

- **Bok Globules** are dark clouds of dense cosmic dust and gas in which star formation sometimes takes place. Bok globules are found within H II regions, and typically have a mass of about 2 to 50 M$_\odot$ contained within a region of about a light year.

Using very general conditions [50–58], we want to show the difference in the Jeans mass value between standard and $f(R)$-gravity. Let us take into account Eq. (117) and Eq.(136):

$$M_J = \frac{\pi}{6}\sqrt{\frac{1}{\rho_0}\left(\frac{\pi\sigma^2}{G}\right)^3}, \tag{137}$$

in which ρ_0 is the ISM density and σ is the velocity dispersion of particles due to the temperature. These two quantities are defined as

$$\rho_0 = m_H n_H \mu, \qquad \sigma^2 = \frac{\mathrm{k_B T}}{\mathrm{m_H}}$$

where n_H is the number of particles measured in m^{-3}, μ is the mean molecular weight, k_B is the Boltzmann constant and m_H is the proton mass. By using these relations, we are able to compute the Jeans mass for interstellar clouds and to show the behavior of the Jeans mass with the temperature. Results are shown in Tab.1 and Fig.4

By using Eq.(136) and by referring to the catalog of molecular clouds in Roman-Duval J. et al. [59], we have calculated the Jeans mass in the Newtonian and $f(R)$ cases. Tab.2 shows the results. In all cases we note a substantial difference between the classical and

Table 1. Jeans masses derived from Eq. (117) **(Newtonian gravity) and** (136) **($f(R)$-gravity)**

Subject	T (K)	n (10^8m^{-3})	μ	M_J ($M_\odot$)	$\tilde{M}_J$ ($M_\odot$)
Diffuse Hydrogen Clouds	50	5.0	1	795.13	559.68
Diffuse Molecular Clouds	30	50	2	82.63	58.16
Giant Molecular Clouds	15	1.0	2	206.58	145.41
Bok Globules	10	100	2	11.24	7.91

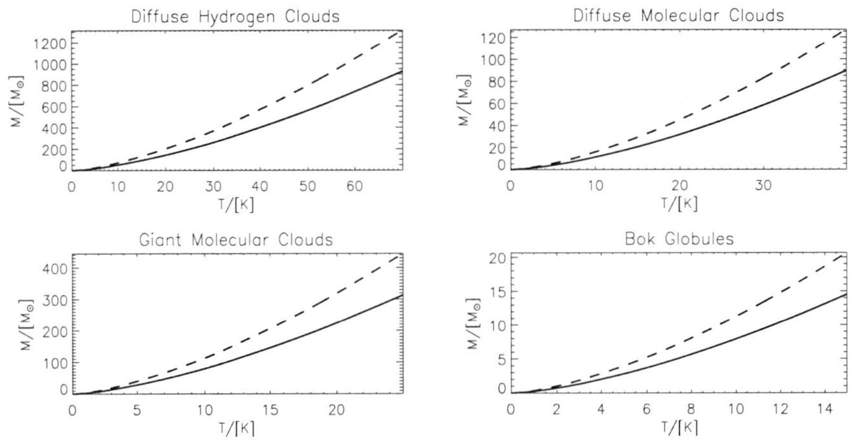

Figure 4. The M_J-T relation. Dashed-line indicates the Newtonian Jeans mass behavior with respect to the temperature. Continue-line indicates the same for $f(R)$-gravity Jeans mass.

$f(R)$ value. We can conclude that, in $f(R)$-scenario, molecular clouds become sites where star formation is strongly supported and more efficient.

13. Discussion and Conclusions

Black hole solutions can be find out, in principle, in any Extended Theory of Gravity. Here, we have shown that exact solutions can be achieved in $f(R)$-gravity starting from symmetry considerations. In particular, the existence of Noether symmetries allow to find exact solutions. Such solutions can be eventually classified as black holes.

In particular, we have shown that the Newman-Janis method, used to derive rotating black hole solutions in GR, works also in $f(R)$-gravity. In principle, it could be consistently applied any time a spherically symmetric solutions is derived. The method does not depend on the field equations but directly works on the solutions that, a posteriori, has to be checked to fulfill the field equations.

The key point of the method is to find out a suitable complex transformation which, from a physical viewpoint, corresponds to the fact that we are reducing the number of independent Killing vectors. From a mathematical viewpoint, it is useful since allows to

Table 2. The name of molecular clouds, the particle number density, the excitation temperature and the value of Jeans mass are reported for Newtonian and $f(R)$ case, respectively. This table is only a part of the catalog of molecular clouds reported in [59]

Subject	T K	n (10^8m^{-3})	M_J ($M_\odot$)	$\tilde{M}_J$ ($M_\odot$)
GRSMC G053.59+00.04	5.97	1.48	18.25	12.85
GRSMC G049.49-00.41	6.48	1.54	21.32	15.00
GRSMC G018.89-00.51	6.61	1.58	22.65	15.94
GRSMC G030.49-00.36	7.05	1.66	22.81	16.06
GRSMC G035.14-00.76	7.11	1.89	28.88	20.33
GRSMC G034.24+00.14	7.15	2.04	29.61	20.84
GRSMC G019.94-00.81	7.17	2.43	29.80	20.98
GRSMC G038.94-00.46	7.35	2.61	31.27	22.01
GRSMC G053.14+00.04	7.78	2.67	32.06	22.56
GRSMC G022.44+00.34	7.83	2.79	32.78	23.08
GRSMC G049.39-00.26	7.90	2.81	35.64	25.09
GRSMC G019.39-00.01	7.99	2.87	35.84	25.23
GRSMC G034.74-00.66	8.27	3.04	36.94	26.00
GRSMC G023.04-00.41	8.28	3.06	38.22	26.90
GRSMC G018.69-00.06	8.30	3.62	40.34	28.40
GRSMC G023.24-00.36	8.57	3.75	41.10	28.93
GRSMC G019.89-00.56	8.64	3.87	41.82	29.44
GRSMC G022.04+00.19	8.69	4.41	47.02	33.10
GRSMC G018.89-00.66	8.79	4.46	47.73	33.60
GRSMC G023.34-00.21	8.87	4.99	48.98	34.48
GRSMC G034.99+00.34	8.90	5.74	50.44	35.50
GRSMC G029.64-00.61	8.90	6.14	55.41	39.00
GRSMC G018.94-00.26	9.16	6.16	55.64	39.16
GRSMC G024.94-00.16	9.17	6.93	56.81	39.99
GRSMC G025.19-00.26	9.72	7.11	58.21	40.97
GRSMC G019.84-00.41	9.97	11.3	58.52	41.19

overcome the problem of a direct search for axially symmetric solutions that, in $f(R)$-gravity, could be extremely cumbersome due to the fourth-order field equations.

We have to stress that the utility of generating techniques is not simply to obtain a new metric, but a metric of a new spacetime with specific properties as the transformation properties of the energy-momentum tensor and Killing vectors. In its original application, the Newman-Janis procedure transforms an Einstein-Maxwell solution (Reissner-Nordstrom) into another Einstein-Maxwell solution (Kerr-Newman). As a particular case (setting the charge to zero) it is possible to achieve the transformation between two vacuum solutions (Schwarzschild and Kerr). Also in case of $f(R)$-gravity, new features emerge by adopting such a technique. In particular, it is worth studying how certain features of spherically simmetric metrics, derived in $f(R)$-gravity, result transformed in the new axially symmetric solutions for black holes. For example, considering the $f(R)$ spherically symmetric solution studied here, the Ricci scalar evolves as r^{-2} and then the asymptotic flatness is recovered. Let us consider now the axially symmetric metric achieved by the Newman-Janis method. The parameter $a \neq 0$ indicates that the spherical symmetry ($a = 0$) is broken. Such a parameter can be immediately related to the presence of an axis of symmetry and then to the fact that a Killing vector, related to the angle θ, has been lost. To conclude, we can say that once the vacuum case is discussed, more general spherical metrics can be transformed in new axially symmetric metrics adopting more general techniques [16], that would allow us to have the best configurations to describe the black hole dynamics.

A part the mathematical interest in finding out new solutions, such results could have remarkable applications, at least as toy models, in the study of self-gravitating systems. We have shown that the stellar theory as well as the Jeans analysis of instability and collapse can be dramatically altered if one adopt $f(R)$-gravity instead of GR.

The study has been performed starting from the Newtonian limit of $f(R)$-field equations. Since the field equations satisfy in any case the Bianchi identity, we can use the conservation law of energy-momentum tensor. In particular adopting a polytropic equation of state relating the mass density to the pressure, we derive the *modified Lané-Emden equation* and its solutions for $n = 0, 1$ which can be compared to the analogous solutions coming from the Newtonian limit of GR. When we consider the limit $f(R) \to R$, we obtain the standard hydrostatic equilibrium theory coming from GR. A peculiarity of $f(R)$-gravity is the non-viability of Gauss theorem and then the *modified Lané-Emden equation* is an integro-differential equation where the mass distribution plays a crucial role. Furthermore the correlation between two points in the star is given by a Yukawa-like term of the corresponding Green function.

These solutions have been matched with those coming from GR and the corresponding density radial profiles have been derived. In the case $n = 0$, we find an exact solution, while, for $n = 1$, we used a perturbative analysis with respect to the solution coming from GR. It is possible to demonstrate that density radial profiles coming from $f(R)$-gravity analytic models and close to those coming from GR are compatible. This result rules out some wrong claims in the literature stating that $f(R)$-gravity is not compatible with self-gravitating systems. Obviously the choice of the free parameter of the theory has to be consistent with boundary conditions and then the solutions are parameterized by a suitable "wave-length" $m = \sqrt{-\frac{1}{3f''(0)}}$ that should be experimentally fixed.

The next step is to derive self-consistent numerical solutions of *modified Lané-Emden equation* and build up realistic star models where further values of the polytropic index n and other physical parameters, e.g. temperature, opacity, transport of energy, are considered. Interesting cases are the non-relativistic limit ($n = 3/2$) and relativistic limit ($n = 3$) of completely degenerate gas. These models are a challenging task since, up to now, there is no self-consistent, final explanation for compact objects (e.g. neutron stars) with masses larger than Volkoff mass, while observational evidences widely indicate these objects [32]. In fact it is plausible that the gravity manifests itself on different characteristic lengths and also other contributions in the gravitational potential should be considered for these exotic objects. As we have seen above, the gravitational potential well results modified by higher-order corrections in the curvature. In particular, it is possible to show that if we put in the gravitational action other curvature invariants also repulsive contributions can emerge [33,46]. These situations have to be seriously taken into account in order to address several issues of relativistic astrophysics that seem to be out of the explanation range of the standard theory.

For this purpose, we have analyzed the Jeans instability mechanism, adopted for star formation, considering the Newtonian approximation of $f(R)$-gravity. The related Boltzmann-Vlasov system leads to modified Poisson equations depending on the $f(R)$-model. In particular, considering Eqs.(79) and (80), it is possible to get a new dispersion relation (129) where instability criterion results modified (see also [30]). The leading parameter is α, i.e. the second derivative of the specific $f(R)$-model. Standard Newtonian Jeans instability is immediately recovered for $\alpha = 0$ corresponding to the Hilbert-Einstein Lagrangian of GR. In Fig. 3, dispersion relations for Newtonian and a specific $f(R)$-model are numerically compared. The modified characteristic length van be given in terms of the classical one.

Both in the classical and in $f(R)$ analysis, the system damps the perturbation. This damping is not associated to the collisions because we neglect them in our treatment, but it is linked to the so called Landau damping [48].

A new condition for the gravitational instability is derived, showing unstable modes with faster growth rates. Finally we can observe the instability decrease in $f(R)$-gravity: such decrease is related to a larger Jeans length and then to a lower Jeans mass. We have also compared the behavior with the temperature of the Jeans mass for various types of interstellar molecular clouds (Fig. 4). In Tables 1 and 2 we show the results given by this new limit of the Jeans mass for a sample of giant molecular clouds. In our model the limit (in unit of mass) to start the collapse of an interstellar cloud is lower than the classical one advantaging the structure formation. Real solutions for the Jean mass can be achieved only for $\alpha < 0$ and this result is in agreement with cosmology [49]. In particular, the condition $\alpha < 0$ is essentials to have a well-formulated and well-posed Cauchy problem in $f(R)$-gravity [49]. Finally, it is worth noticing that the Newtonian value is an upper limit for the Jean mass coinciding with $f(R) = R$.

This analysis is intended to indicate the possibility to deal with ISM collapsing clouds under different assumptions about gravity. It is important to stress that we fully recover the standard collapse mechanisms but we could also describe proto-stellar systems that escape the standard collapse model. On the other hand, this is the first step to study star formation and physical black holes in alternative theories of gravity (see also [28–31]). From an observational point of view, reliable constraints can be achieved from a careful analysis of the proto-stellar phase taking into account magnetic fields, turbulence and collisions. Finally, addressing stellar systems by this approach could be an extremely important to test observationally $f(R)$-gravity.

Moreover, the approach developed here admits direct generalizations for other modified gravities, like non-local gravity, modified Gauss-Bonnet theory, string-inspired gravity, etc. In these cases, the constrained Poisson equation may be even more complicated due to the presence of extra scalar(s) in non-local or string-inspired gravity. Developing further this approach gives, in general, the possibility to confront the observable dynamics of astrophysical objects (like stars) with predictions of alternative gravities.

Acknowledgments

The Authors wish to thank I. De Martino, M. Formisano, S.D. Odintsov, A. Stabile for useful comments and discussions on the topics discussed here.

References

[1] S. Capozziello, *Int. Jou. Mod. Phys.* **D 11**, 483 (2002); S. Nojiri, S.D. Odintsov, *Phys. Lett.* **B 576**, 5, (2003); S. Nojiri, S.D. Odintsov, *Phys. Rev.* bf D 68, 12352, (2003); S. M. Carroll, V. Duvvuri, M. Trodden and M. S. Turner, *Phys. Rev.* **D70**, 043528 (2004); G. Allemandi, A. Borowiec, M. Francaviglia, *Phys. Rev.* **D 70**, 103503 (2004); S. Capozziello, V. F. Cardone and A. Troisi, *Phys. Rev.* **D 71**, 043503 (2005); S. Carloni, P.K.S. Dunsby, S. Capozziello, A. Troisi, *Class. Quant. Grav.* **22**, 4839 (2005).

[2] S. Capozziello S., V.F. Cardone, A. Troisi, *JCAP* 08, 001(2006)

[3] S. Capozziello and M. Francaviglia, *Gen. Rel. Grav.* **40**, 357 (2008).

[4] S. Capozziello, M.F. De Laurentis, M. Francaviglia, S. Mercadante, *Found. of Physics* **39**, 1161 (2009).

[5] S. Capozziello, S. Tsujikawa, *Phys. Rev. D*, **77**, 10750, (2008).

[6] S. Capozziello, M. De Laurentis, S. Nojiri, S.D. Odintsov, *Gen. Rel. Grav.* **41**, 2313, (2009); *Physical Review D* **79**, 124007, (2009); M. De Laurentis, S. Capozziello, *Nuclear Physics B (Proceedings Supplements)*, 212, (2009); M. De Laurentis, S. Capozziello, S. Nojiri, S.D. Odintsov, *Annalen der Physik* **19**, 347, (2009); S. Capozziello, N. Carlevaro, M. De Laurentis, M. Lattanzi, G. Montani, arXiv:1104.2169v1 [astro-ph.CO] (2011); O.M. Lecian, G. Montani, *Class. Quant. Grav.* **26**, 045014, (2009).

[7] L. Pogosian, A. Silvestri, *Phys. Rev. D* **77** , 023503, (2008).

[8] S. Capozziello, S. Vignolo *Class Quantum Grav*, **26** 168001 2009; V. Faraoni *Class. Quantum Grav.* **26** 168002 (2009); G.J. Olmo, P. Singh, *JCAP* 030, 0901 (2008); V. Faraoni, N. Lanahan-Tremblay , *Phys. Rev.* **D 78**, 064017 (2008); S. Capozziello, S. Vignolo, *Class. Quantum. Grav.* **26** 175013 (2009).

[9] S. Capozziello, A. Stabile, A. Troisi *Class. Quant. Grav.* **24**, 2153 (2007).

[10] T. Multamaki, I. Vilja, *Phys. Rev.* **D 74**, 064022 (2006).

[11] T. Multamaki, I. Vilja *Phys. Rev.* **D 76**, 064021 (2007); K. Kainulainen, J. Piilonen, V. Reijonen, D. Sunhede, *Phys. Rev.* **D 76**, 024020 (2007).

[12] S. Capozziello, A. Stabile, A. Troisi, *Class. Quant.Grav.*, **25**, 085004 (2008).

[13] R. P. Kerr, *Phys. Rev. Lett.* **11** 237 (1963).

[14] D. McManus, *Class. Quant. Grav.* **8** 863 (1991).

[15] A. Kransinski, *Ann. Phys.* **112** 22 (1978).

[16] H. Stephani, D. Kramer, M. MacCallum, C. Hoenselaers, E. Herlt *Exact Solutions to Einstein's Field Equations, Second Edition*, Cambridge University Press, Cambridge (2003).

[17] E. T. Newman, A. I. Janis, *J. Math. Phys.* **6** 915 (1965).

[18] E. T. Newman, E. Couch, K. Chinnapared, A. Exton, A. Prakash and R. Torrence, *J. Math. Phys.* **6** 918 (1965).

[19] R. D'Inverno, *Introducing Einstein's Relativity* Oxford University Press, New York, (1992).

[20] M. M. Schiffer, R. J. Adler, J. Mark and C. Sheffield, *J. Math. Phys.* **14** 52 (1973).

[21] G. C. Debney, R. P. Kerr and A. Schild, *J. Math. Phys.* **10** 1842 (1969).

[22] S. Capozziello, M. De Laurentis, and A. Stabile, *Class. Quantum Grav.* **27**, 165008 (2010); M. De Laurentis, arXiv:1111.2071v1 [gr-qc] (2011).

[23] D. Psaltis, *Living Rev. Relativity* **11**, (2008).

[24] F. Briscese, E. Elizalde, S. Nojiri, S.D. Odintsov, *Phys. Lett. B* **646**, 105 (2007).

[25] J. Khoury, A. Weltman, *Phys. Rev. D* **69**, 044026 (2007).

[26] W. Hu and I. Sawicki, *Phys. Rev. D* **76**, 064004 (2007);
S. Capozziello and S. Tsujikawa *Phys. Rev. D* **77**, 107501 (2008);
S. Tsujikawa, T. Tamaki, R. Tavakol, *JCAP* 0905, 020 (2009);
A. Upadhye and W. Hu, *Phys. Rev. D* **80**, 064002 (2009).

[27] E. Babichev and D. Langlois, *Phys. Rev. D* **80**, 121501 (2009);
E. Babichev and D. Langlois, *Phys. Rev. D* **81** ,124051 (2010).

[28] A. Cooney, S. DeDeo, D. Psaltis, *Phys. Rev. D*, **82**, 064033 (2010).

[29] A. Upadhye, W. Hu, *Phys. Rev. D*, **82**, 064002 (2010).

[30] S. Capozziello, M. De Laurentis, S.D. Odintsov, A. Stabile, *Phys. Rev. D*, **83**, 064004, (2011).

[31] Ph. Chang, L. Hui, 2011, ApJ, **732**, 25

[32] M.P. Muno, at al., *The Astrophysical Journal* **636**, L41 (2006).

[33] A. Stabile, *Phys. Rev. D* **82**, 124026 (2010).

[34] S. Capozziello and M. De Laurentis, *Phys. Repts.* **509**, 167 (2011)

[35] G. Allemandi, M. Capone, S. Capozziello, M. Francaviglia, *Gen. Rel. Grav.* **38**, 33 (2006).

[36] E.L. Ince, *Ordinary Differential Equations*, Dover, New York (1956).

[37] J.D. Barrow, A.C. Ottewill, *J. Phys.* **A 16** 2757 (1983).

[38] V.I. Arnold, *Mathematical Methods of Classical Mechanics*, Springer–Verlag, Berlin (1978).

[39] G. Marmo, E.J. Saletan, A. Simoni and B. Vitale, *Dynamical Systems. A Differential Geometric Approach to Symmetry and Reduction*, Wiley, New York (1985).

[40] G. Morandi, C. Ferrario C., G. Lo Vecchio, G. Marmo, C. Rubano *Phys. Rep.* **188** 149 (1990).

[41] S.Weinberg *"Gravitation and Cosmology"*, John Wiley & Sons, Inc., New York, (1972); Ch. W. Misner, K.S. Thorne, J.A. Wheeler, *Gravitation*, ed. W.H. Freeman, San Francisco (1973).

[42] Chandrasekhar S. *The mathematical theory of black holes*, Oxford, (1983).

[43] Kippenhahn R., Weigert A., *Stellar structures and evolution*, Springer-Verlag, Berlin, (1990).

[44] Capozziello S., Stabile A., Troisi A., *Phys. Rev. D* **76**, 104019 (2007).

[45] Stabile A., *Phys. Rev. D* **82**, 064021 (2010).

[46] S. Capozziello and A. Stabile, *Class. Quant. Grav.* **26**, 085019 (2009).

[47] Capozziello S., Stabile A., Troisi A., *Phys. Lett. B* **686**, 79 (2010).

[48] J. Binney, S. Tremaine, *Galactic Dynamics*, Princeton University Press, Princeton (1994).

[49] S. Capozziello, S. Vignolo, *Class. Quant. Grav.* **26**, 175013, (2009).

[50] C. F. McKee, E.C. Ostriker, *Annu. Rev. Astron. Astrophys.* **45**, 565, (2007).

[51] B.W. Carroll, D.A Ostlie., *An Introduction to Modern Astrophysics*, Pearson (2007).

[52] R. Kippenhahn, A. Weigert, *Stellar structures and evolution, Springer-Verlag*, Berlin, (1990).

[53] M.A. Dopita, R.S. Sutherland , *Astrophysics of the Diffuse Universe* Springer, New York (2003).

[54] H. Scheffler, H. Elsasser, *Physics of the Galaxy and Interstellar Matter* Springer, New York (1982).

[55] C. F. McKee *The Origin of Stars and Planetary Systems*, ed. C. J. Lada & N.D. Kylafis (Dordrecht: Kluwer), 29, (1999).

[56] L. Blitz, *Giant molecular clouds. In Protostars and Planets III*, eds. E.H. Levy and J.I. Lunine, (Tucson: Univ. of Arizona Press), pp. 125-161 (1993).

[57] L. Blitz, L. Magnani, L. Mundy, *Astrophys. J. Lett.* , **282**, L9, (1984).

[58] L.D. Anderson, T.M. Bania, J.M. Jackson et al. , *ApJS* **181**, 255A, (2009).

[59] J. Roman-Duval, J.M. Jackson, M. Heyer, HJ. Rathborne, R. Simon *ApJ* **723**, 492, (2010).

In: Black Holes: Evolution, Theory and Thermodynamics ISBN: 978-1-61942-929-1
Editors: A. J. Bauer and D. G. Eiffel, pp. 81-102

Chapter 3

PRIMORDIAL BLACK HOLES IN HIGHER ORDER CURVATURE GRAVITY AND BEYOND

S. O. Alexeyev and D. A. Tretyakova
Sternberg Astronomical Institute (MSU),
Universitetsky Prospekt,
Moscow, Russia

Abstract

We summarize different aspects of dilatonic black hole in 4D Gauss-Bonnet gravity that were hardly investigated not so long ago. Such study has been become real because of a new numerical integration method, based on the analysis of the main system determinant. The most interesting result is the existence of a minimal black hole in Gauss-Bonnet gravity, and, so, a new approach to the last stages of black hole evaporation is discussed. Such a black hole does not evaporate comletely, but only to some relic remnant. If this remnant is stable, being the endpoint of the evaporation, it can represent the relic remnant of the black holes formed during the initial stages of our Universe formation. The coupling with early Universe parameters is also discussed.

1. Introduction

At the current moment one of the ways to clarify important unsolved problems of modern theoretical physics and astrophysics (dark energy or the endpoint of Hawking evaporation, for example) is to extend the General Relativity. That is why dilatonic black holes in the 4D low-energy limit of string theory have attract so much attention. As the differential equations in such class of models have a very complicated and special structure, there is no analyticall solution, hence, the researchers have to use perturbative [1]-[3] or numerical [4]-[6] methods.

String theory as one of the approaches of "theory of everything" predicts the Einstein equations to be modif ed by higher order curvature corrections in the range where the curvature of space-time has the near-Planckian values. The form of these corrections is not investigated completely [7] yet. Since the general structure of the expansion is not known yet, the direct summing up is impossible. However, as we deal with the expansion, the most important correction is the second order one which is the product of the Gauss-Bonnet and dilatonic terms:

$$S = \frac{1}{16\pi}\int d^4x\sqrt{-g}\Big[m_{Pl}^2(-R+2\partial_\mu\phi\partial^\mu\phi)+\lambda e^{-2\phi}S_{GB}\Big]. \qquad (1)$$

Here ϕ is the dilatonic f eld, λ is the string coupling constant showing the inf uence of the Gauss-Bonnet term $S_{GB} = R_{ijkl}R^{ijkl} - 4R_{ij}R^{ij} + R^2$, R is Ricci scalar. Being the full derivative, the Gauss-Bonnet term does not contribute to the f eld equations by itself, so it should be coupled with dilatonic f eld. The order of the differential equations does not increase and remains the second one, so the presence of the dilatonic term makes the Gauss-Bonnet term contribution dynamical.

As it is known the uniqueness theorem is violated for black hole solutions in particular points [8] in second order curvature gravity, so it was important to develop the methods allowing us to work with multiple solutions during calculations. The existence of multiple solutions at some point corresponds to a particular point of the differential equations system, so it is better to use the most strong method for a numerical integration to pass these points successfully and to obtain the solution in the widest range. Such a method was proposed by some of us in [9]; the integration over the additional parameter, that allows to investigate the linear system of the second order differential equations with branching points and singularities.

2. Field Equations

The purpose was to study spherically symmetric black hole solutions, so, the most convenient choice of metric has been chosen as

$$ds^2 = \Delta(r)dt^2 - \frac{\sigma^2(r)}{\Delta(r)}dr^2 - f^2(r)(d\theta^2 + \sin^2\theta d\varphi^2). \qquad (2)$$

Metric functions Δ, σ and f depend only upon a radial coordinate r. One can use various types of the metric gauge. The most convenient of them are, so-called, "curvature gauge" ($f = r$, used in [9]) and GHS or GM-GHS (Gibbons-Maeda-Garf ncle-Horowits-Strominger) gauge ($\sigma = 1$, used in [10]). Field equations in the curvature gauge have the following form [11]:

$$\begin{aligned}&\sigma^2\left[-r\sigma' + \sigma r^2(\phi')^2\right] + 4\mathrm{e}^{-2\phi}\lambda\sigma\left(\Delta - \sigma^2\right)\left[\phi'' - 2(\phi')^2\right]\\ &+4\mathrm{e}^{-2\phi}\phi'\lambda\sigma'\left(\sigma^2 - 3\Delta\right) = 0,\end{aligned} \qquad (3)$$

$$\sigma^2 = \frac{-k+\sqrt{k^2-4n}}{2}, \qquad (4)$$

where

$$k = \left(\Delta r^2(\phi')^2 - \Delta' r - \Delta\right) + 4\mathrm{e}^{-2\phi}\lambda\phi'\Delta',$$
$$n = -4\mathrm{e}^{-2\phi}\lambda\phi'\left(3\Delta\Delta'\right).$$

$$\begin{aligned}&\sigma^2\left[\Delta''r\sigma-\Delta'r\sigma'+2\Delta'\sigma-2\Delta\sigma'+2\Delta r\sigma(\phi')^2\right]\\&\quad+4\mathrm{e}^{-2\phi}\lambda 2\sigma\Delta\Delta'\left[\phi''-2(\phi')^2\right]\\&\quad+4\mathrm{e}^{-2\phi}\phi'\lambda 2\left[(\Delta')^2\sigma+\Delta\Delta''\sigma-3\Delta\Delta'\sigma'\right]=0,\end{aligned} \tag{5}$$

$$\begin{aligned}&-2\sigma^2\left[\Delta'r^2\sigma\phi'+2\Delta r\phi'\sigma-\Delta r^2\phi'\sigma'+\Delta r^2\phi''\sigma\right]\\&\quad+4\mathrm{e}^{-2\phi}\lambda\big[(\Delta')^2\sigma+\Delta\Delta''\sigma-3\Delta\Delta'\sigma'-\Delta''\sigma^3\\&\qquad+\Delta'\sigma'\sigma^2\big]=0.\end{aligned} \tag{6}$$

Here and furter we f x $m_{Pl} = 1$ for simplitity. The Arnowitt-Diezer-Misner (ADM) mass M and the new value named "dilaton charge" D are def ned by the asymptotic expansions for Δ, f and ϕ [11] at the inf nity:

$$\Delta(r\to\infty) = 1-\frac{2M}{r}+O\left(\frac{1}{r}\right), \tag{7}$$

$$\sigma(r\to\infty) = 1-\frac{1}{2}\frac{D^2}{r^2}+O\left(\frac{1}{r^2}\right), \tag{8}$$

$$\phi(r\to\infty) = \frac{D}{r}+O\left(\frac{1}{r}\right), \tag{9}$$

3. The Numerical Method: Brief Overview

A system of ordinary differential equations with unknown functions $x_i(t) \in R^n$ can be established in the following form [9]:

$$\begin{aligned}&\sum_{j=1}^{n} x_j'(t)a_{1j}(x,t) = b_1(x,t),\\&\sum_{j=1}^{n} x_j'(t)a_{2j}(x,t) = b_2(x,t),\\&\qquad\ldots\\&\sum_{j=1}^{n} x_j'(t)a_{nj}(x,t) = b_n(x,t),\end{aligned} \tag{10}$$

where the coeff cients b_i, $a_{ij} \in C^1[R^n \times R^1]$. According to the existence theorem, the solution of (10) would exist only in the case when the main system discriminant not to be equal to zero, i.e.:

$$\det A(x(t),t) \neq 0. \tag{11}$$

If in any point t^* $\det A(x(t^*),t^*) = 0$, the system (10) has several solutions and a numerical integration performed by the usual (Runge-Kutta, Adams, etc.) method would stop before the point t^*. So, it is necessary to choose such a numerical integration strategy which allows us to pass the point t^*. This method was developed in [9] (and now we follow this paper), its applications for other problems were discussed earlier in [12].

For extending the system phase-space one can introduce a new independent variable s and consider x and t to be functions of s: $x = x(s)$, $t = t(s)$. Therefore,

$$\frac{dx}{dt} = \frac{dx}{ds}\frac{ds}{dt} = x'_s(t'_s)^{-1}. \tag{12}$$

Putting (12) to the system (10), one obtains a new homogeneous system which consists of n equations with $(n+1)$ unknowns. One needs to add $(n+1)$th equation representing the normalisation condition for the system, which leads to the following system:

$$\begin{aligned}
&\sum_{j=1}^{n} \frac{dx_j}{ds} a_{1j}(x,t) - \frac{dt}{ds} b_1(x,t) = 0,\\
&\sum_{j=1}^{n} \frac{dx_j}{ds} a_{2j}(x,t) - \frac{dt}{ds} b_2(x,t) = 0,\\
&\qquad \cdots\\
&\sum_{j=1}^{n} \frac{dx_j}{ds} a_{nj}(x,t) - \frac{dt}{ds} b_n(x,t) = 0,\\
&\sum_{j=1}^{n} \frac{dx_j}{ds} \tau_j(s) - \frac{dt}{ds} \tau_t(s) = 1,
\end{aligned} \tag{13}$$

where new functions $\tau(s) \in R^{n+1}$. These new functions $\tau(s)$ are fxed by the following normalisation conditions:

$$\begin{aligned}
\tau_i(s) &= \frac{dx_i/ds}{\sqrt{\sum\limits_{j=1}^{n} (dx_i/ds)^2 + (dt/ds)^2}},\\
\tau_t(s) &= \frac{dt/ds}{\sqrt{\sum\limits_{j=1}^{n} (dx_i/ds)^2 + (dt/ds)^2}}.
\end{aligned} \tag{14}$$

So, $\vec{\tau}(s)$ is the tangent vector to the solution trajectory in $(n+1)$ phase-space, therefore, the integration is proceeded along the solution trajectory, not along the radial coordinate r, as usually performed. At a frst step the most convenient choice of $\vec{\tau}$ is: $\tau_i = 0$, $\tau_t = \pm 1$, where the sign depends upon the solution direction moving. At the next step $\vec{\tau}(s)$ are extrapolated by the Legendre polynomials with $\vec{\tau}(s)$ values from the previous steps according to formulae (14). The system (13) remains the linear one. One can argue [9] that if the integrating step vanishes the solution of the system (13) with the extrapolating functions $\vec{\tau}(s)$ reduces to the solution of the system (13) with the functions $\vec{\tau}(s)$ defned in (14). In such a case our system (3), (5), (6) could be rewritten as follows:

$$\begin{aligned} \frac{d\Delta}{ds} + 0 + 0 + 0 + 0 - \Delta' \frac{dr}{ds} &= 0, \\ 0 + \frac{d\phi}{ds} + 0 + 0 + 0 - \phi' \frac{dr}{ds} &= 0, \\ 0 + 0 + a_{11}\frac{d\Delta'}{ds} + a_{12}\frac{d\sigma}{ds} + a_{13}\frac{d\phi'}{ds} - b_1\frac{dr}{ds} &= 0, \\ 0 + 0 + a_{21}\frac{d\Delta'}{ds} + a_{22}\frac{d\sigma}{ds} + a_{23}\frac{d\phi'}{ds} - b_2\frac{dr}{ds} &= 0, \\ 0 + 0 + a_{31}\frac{d\Delta'}{ds} + a_{32}\frac{d\sigma}{ds} + a_{33}\frac{d\phi'}{ds} - b_3\frac{dr}{ds} &= 0, \\ \frac{d\Delta}{ds}\tau_\Delta + \frac{d\phi}{ds}\tau_\phi + \frac{d\Delta'}{ds}\tau_{\Delta'} + \frac{d\sigma}{ds}\tau_\sigma + \frac{d\phi'}{ds}\tau_{\phi'} - \tau_r\frac{dr}{ds} &= 1, \end{aligned} \tag{15}$$

where stroke denotes $\partial/\partial r$ and a_{ij}, b_i have the form:

$$\begin{aligned} a_{11} &= 0, \\ a_{12} &= -\sigma^2 r + 4\mathrm{e}^{-2\phi}\lambda\phi'(\sigma^2 - 3\Delta), \\ a_{13} &= 4\mathrm{e}^{-2\phi}\lambda\sigma(\Delta - \sigma^2), \\ a_{21} &= \sigma^3 r + 4\mathrm{e}^{-2\phi}\lambda\phi' 2\Delta\sigma, \\ a_{22} &= -\sigma^2(\Delta' r + 2\Delta) - 4\mathrm{e}^{-2\phi}\lambda\phi' 2\Delta 3\Delta', \\ a_{23} &= 4\mathrm{e}^{-2\phi}\lambda 2\Delta\Delta'\sigma, \\ a_{31} &= 4\mathrm{e}^{-2\phi}\lambda(\Delta\sigma - \sigma^3), \\ a_{32} &= 2\sigma^2\Delta r^2\phi' + 4\mathrm{e}^{-2\phi}\lambda(-3\Delta\Delta' + \Delta'\sigma^2), \\ a_{33} &= -2\sigma^2\Delta r^2\sigma, \end{aligned}$$

$$\begin{aligned} b_1 &= -\sigma^3 r^2(\phi')^2 + 4\mathrm{e}^{-2\phi}\lambda\sigma(\Delta - \sigma^2)2(\phi')^2, \\ b_2 &= -\sigma^2(2\Delta'\sigma + 2\Delta r\sigma(\phi')^2) + 4\mathrm{e}^{-2\phi}\lambda 2\sigma\Delta\Delta' 2(\phi')^2 \\ &\quad -4\mathrm{e}^{-2\phi}\lambda\phi' 2(\Delta')^2\sigma, \\ b_3 &= 2\sigma^2(\Delta' r^2\sigma\phi' + 2\Delta r\sigma\phi') - 4\mathrm{e}^{-2\phi}\lambda(\Delta')^2\sigma. \end{aligned}$$

Equation (4) is used as a constraint for the addition accuracy control at every integration step. The integration is proceeded by the independent variable quantity s and starts from the point of $s = 0$. At the f rst step it is necessary the following condition to be fulf lled: det B=det $A \neq 0$, where B is the equation matrix def ned as follows:

$$B = \begin{pmatrix} 1 & 0 & 0 & 0 & 0 & -\Delta' \\ 0 & 1 & 0 & 0 & 0 & -\phi' \\ 0 & 0 & a_{11} & a_{12} & a_{13} & -b_1 \\ 0 & 0 & a_{21} & a_{22} & a_{23} & -b_2 \\ 0 & 0 & a_{31} & a_{32} & a_{33} & -b_3 \\ \tau_\Delta & \tau_\phi & \tau_{\Delta'} & \tau_\sigma & \tau_{\phi'} & \tau_r \end{pmatrix} \tag{16}$$

At every integration step one calculates $\vec{\tau}$ by the formulae (14). The 8th order Runge-Kutta code for integration by one step was used, it divides the given step automatically to reach the accuracy required.

One can argue that if the integration step becomes quite small, the matrix B (16) will not be degenerated at that trajectory point and rang$A \geq (n-1)$. When rang$A < (n-1)$, a ramifying point for two or more branches could exist in that point that can not be passed by the describing strategy.

Matrix B was turned out by the Singular Value Decomposition (SVD) method [13]. It decomposes the given matrix B to the product of three matrices $U\Sigma V^{\dagger}$, where U and V are the unitary matrices $(n+1) \times (n+1)$ dimension and Σ is the diagonal matrix of absolute own values of B, i.e. $\Sigma = \{\sigma_{ii}\}$, $\sigma_{ii} \geq 0$. Therefore, $B^{-1} = V\Sigma^{-1}U^{\dagger}$ where $\Sigma^{-1} = \{1/\sigma_{ii}\}$. This allows the method to pass the degenerated point.

4. Minimal Mass Black Hole

The results of numerical integration are shown in the Fig. 1. It presents the dependence of the metric functions Δ (a), σ (b) and the dilatonic function $e^{-2\phi}$ (c) against the radial coordinate r. The behaviour of the solution outside the horizon has the usual Schwarzchild form. Under the inf uence of the Gauss-Bonnet term the structure of the solution inside the horizon changes: new singular point, so-called r_s, appears (see Fig. 1). The solution turns to the other branch when reaching this point. Only two branches exist near the position r_s and there are no any branches between r_s and the origin. Therefore, a new singular surface with the topology $S^2 \times R^1$ appears (it is an inf nite "tube" in the t direction). Such a singularity is absent in the f rst order curvature gravity and is present for any choice of metric paramtrization [9].

Fig. 2 shows the graph of the metric function Δ versus the radial coordinate r at the different values of the event horizon r_h. The curve (a) represents the case where r_h is rather large and is equal to 30.0 Planck unit values (P.u.v.). The curve (b) shows the changes in the behaviour of $\Delta(r)$ when r_h is equal to 7.5 P.u.v. The curve (c) represents the boundary case with $r_h = r_{hmin}$ where all the particular points merge and the internal structure disappears. The curve (d) shows the case where $2M \ll r_{hmin}$ and any horizon is absent.

5. Black Hole Evaporation at Last Stages

The most interesting consequence of the second order curvature corrections is the existence of a minimal black hole [9]. This effect was found both in the numerical calculations and in the analytical ones [15] and is independent from the metric parametrisation. Using Green-Witten-Schwartz formula [26], one can calculate the value of the string coupling constant λ, therefore, one f nds the limit value on this minimal black hole mass. In Planckian values it has the form

$$r_h^{min} = \sqrt{\lambda}\sqrt{\frac{4\sqrt{6}}{m_{Pl}^2}}, \tag{17}$$

where r_h^{min} is the lower limit value of the horizon radius [9]. This restriction appears in the second order curvature gravity and is absent in the minimal Einstein-Schwarzschild grav-

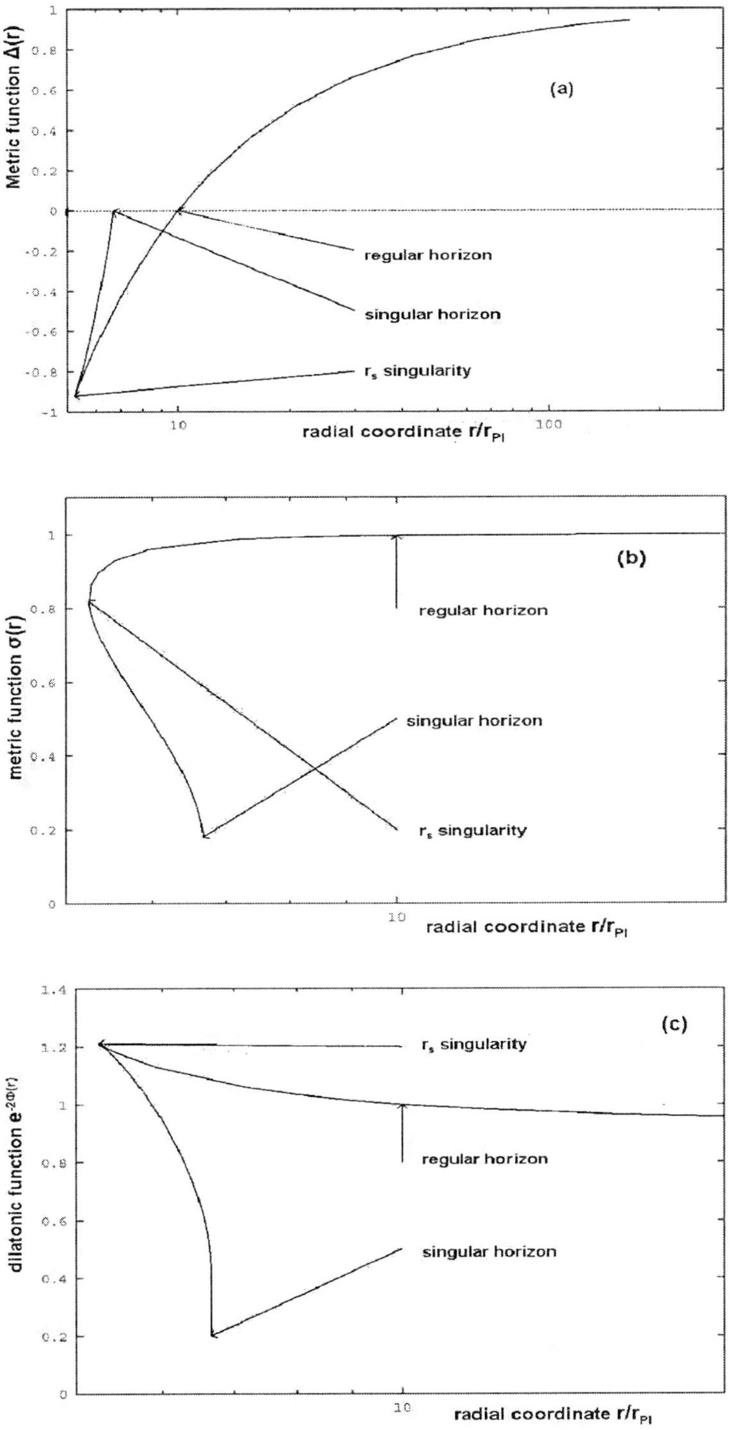

Figure 1. Dependence of the metric functions Δ (a), σ (b) and the dilaton function $e^{-2\phi}$ (c) against the radial coordinate r.

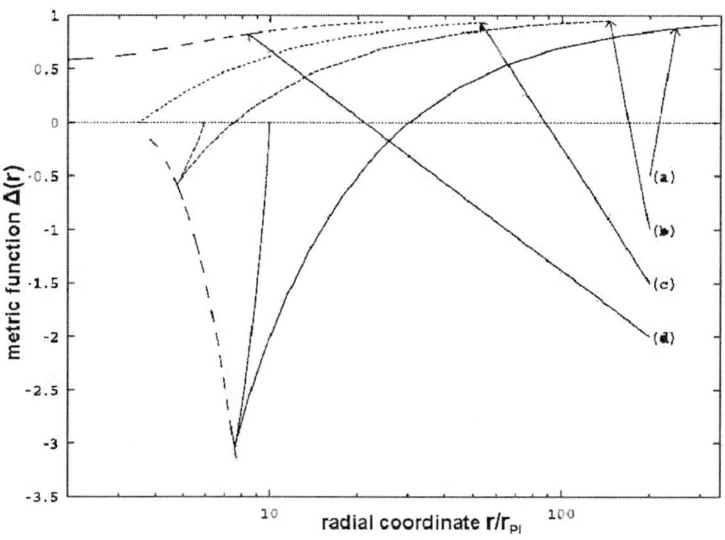

Figure 2. The dependence of the metric function Δ versus the radial coordinate r.

ity. From the analysis of [28] one can conclude that when the third curvature correction in addition to Gauss-Bonnet term is taken into account, the restriction to the minimal black hole size does retain. It's numerical value slightly differs from the second order one, but the effect retains principally. Since in low energy string gravity the curvature corrections starting from the fourth order one do not produce higher derivatives of metric functions, the effect should exist in all perturbative orders. Thus, the discussed restriction to the minimal black hole mass is a fundamental restriction of the string theory (as it takes place in all perturbative orders). So, there is no static, asymptotically f at, spherically symmetric dilatonic black hole with a mass smaller than the minimum mass M_{min}. This state $(M = M_{min})$ is called the ground one for a quasi-classical model. We consider a quasi-classical state that becomes the ground state when the model is quantized [23].

The question of whether the transition from the f rst excited state to the ground state (Fig. 2(c)) is possible in principle must be clarif ed. We consider a diagonal quasi-Schwarzschild metric of the form (2). At position r_{hmin} (see Fig. 2), the asymptotic form of metric (2) is

$$\Delta = \text{Const}_1 \times \sqrt{r - r_{hmin}}, \tag{18}$$

$$\sigma = \text{Const}_2 \times \sqrt{r - r_{hmin}}. \tag{19}$$

Consequently, $R_{ijkl}R^{ijkl} \approx \text{Const}_3 \times (r - r_{hmin})^{-6}$; i.e., the curvature invariant diverges, and this is a nonintegrable singularity. At the same time, the asymptotics on the horizon is regular in all the remaining states and shows a quasi-Schwarzschild behavior:

$$\Delta = d_1(r - r_h) + d_2(r - r_h)^2 + \cdots, \tag{20}$$

$$\sigma = s_0 + s_1(r - r_h) + \cdots, \tag{21}$$

where $(r - r_h) \ll 1$, s_0 and r_h are free parameters. The probability of the transition from the f rst excited state to the ground state with a minimum mass is [33]

$$P = \text{Const} \times e^{S_{r_h} - S_{r_{hmin}}} = \text{Const} \times e^{-S_{r_{hmin}}} \tag{22}$$

$$\propto \text{Const} \times e^{-\frac{1}{(r-r_{hmin})^5}} = \text{Const} \times e^{-\infty} = 0. \tag{23}$$

Thus, this transition is forbidden and the black hole will never reach the ground state during its evaporation.

6. The Deceleration of Evaporation

To f nish the evaporation before reaching the ground state M_{min} requires to add the condition for stopping the evaporation in the classical emission model. Since the mass of the relic remnant of a black hole becomes comparable to the mass of its emitted particles, this condition can be extracted from the fact that the black hole cannot emit more matter than it contains. In the standard approach, the mass of the emitted particles is much smaller than M of the emitting object itself. The smaller is the black-hole mass, the more intense is the emission. The evaporation rate is inversely proportional to M^2:

$$-\frac{dM}{dt} \approx 4 \times 10^{-5} \left(\frac{m_{Pl}}{M}\right)^2 \frac{m_{Pl}}{t_{Pl}} f, \tag{24}$$

where $f = 1.023h(1/2) + 0.420h(1) + 0.048h(2)$ is a function of the number of degrees of freedom for particles with spins 1/2, 1, and 2, respectively [34]. Let us consider the toy model of the the emission process (24) at last stages[1] by taking into account the fact that the mass of the emitted particle for a black hole with a mass of the order of the minimum mass becomes comparable to that of the emitting object itself. In this case, it is natural to impose the condition that during its evaporation the black hole cannot emit more matter than it contains. This condition follows from the existence of a forbidden transition between the ground state of the black hole and its excited states (see previous part). In the standard model of Hawking emission, the mass-loss rate of the black hole increases to inf nity as its mass tends to zero (24). If conservation of energy is taken into account, the mass of the black hole decreases as it evaporates (in the simplest way) and its temperature increases, but the mean energy of the emitted particle and its frequency decreases. Thus, the classical Planck spectrum becomes cut off by the condition that the energy E of the emitted particle should not exceed $M - M_{min}$. When emitting particles, the system successively passes to new states without reaching the forbidden ground state $E = M - M_{min} = 0$ (Fig. 3). By the simplest way this condition can be taken into account by inserting the Heaviside function (H) into the emission equation:

$$\frac{d^2N}{dEdt} = \frac{\Gamma_s(M \times E)}{2\pi} \frac{H(M - M_{min} - E)}{e^{8\pi ME} - (-1)^{2s}}, \tag{25}$$

where $\Gamma_s(M \times E)$ is the Page function [36] depending on the mass, energy, and spin of the corresponding particle emitted by the black hole as:

$$\Gamma_{sboson} = \left[\frac{(l-s)!(l+s)!}{(2l)!(2l+1)!!}\right]^2$$

[1]The complete description of Gauss-Bonnet black hole evaporation was presented in [16] and, earlier, in [15], here we restrict ourselves with the toy model for clarifying the approach

$$\times \prod_{n=1}^{l} \left[1 + \frac{16}{n^2}(ME)^2\right] 8(ME)\,[2ME]^{2l+1}, \tag{26}$$

$$\Gamma_{sfermion} = \left[\frac{(l-s)!(l+s)!}{(2l)!(2l+1)!!}\right]^2$$

$$\times \prod_{n=1}^{l+1/2} \left[1 + \frac{64}{(2n-1)^2}(ME)^2\right] [2ME]^{2l+1}, \tag{27}$$

where l and s are the quantum numbers and $(M \times E) \ll 1$. Below, we take into account the contribution of the $l = s$ mode alone, which is dominant. Using given expression (25), the emission law (24) is modif ed as follows:

$$-\frac{dM}{dt} = \frac{1}{2\pi}\int_0^{M-M_{min}} dE \frac{\Gamma_s(M \times E)E}{e^{8\pi ME} - (-1)^{2s}}, \tag{28}$$

where the integration is no longer over all particle energies but only up to a f nite limit. Let us consider the emission law separately for particles of different spins at $M - M_{min} \ll 1$. At f nal evaporation stages, the corresponding rates of decrease in the black-hole mass can be represented as the joining of analytic asymptotics and series with the corresponding domains of convergence (Fig. 4).

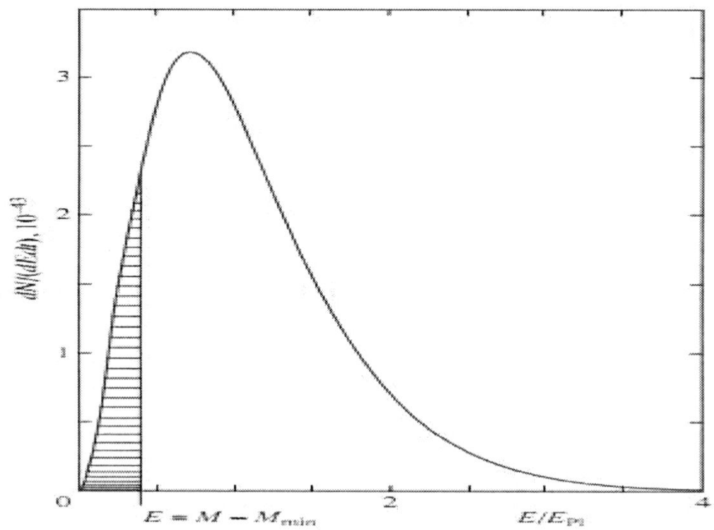

Figure 3. A simple model for the deceleration and stoping of black hole evaporation at f nal stages involves introducing a cutoff of the Planck spectrum by the law of energy conservation. A particle of the corresponding energy can be emitted if the particle energy lies to the left of the cutoff line $E = M - M_{min}$ (hatched region) and can not be emitted if it lies to the right, becaurse it's energy exceeds the total energy of the system. In this case, the black hole can pass into the forbidden region, thus, athough the black hole temperature increases, the energy of the emitted particle decreases until the evapration is completely stopped.

For particles of zero spin:

$$
\begin{aligned}
-\frac{dM}{dt} &= \frac{1}{512\pi^5}\frac{1}{M^2}\times\int_0^{8\pi M(M-M_{min})} dx\frac{x^3}{e^x-1} \\
&= \frac{1}{\pi^2}M(M-M_{min})^3\left(\frac{1}{3}-\pi M(M-M_{min})\right) \\
&\times\left(\sum_{k=1}^{\infty}\frac{B_{2k}(8\pi M(M-M_{min}))^{2k}}{(2k+3)(2k)!}\right).
\end{aligned}
\tag{29}
$$

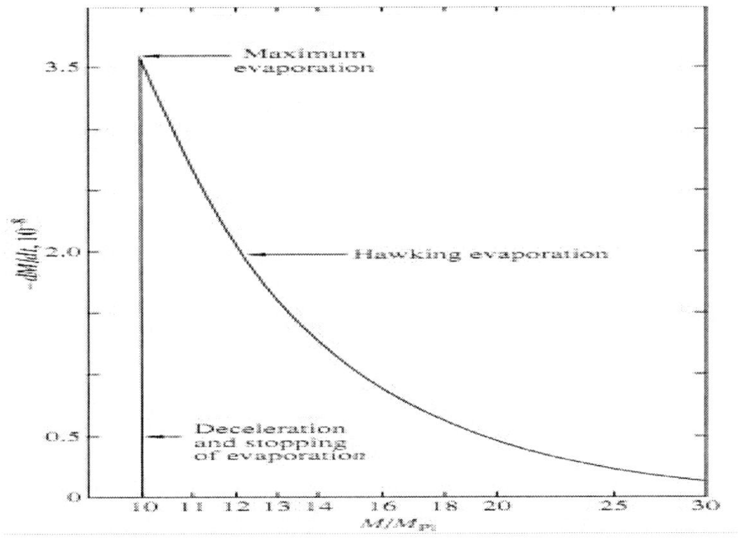

Figure 4. The black hole evaporation law with energy conservation: the standart Hawking part where $-dM/dt \sim 1/M^2$; the stopping of evaporation at f nal stages upon reaching the state with a minimum mass.

For particles of spin $1/2$:

$$
\begin{aligned}
-\frac{dM}{dt} &= \frac{1}{819\pi^5}\frac{1}{M^2}\times\int_0^{8\pi M(M-M_{min})} dx\frac{x^3}{e^x+1} \\
&= \frac{1}{1024\pi^2}\frac{1}{M^2}e^{-4\pi(M-M_{min})} \\
&\times\left(\sum_{k=0}^{\infty}\frac{E_{2k}(4\pi M(M-M_{min}))^{2k+4}}{(2k+4)(2k)!}\right) \\
&\times\left(\sum_{k=0}^{\infty}\frac{E_{2k}\gamma(2k+4,4\pi M(M-M_{min}))}{(2k+4)(2k)!}\right).
\end{aligned}
\tag{30}
$$

For particles of spin 1,

$$
-\frac{dM}{dt} = \frac{1}{73728\pi^7}\frac{1}{M^2}\times\int_0^{8\pi M(M-M_{min})} dx\frac{x^5}{e^x-1}
$$

$$= \frac{4}{9\pi^2} M^3 (M - M_{min})^5 \left(\frac{1}{5} - \frac{2}{3}\pi M (M - M_{min}) \right) \times \left(\sum_{k=1}^{\infty} \frac{B_{2k} (8\pi M (M - M_{min}))^{2k}}{(2k+5)(2k)!} \right). \quad (31)$$

For particles of spin 2,

$$-\frac{dM}{dt} = \frac{1}{29491200\pi^9} \frac{1}{M^2} \times \int_0^{8\pi M(M-M_{min})} dx \frac{x^9}{e^x - 1}$$
$$= \frac{16}{225\pi^2} M^5 (M - M_{min})^7 \left(\frac{1}{7} - \frac{1}{2}\pi M (M - M_{min}) \right) \times \left(\sum_{k=1}^{\infty} \frac{B_{2k} (8\pi M (M - M_{min}))^{2k}}{(2k+7)(2k)!} \right). \quad (32)$$

In the high-mass limit $(M(M - M_{min}) \gg 1)$ the evaporation rates of particles with different spins are virtually indistinguishable. Indeed, using an asymptotic representation of the integrals and taking into account the fact that the Page function in Eq. (25) is simply $\Gamma_s(M \times E) = M^2 E^2$ in the high-mass limit, irrespective of the spin of the emitted particles, we obtain for bosons ($s = 0, 1, 2$):

$$-\frac{dM}{dt} = \frac{1}{128\pi^3} \frac{1}{M^2} \times \int_0^{8\pi M(M-M_{min})} dx \frac{M^2 x^3}{e^x - 1}$$
$$\approx \frac{1}{128\pi^3} \times \left(\frac{\pi^4}{15} - e^{-A}(A^3 + 3A^2 + 6A + 6) \right) \quad (33)$$

and for fermions

$$-\frac{dM}{dt} = \frac{1}{128\pi^3} \frac{1}{M^2} \times \int_0^{8\pi M(M-M_{min})} dx \frac{M^2 x^3}{e^x + 1}$$
$$\approx \frac{1}{128\pi^3} \times \left(\frac{7\pi^4}{120} - e^{-A}(A^3 + 3A^2 + 6A + 6) \right) \quad (34)$$

where $A = 8\pi M(M - M_{min})$. Thus, at the inf nity, i.e., in the classical high-mass approximation, the ratio of the boson and fermion emission rates is 8/7. For $M(M - M_{min}) \ll 1$, it would be natural to use a different approximation:

$$-\frac{dM}{dt}|_{s=0} \approx \frac{1}{3\pi^2} M (M - M_{min})^3, \quad (35)$$
$$-\frac{dM}{dt}|_{s=1/2} \approx \frac{1}{16\pi} M^2 (M - M_{min})^4, \quad (36)$$
$$-\frac{dM}{dt}|_{s=12} \approx \frac{4}{45\pi^2} M^3 (M - M_{min})^5, \quad (37)$$
$$-\frac{dM}{dt}|_{s=2} \approx \frac{16}{1575\pi} M^5 (M - M_{min})^7, \quad (38)$$

The evaporation behavior at f nal stages for bosons and fermions is shown in Fig. 5. Relations (35)-(38) lead us to conclude that particles with an integer spin will be emitted at f nal stages of black hole evaporation with a high probability.

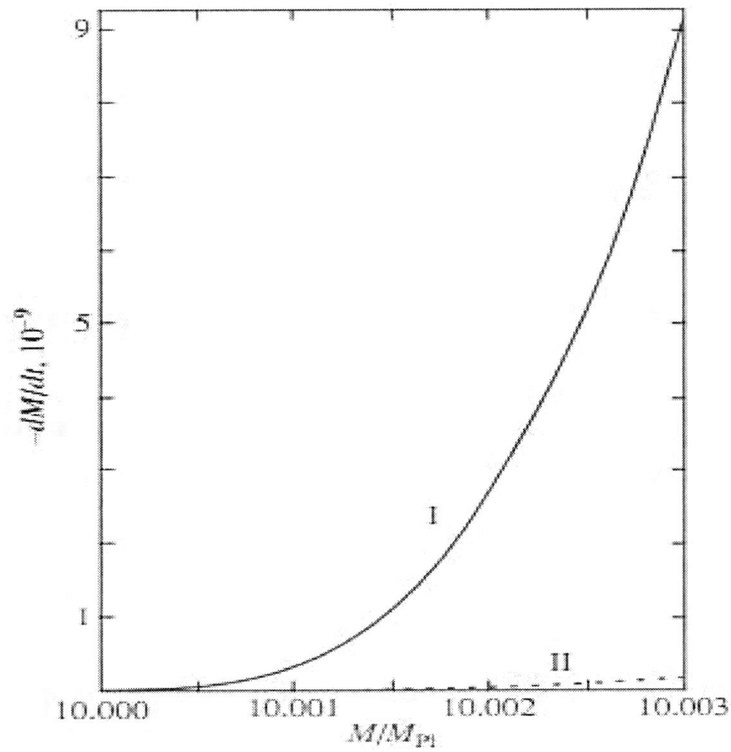

Figure 5. The black hole evaporation law with energy conservation at f nal stage when the difference between the emission probabilities of particles with different spins shows up most clearly: the emission rates of bosons (I) and fermions (II).

So, we have presented a simple model to describe the evaporation of primordial black holes at f nal stages with the formation of a relic remnant with a mass of $1 - 10^3 m_{Pl}$. These relic remnants may account for a substantial fraction of dark matter in the Universe.

It is belived that the most realistic models describing the black holes in our Universe are Schwarzschild one and Kerr one. It was established in papers [13, 22] that spinning black hole looses its angular momentum during the Hawking radiation. So, if this object is stable, the minimal (quasi-Schwarzschild) black hole, being the endpoint of the evaporation, can represent the relic remnant of the black holes formed during the initial stages of our Universe formation. This means that the evaporation of such black holes requires additional investigations.

7. Primordial Black Holes and Early Universe Parameters

As we discussed previously according to the high-order curvature gravity [28] a PBH does not evaporate completely but only to some relic remnant. This relic remnant can exist at the present time t_0 in the Universe, as the primordial black hole (PBH) mass did not exceed the critical value of $M = M^* \approx 10^{15} g$ [24] at the reheating time $t = t_{rh}$. For $M > M^*$, PBHs would continue evaporation until $t = t_0$ and would have no time to evolve into their relic remnant state. Here we discuss the conditions under which the relic remnants of PBHs have time to be formed by now, following our paper [25]. At the present time the inf ation theory [29] is widelly accepted theory for the Early Universe. According to this theory, the Early Universe exponentially expands till the reheating phase, when the matter temperature rises to $t = t_{rh}$ [30]. The reheating temperature is determined by the energy of the scalar-f eld particles providing inf ation [31]-[32]. Only the range of possible reheating temperatures is known now. Its upper limit is $m_{up} \leq 3 \times 10^{16} GeV$, as estimated by [39]. At the same time, some authors believe that $m_{up} \approx 1TeV$ [40].

Our aim is to f nd a condition on the reheating temperature at which the evaporation of PBHs has f nished by now, and the relic remnants of such objects have been formed. The temperature range is f xed by the critical PBH mass. In Early Universe black holes are formed from an enhanced overdensity in the range of the comoving particle horizon. Therefore, the characteristic (most probable) BH mass at time t depends upon the horizon size, $l \sim t$ [35]. Thus, the characteristic PBH mass is determined by the total mass within the particle horizon [35]:

$$M(t) = \frac{4}{3}\pi \left(a(t) \int_0^t \frac{d\tilde{t}}{a(\tilde{t})} \right)^3 \frac{3}{32\pi G t^2} = \gamma^{3/2} \frac{t}{t_{Pl}} m_{Pl}, \tag{39}$$

where $a(t)$ is the scale factor of the radiation-dominated Universe according to the law $a(t) \sim t^{1/2}$, and $\gamma^{3/2}$ denotes the following:

$$\gamma^{3/2} = \frac{t_{Pl}}{m_{Pl}} \frac{c^3}{G}. \tag{40}$$

According to [35], the BH mass at time $t = t_{rh}$ is

$$M(t_{rh}) = \gamma^{3/2} \frac{t_{rh}}{t_{Pl}} m_{Pl}. \tag{41}$$

There is a set of reheating models [37] now. For a fast transition rate the temperature can be used as a characteristic of the epoch. It is possible to calculate the time t_{rh} using the condition of same entropy for two times (t_{rh} and $t_\gamma = t_0$). This means that there were no signif cant phase transitions with a entropy jump after reheating, as it was pointed out by [31] and [38]. The entropy S is equal to

$$S = N_f(t, T) T^3 a^3, \tag{42}$$

where $N_f(t, T)$ is the number of degrees of freedom of the particles existing at some time and at the Universe temperature (T). Equating the entropies for the current and reheating times:

$$N_f|_{rh} T_{rh}^3 a(t_{rh})^3 = N_f|_\gamma T_\gamma^3 a(t_\gamma)^3, \tag{43}$$

where $T_f|_\gamma$ is the current microwave background radiation (CMBR) temperature and taking into account the relationship between the scale factor and redshift, formula (43) can be rewritten as

$$N_f|_{rh} T_{rh}^3 = N_f|_\gamma T_\gamma^3 (1 + z_{rh})^3. \tag{44}$$

The absence of signif cant phase transitions allows to derive a relationship between the temperature of the particle horizon and the redshift. Hence it is possible to calculate the redshift z_{rh} as a function of the current CMBR temperature T_γ, the number of degrees of freedom at the reheating time, $N_f|_{rh}$, and the number of degrees of freedom of the photons at present, $N_f|_\gamma (N_f|_\gamma = 2)$:

$$1 + z_{rh} = \left(\frac{N_f|_{rh}}{2}\right)^{1/3} \frac{T_{rh}}{T_\gamma}. \tag{45}$$

To calculate the PBH mass at the reheating time $t = t_{rh}$ using formula (7), one has to derive a relation between time t_{rh} and redshift z_{rh}. Reheating is known [35] to take place at the radiation-dominated stage characterized by the following density dependence upon the pressure: $p = \epsilon/3$. Thus, the scale factor varies as $a(t) \sim t^{1/2}$. Hence,

$$\frac{a(t_0)}{a(t_{rh})} = \left(\frac{t_0}{t_{rh}}\right)^{1/2}, \tag{46}$$

and, consequently, the redshift at the reheating time is related with the initial time as

$$1 + z_{rh} = \left(\frac{t_0}{t_{rh}}\right)^{1/2}, \tag{47}$$

where $t_0 = (2/3)H_0^{-1}$ and H_0 is the present value of Hubble constant. According to Eqs (45) and (47) we arrive to the following relation between t_{rh} and T_{rh}:

$$t_{rh} = \frac{2}{3} H_0^{-1} \left(\frac{2}{N_f|_{rh}}\right)^{2/3} \left(\frac{T_\gamma}{T_{rh}}\right)^2. \tag{48}$$

Using Eqs (39) and (48) the temperature dependence of the PBH mass at the reheating time can be obtained as

$$M(t_{rh}) = \frac{2}{3} H_0^{-1} \gamma^{3/2} \frac{m_{Pl}}{t_{Pl}} \left(\frac{2}{N_f|_{rh}}\right)^{2/3} \left(\frac{T_\gamma}{T_{rh}}\right)^2. \tag{49}$$

One can use the following designation to make the calculations more easy:

$$T_0 = \frac{2}{3} \gamma^{3/4} \left(\frac{2}{N_f|_{rh}}\right)^{1/3} \frac{1}{(t_{Pl} H_0)^{1/2}} T_\gamma. \tag{50}$$

From Eq (49) it follows that $T_{rh} = \sqrt{m_{Pl}/M(t_{rh})} T_0$. We consider $M(t_{rh}) = M^* \approx 10^{15} g$. For $M > M^*$ no relic remnants of PBHs have been formed by the present time $t = t_0$. The corresponding inequality for the reheating temperature is

$$T_{rh} < \left(\frac{m_{Pl}}{M^*}\right) T_0. \tag{51}$$

The estimation for T_{rh} is: $T_{rh} < \sqrt{m_{Pl}/M^*} T_0 \approx \alpha \times 5.6 \times 10^7 GeV$, where

$$\alpha = \sqrt{2/3} \gamma^{3/4} \left(\frac{2}{N_f|_{rh}}\right)^{1/3} \approx 3.3 \left(\frac{2}{N_f|_{rh}}\right)^{1/3}. \tag{52}$$

This relation implies that if the reheating temperature was less than $\sim 10^8 GeV$, the relic remnants of PBHs have not yet formed till now and such BHs are still evaporating. Note, that our estimation of the reheating temperature at which PBHs can produce relic remnants at present is correct in order of magnitude. However, we ignored details of the Universe evolution from $t = t_{rh}$ to the current epoch.

For more accurate estimation it is usefull to consider the transition from the radiation-dominated stage to the dust stage. As it was mentioned above, in the radiation-dominated Universe the scale factor depends upon time as $a(t) \sim t^{1/2}$. In the dust Universe, the pressure is equal to $p = 0$ and the scale factor depends on time as $a(t) \sim t^{2/3}$. The transition from one stage to the other one takes place at the so-called decoupling time. Now one can estimate the reheating temperature more accurately by taking into account the parameters of the Universe at the decoupling time:

$$\frac{a(t_0)}{a(t_{rh})} = \frac{a(t_d)}{a(t_{rh})}\frac{a(t_0)}{a(t_d)}. \tag{53}$$

The redshift z_{rh} at $t = t_{rh}$ is

$$1 + z_{rh} = \frac{a(t_0)}{a(t_{rh})} = \frac{a(t_d)}{a(t_{rh})}\frac{a(t_0)}{a(t_d)}. \tag{54}$$

According to the def nition, the scale factor is related to the redshift by

$$\frac{a(t_0)}{a(t_d)} = 1 + z_d. \tag{55}$$

Considering the radiation-dominated stage, one can write a relation similar to formula (46):

$$\frac{a(t_d)}{a(t_{rh})} = \left(\frac{t_d}{t_{rh}}\right)^{1/2}. \tag{56}$$

The last three formulas reveal a relation between t_{rh}, t_d, z_d and z_{rh}:

$$t_{rh} = t_d \left(\frac{1 + z_d}{1 + z_{rh}}\right)^2. \tag{57}$$

Thus, the reheating time now also depends on the conditions in the Universe at the decoupling time. Similarly to calculate the PBH mass using Eq () it is necessary to determine the redshift z_d at the decoupling time $t = t_d$ and the time t_d by itself. The dust density for some f xed redshift z is $\rho_m = 0.3(3H_0^2)/(8\pi G)$, $\rho_m(z) = (1 + z)^3 \rho_{m0}$, and the radiation density is $\rho_\gamma = 10^{-4} \times (3H_0^2)/(8\pi G)$, $\rho_\gamma(z) = (1 + z)^4 \rho_{\gamma 0}$ [31]. So,

$$\frac{\rho_\gamma(z)}{\rho_m(z)} = (1 + z)\frac{\rho_{\gamma 0}}{\rho_{m0}}. \tag{58}$$

At the decoupling time $t = t_d$,

$$\frac{\rho_\gamma(z)}{\rho_m(z)} = 1, \frac{\rho_{\gamma 0}}{\rho_{m0}} = \Omega_m \Omega_\gamma, z = z_d. \tag{59}$$

Therefore, the redshift at $t = t_d$ is

$$z_d = \frac{\Omega_m}{\Omega_\gamma} - 1 \approx 1.1 \times 10^4, \tag{60}$$

where

$$\Omega_\gamma = \frac{4\sigma}{c} T^4 \frac{8\pi G}{3H_0^2}. \tag{61}$$

Here, $\Omega_m \approx 0.2$ and σ is the Boltzmann constant [41]. The corresponding time t_d for the dust stage is

$$t_d = \frac{t_0}{(1+z_d)^{3/2}} = \frac{2}{3} H_0^{-1} \frac{1}{(1+z_d)^{3/2}}. \tag{62}$$

Thus, the reheating time from Eq (57) depends upon the redshift at the decoupling time as

$$t_{rh} = \frac{2}{3} H_0^{-1} \frac{1}{(1+z_d)^{3/2}} \left(\frac{1+z_d}{1+z_{rh}} \right)^2. \tag{63}$$

Estimating T_{rh} relative to M^* similar to previous calculations (see Eq (51)), one obtains

$$T_{rh} < T_0 \left(\frac{m_{Pl}}{M^*} \right)^{1/2} (1+z_d)^{1/4} \approx \alpha \times 5.7 \times 10^8, \tag{64}$$

where T_0 and α are equal to

$$T_0 = \frac{2}{3} \gamma^{3/4} \left(\frac{2}{N_f|_{rh}} \right)^{1/3} \frac{1}{(t_{Pl} H_0)^{1/2}} T_\gamma \tag{65}$$

and

$$\alpha = \sqrt{2/3} \gamma^{3/4} \left(\frac{2}{N_f|_{rh}} \right)^{1/3} \approx 3.3 \left(\frac{2}{N_f|_{rh}} \right)^{1/3}, \tag{66}$$

respectively, and z_d is determined by condition (60). The value of N_f at the reheating time depends on the number of particles at given energies (e.g., $N_f \approx 115$ for $E \sim 100 GeV$ [42]). When additional conditions at $t = t_d$ are taken into account, the lower limit for T_{rh} increases. Thus, the relic remnants of PBHs can exist at present when the reheating temperatures is higher than

$$T_{rh} \sim 2 \left(\frac{2}{N_f|_{rh}} \right)^{1/3} \times 10^9 GeV \tag{67}$$

So, from this simple analysis one arrives to a conclusion that if the reheating temperature exceeds $\approx 10^9 GeV$, the relic remnants of PBHs can exist at present epoch and be the dark-matter candidates.

8. Tidal-Charged Black Hole

Some time ago models with non-compact extra dimensions [43, 44] became very popular. They help to avoid a set of problems in $D > 4$ dimensional gravity. One of the f rst attempts to apply these models to astronomy was made in [26]. In this paper a new black hole metric originating in Randall-Sundrum models was suggested. This tidal-charged black hole is described by the following metric [26]:

$$ds^2 = A(r)dt^2 - A^{-1}(r)dr^2 - r^2(d\theta^2 + sin^2\theta d\phi^2), \quad (68)$$
$$\text{where} \quad A(r) = 1 + \frac{\alpha}{r} + \frac{\beta}{r^2}, \qquad \alpha = -2M/M_{Pl}^2,$$

β is the "tidal charge" parameter ($\beta < 0$) describing the contribution of extra dimension geometry. The restrictions on "tidal charge" value from the Solar System stability condition were found in [26] and more accurately in [27]. The most interesting feature is that opposite to the previous results the presence of "tidal charge" decreases the temperature of black hole and, therefore, it's evaporation rate. Non-zero tidal-charge corresponds to lower temperature (see Fig.6).

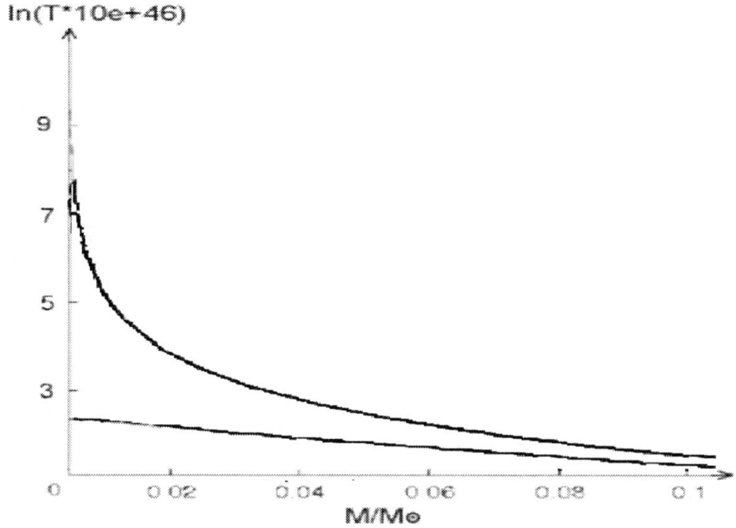

Figure 6. Temperature vs. black hole mass. The upper curve corresponds to $\beta = 0$, the lower one to $\beta = -0.1$

One can see from Fig. 7 that the evaporation rate is also less for non-zero tidal charge and is equal to:

$$\frac{d^2N}{dEdt} = \frac{\Gamma_s}{e^{-\xi} - 1}, \quad (69)$$
$$\text{where} \quad \xi = -\frac{\pi E(\alpha - \sqrt{\alpha^2 - 4\beta})}{\sqrt{\alpha^2 - 4\beta}}$$

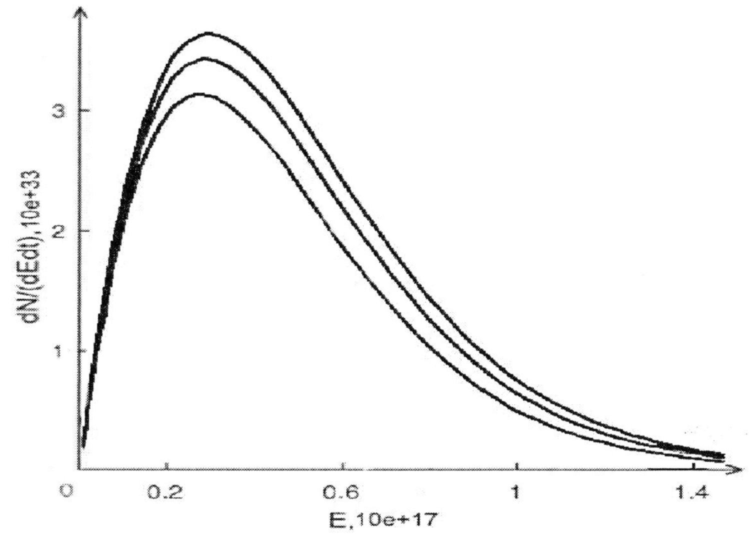

Figure 7. The number of particles emitted vs. energy (l=s=0, so $\Gamma_0 = 16M^2E^2$). The curves upside down: $\beta = 0$, $\beta = -1$, $\beta = -2$. This tidal charge values are chosen to make the craph representational.

Figure 7 shows that the presence of the tidal charge results in desreasing of the maximum value of (69) and shifting it to lower energies. This leads to decreasing in the mass loss rate. Unfortunately, the limit on "tidal charge" cuts off all the deviations from Schwarzshild black hole for astrophysical masses.

Conclusion

We discussed analytical and numerical properties of dilatonic black holes in 4D Gauss-Bonnet gravity and their inf uence on the early Universe. The combination of standard methods of gravity with the high energy physics ones allowed to clarify different problems, as black hole evaporation in higher order curvature gravity and related models. The most interesting consequence of the Gauss-Bonnet term existence is the limit on minimal black holes that gives new breathe to related models study. Such black holes do not evaporate comletely, but only to some relic remnant. If this object is stable, the minimal black hole, being the endpoint of the evaporation, can represent the relic remnant of the black holes formed during the initial stages of our Universe formation. These objects can pretend to be dark matter candidates.

Acknowledgments

This work was supported by Federal Agency on science and innovations of Russian Federation via State Contract No. 02.740.11.0575. S.A. would like to thank his co-authors of original papers M.Pomazanov, M.Sazhin, O.Sazhina, A.Barrau, G.Boudoul.

References

[1] S. Mignemi and N.R. Stewart, *Phys. Rev.* **D47**, 5259 (1993);
S. Mignemi, *Phys. Rev.* **D51**, 934 (1995).

[2] A. Economou and C.O.Luosto, *Phys. Rev.* **D49**, 5278 (1994);
M. Campanelli and C.O.Luosto, *Phys. Rev.* **D49**, 5188 (1994), *Phys. Rev.* **D51**, 6810 (1995); *Phys. Rev.* **D54**, 3854 (1996).

[3] M. Natsuume, *Phys. Rev.* **D50**, 3945 (1994).

[4] P. Kanti, N.E. Mavromatos, J. Rizos, K. Tamvakis and E. Winstanley, *Phys. Rev.* **D54**, 5049 (1996);
P. Kanti and K. Tamvakis, *Phys. Lett.* **B392**, 30 (1997);
P. Kanti, N.E. Mavromatos, J. Rizos, K. Tamvakis and E. Winstanley, *Phys.Rev.* **D57**, 6255 (1998).

[5] T. Torii, H. Yajima and K. Maeda, *Phys. Rev.* **D55**, 739 (1997).

[6] E.E. Donets and D.V. Gal'tsov, *Phys. Lett.* **B352**, 261 (1995);
E.E. Donets, D.V. Gal'tsov, M.Yu. Zotov, [gr-qc/9612067];
P. Breitenlohner, G. Lavrelashvili and D. Maison, [gr-qc/9703047].

[7] M.C. Bento and O. Bertolani, *Phys.Lett.* **B368** (1996) 198.

[8] H.S. Reall, *Phys. Rev.* **D68**, 024024 (2003).

[9] S.O. Alexeyev, M.V. Pomazanov, *Phys.Rev.* **D55** (1997) 2110-2118.

[10] S.O. Alexeyev, *Grav.Cosmol.* **3**, 161 (1997).

[11] S.O. Alexeyev, M.V. Sazhin, *Gen. Rel. Grav.* **V.30** (1998) 1187-1201.

[12] Egorov V.A., Sazonov V.V., Egorov M.A. and Smirnov V.V., *Cosmic Researchs,* **32/6**, 77 (1994).

[13] Golub G.N. and Reinsch C., *Numer. Math.* **14**, 403 (1970).

[14] M.B.Green, J.H.Schwarz and E.Witten, *Superstring theory,* Cambridge University Press, Cambridge, (1987).

[15] S. Alexeyev, A. Barrau, G. Boudoul, O. Khovanskaya, M. Sazhin, *Class.Quant.Grav.* **19** (2002) 4431-4444.

[16] S.Alexeyev, Black Holes in Higher Order Curvature Gravity, in "Black Holes and Galaxy Formation" (serie Space Science, Exploration and Policies), ed. Adonis D. Wachter, Raphael J. Propst, ISBN: 978-1-60741-703-3, NovaPublishers (New-York, 2009).

[17] S. Massar & R. Parentani, *Nucl.Phys.* **B575**, 333-356 (2000).

[18] M.K.Parikh, F.Wilczek, *Phys. Rev. Lett.* **85**, 24 (2000).

[19] K.Srinivasan, T. Padamanabhan, *Phys. Rev.* **D60**, 24007 (1999).

[20] T.Damour, R.Ruff ni, *Phys. Rev.* **D14**, 332 (1976).

[21] D.N.Page, *Phys.Rev.* **D13**, 198 (1976); **D14**, 3260 (1976); **D16**, 2402 (1977).

[22] C.M.Chambers, W.A.Hiscock, B.E.Taylor, The UPS and DOWNS of a spinning black hole, Proceedings of the 8th Marcel Grossman Meeting on Recent Developments in Theoretical and Experimental General Relativity, Gravitation and Relativistic Field Theory (MG8), Jerusalem, Israel, 22 May - 27 Jun 1997; [gr-qc/9710013].

[23] S. O. Alexeyev, A. Barrow, G. Boudoul, M. V. Sazhin, O. S. Khovanskaya, *Astron. Lett.*, **V. 28**, N. 7, (2002) 428-433.

[24] S.O. Alexeyev, O.S. Khovanskaya, *Grav.Cosmol.* **6** (2000) 14-18.

[25] S. O. Alexeyev, M. V. Sazhin, O. S. Khovanskaya, *Astron. Lett.,* **V. 28**, N. 3, (2002) 139-142.

[26] N. Dadhich, R. Maartens, P. Papadopoulos and V. Rezania, *Phys. Lett.* **B 487**, 1 (2000).

[27] S.O. Alexeyev, D.A. Starodubtseva, *J.Exp.Theor.Phys.* **111** (2010) 576-581.

[28] S.O. Alexeev, M.V. Sazhin, M.V. Pomazanov, *Int.J.Mod.Phys.* **D10** (2001) 225-230.

[29] A.Guth, *Phys.Rev.D* **23**, 347 (1981).

[30] L.Kofman, A.Linde and A.Starobinsky, *Phys.Rev.Lett.,* **73**, 3195 (1994) L.Kofman, A.Linde and A.Starobinsky, *Phys.Rev.Lett.,* **56**, 3528 (1997).

[31] A.D. Dolgov, Ya. B. Zel'dovich and M. V. Sazhin, *Cosmlogy of early universe* (Mosk.Gos. Univ., Moscow, 1998).

[32] A.Linde, *Phys. Rep.,* **333** 575 (2000).

[33] P. R. Branoff and D. R. Brill, gr-qc/9811079.

[34] I. D. Novikov and V. P. Frolov, *Physics of Black Holes* (Nauka, Moscow, 1986).

[35] I. Novikov, A. Polnarev, A. Starobinsky, and Ya. Zeldovich, *Astron. Astrophys.* **80**, 104 (1979).

[36] A. A. Starobinsky, *Zh. Eksp. Teor. Fiz.* **64**, 48 (1973) [*Sov. Phys. JETP* **37**, 28 (1973)]. A. A. Starobinsky and S. M. Churilov, *Zh. Eksp. TeFiz.* **65**, 3 (1973) [*Sov. Phys. JETP* **38**, 1 (1974)]. D. N. Page, *Phys. Rev. D* **13**, 198 (1976). D. N. Page, *Phys. Rev. D* **14**, 3260 (1976).

[37] L. Kofman, A. Linde, and A. Starobinsky, *Phys. Rev.D* **56**, 3258 (1997).

[38] B. Carr and J. Gilbert, *Phys. Rev. D* **50**, 4853 (1994).

[39] A. Melchiorri, M. Sazhin, V. Shulga, and N. Vittorio, *Astrophys. J.* **518**, 562 (1999).

[40] M.Libanov, V.Rubakov, P.Tinyakov, *Phys. Lett. B* **442**, 63 (1998).

[41] L. D. Landau and E. M. Lifshitz, Course of Theoretical Physics, Vol.5: *Statistical Physics* (Nauka, Moscow, 1988; Pergamon, Oxford, 1980).

[42] V. A. Rubakov, *Mon. Not. R. Astron. Soc.* **229**, 357 (1987).

[43] N.Arkani-Hamed, S.Dimopoulos, G.Dvali, *Phys. Lett.* **B429**, 263 (1998).

[44] L.Randall, R.Sundrum, *Phys. Rev. Lett.* **83**, 4690 (1999).

In: Black Holes: Evolution, Theory and Thermodynamics ISBN: 978-1-61942-929-1
Editors: A. J. Bauer and D. G. Eiffel, pp. 103-116

Chapter 4

The Relationship between the Tunneling Rate of a Rotating Black Hole and the First Law of Black Hole Thermodynamics

Zheng Ze Ma*
Department of Physics, Southeast University,
Nanjing, P. R. China

Abstract

We study the relationship between the tunneling rate of a rotating black hole and the f rst law of black hole thermodynamics. We show that the tunneling rate of a rotating black hole $e^{-2\mathrm{Im}\mathcal{I}}$ can be set up from the emission rate $e^{\Delta S}$ and the f rst law of black hole thermodynamics. Besides, we give an analysis of the one loop corrected tunneling rate of a four-dimensional Kerr black hole from the corrected Bekenstein–Hawking entropy of the Kerr black hole.

PACS: 04.70.Dy, 04.70.-s, 04.50.+h, 03.65.Xp

1. Introduction

As a special state of spacetime, black hole occurs in any gravitational theories. Black holes have many intriguing properties, such as the Hawking radiation, entropy, and the f rst law thermodynamics [1–6]. Many different explanations and derivations of black hole radiation have been given in the literature [7–13]. In [14–17], a semi-classical description of black hole radiation has been proposed by Parikh and Wilczek which is called tunneling. In this method, Hawking radiation of a black hole is interpreted as radiated particles tunneling across a potential barrier on the horizon. The barrier which a radiated particle tunnels across is generated by the contraction of the horizon when the black hole loses its total energy during the radiation.

The tunneling method of Parikh and Wilczek has been used in the research of Hawking radiation of many different types of black holes by many authors [18–46]. Using

*E-mail address: z.z.ma@seu.edu.cn

this method, one can derive the Hawking temperatures of many different types of black holes and obtain the Boltzmann distribution of radiated particles, including the radiation of fermion particles. Some recent studies and applications of this method have been carried out in [47–56].

In the tunneling picture, the tunneling rate of a particle emitted from a black hole horizon is given by [14–17]

$$\Gamma = \Gamma_0 \exp(-2\mathrm{Im}\mathcal{I}), \tag{1}$$

where $\mathcal{I}$ is the classical action of the tunneling particle, Γ_0 is a normalization factor. Because the radiation is related with the horizon singularity of a black hole, one can consider that the tunneling particle is moving in a complex path which is permissive in a spacetime manifold with singularities, and thus, the classical action of the particle becomes complex and the tunneling amplitude is determined by the imaginary part of the action, while the real part of the action can be absorbed in a normalization factor of the amplitude.

On the other hand, in [57], from the approach of geometric dynamics, Massar and Parentani has obtained that the probability for a black hole to emit a particle is given by

$$\Gamma = \Gamma_0 e^{\Delta A/4} = \Gamma_0 e^{\Delta S}, \tag{2}$$

where ΔA is the change of the horizon area induced by the emission ($\Delta A < 0$), ΔS is the change of the entropy of the black hole induced by the emission ($\Delta S < 0$). Equation (2) is also in consistent with the analysis of statistical mechanics. At the same time, we know that in the radiation process of a black hole, it satisf es the f rst law of thermodynamics

$$\delta M = T_H \delta S + \Omega_i \delta J_i + \Phi_j \delta Q_j, \tag{3}$$

where δM, δJ_i, and δQ_j are the changes of mass, angular momenta, and charges of the black hole.

As we can see in that [14–17], the tunneling rate of formula (1) was f rst set up for a spherically symmetric black hole. However, in the references, many authors have used such a formula directly to the tunnelings of rotating black holes. Therefore we consider that it is necessary for us to search for a basis that formula (1) can be used to the tunneling of a rotating black hole. We f nd that there is a relationship between the tunneling rate of a rotating black hole and the f rst law of black hole thermodynamics. Therefore in Sec. 2 of this paper, we will show that the tunneling rate (1) can be derived from the emission rate (2) and formula (3) for a rotating black hole. In this paper, in studying the tunneling of a general higher-dimensional rotating black hole, we will use the rotating coordinate transformation method that was f rst introduced in [24, 27].

In [47–51], it was shown that the tunneling method had also offered an approach for the calculation of entropy correction of a black hole. The corrected Bekenstein–Hawking entropy of Kerr black hole has been obtained in [50] from the tunneling approach. Using formula (1) and (2) and the corrected Bekenstein–Hawking entropy of Kerr black hole, we can calculate the corrected tunneling rate of Kerr black hole. These results are given in Sec. 3. Section 4 is the conclusion.

2. Tunneling Rate of a Higher-Dimensional Rotating Black Hole from Black Hole Thermodynamics

To consider a stationary rotating black hole in D-dimensional spacetime, its metric can generally be written in the form [58–60]

$$\begin{aligned} ds^2 &= -g_{tt}dt^2 + g_{rr}dr^2 + g_{\theta\theta}d\theta^2 + g_{\psi_i\psi_i}d\psi_i^2 + 2g_{\theta\psi_i}d\theta d\psi_i + \sum_{i<j} 2g_{\psi_i\psi_j}d\psi_i d\psi_j \\ &\quad + g_{\phi_i\phi_i}d\phi_i^2 + \sum_{i<j} 2g_{\phi_i\phi_j}d\phi_i d\phi_j - 2g_{t\phi_i}dtd\phi_i, \end{aligned} \tag{4}$$

where in metric (4), i and j on ϕ run from 1 to $[(D-1)/2]$; for even-dimensional case, i and j on ψ run from 1 to $[(D-3)/2]$; for odd-dimensional case, i and j on ψ run from 1 to $[(D-5)/2]$. The black hole is rotating in each ϕ_i-direction, so there are $[(D-1)/2]$ angular momentum components. Due to the higher-dimensional axial symmetry of the metric, all metric components are functions of the coordinates (r, θ, ψ_i), but are not related with the coordinates ϕ_j. The horizon of the black hole is determined by the largest positive root of the equation $g^{rr}|_{r=r_h} = g_{rr}^{-1}|_{r=r_h} = 0$.

For a rotating black hole, generally there exists the radiation of angular momenta, and thus, its radiation not only contains the s-wave components whose orbital angular momenta are zero, but also contains higher rank partial wave components whose orbital angular momenta are not zero, although the contraction of the horizon is still spherically symmetric in the process of the radiation. For a radiated particle with zero orbital angular momentum, i.e. an s-wave, its motion can be described by a radial movement. However, for a radiated particle with nonzero orbital angular momenta, i.e. a higher rank partial wave, it not only contains the motion in the radial direction, but also contains the motion in the ϕ_i-directions, i.e., in general, the tunneling particles from a rotating black hole does also take part in the motions on the rotating degrees of freedom.

To consider an observer located at inf nity, a black hole, no matter whether it is rotating or not, can always be regarded as a point radiation source, the f nite scale of its horizon can be neglected always. Thus, for an observer at inf nity, the radiation of a rotating black hole can still be regarded as radial waves. However, the rotation of the black hole may have certain inf uences on the radiated particles. Because the radiated particles of a black hole are generated near the horizon and the tunneling has occurred on the horizon, we can consider that the inf uences of the rotation of the black hole on its radiated particles are reduced to a dragging effect on them on the horizon. Thus, all of the radiated particles are dragged by the rotation of the black hole when they tunnel across the horizon, and then radiate to inf nity in radial directions. Therefore, for an observer at inf nity, the effective motion of a radiated particle from a rotating black hole including s-waves and higher rank partial waves can be regarded as a decomposition of a dragging motion on the rotating degrees of freedom on the horizon and a radial motion from horizon to inf nity. That is to say, for a rotating black hole, for an observer located at inf nity, its radiated particles can still be regarded as moving in radial directions; but because of the rotation of the black hole, the radiated particles have been dragged by the rotation when they tunnel across the horizon.

The dragging motion on the rotating degrees of freedom of a tunneling particle can be eliminated in a reference frame co-rotating with the black hole horizon. Thus we make the following rotating coordinate transformation to metric (4)

$$\phi_i' = \phi_i - \Omega_i t, \tag{5}$$

where Ω_i are angular velocities of the horizon in ϕ_i-directions respectively which are constants. Under this coordinate transformation, metric (4) turns to the form

$$\begin{aligned} ds^2 &= -G_{tt}dt^2 + g_{rr}dr^2 + g_{\theta\theta}d\theta^2 + g_{\psi_i\psi_i}d\psi_i^2 + 2g_{\theta\psi_i}d\theta d\psi_i + \sum_{i<j} 2g_{\psi_i\psi_j}d\psi_i d\psi_j \\ &\quad + g_{\phi_i\phi_i}d\phi_i'^{\,2} + \sum_{i<j} 2g_{\phi_i\phi_j}d\phi_i' d\phi_j' - 2g_{t\phi_i'}dtd\phi_i', \end{aligned} \tag{6}$$

where

$$\begin{aligned} G_{tt} &= g_{tt} - g_{\phi_i\phi_i}\Omega_i^2 - \sum_{i<j} 2g_{\phi_i\phi_j}\Omega_i\Omega_j + 2g_{t\phi_i}\Omega_i, \quad (7) \\ g_{t\phi_i'} &= g_{t\phi_i} - g_{\phi_i\phi_j}\Omega_j. \quad (8) \end{aligned}$$

G_{tt} and $g_{t\phi_i'}$ satisfy

$$G_{tt}\big|_{r=r_h} = 0, \qquad g_{t\phi_i'}|_{r=r_h} = 0. \tag{9}$$

The horizon's radius of metric (6) is still determined by $g^{rr}|_{r=r_h} = g_{rr}^{-1}|_{r=r_h} = 0$.

For a higher-dimensional rotating black hole, because its spacetime is axially symmetric, its radiation must also be axially symmetric. For the spacetime of metric (4), its radial directions are determined by the azimuthal angles (θ, ψ_i, ϕ_j); for the spacetime of metric (6), its radial directions are determined by the azimuthal angles $(\theta, \psi_i, \phi_j')$. Because of the axial symmetry of the radiation, the tunneling motions of the radiated particles should be same for a constant set of azimuthal angles (θ_0, ψ_{i0}), despite the changes of the azimuthal angles ϕ_j or ϕ_j'. For the radiation of a zero-mass particle, its motion is determined by the null-geodesic equation. Because the tunneling motions of the radiated particles are the same for a constant set of azimuthal angles (θ_0, ψ_{i0}), we can set the azimuthal angles (θ, ψ_i) to be a set of constant values in metric (6) for the purpose of determining the null-geodesic equation of a zero-mass particle.

Thus, to let the azimuthal angles (θ, ψ_i) be a set of constant values (θ_0, ψ_{i0}) in metric (6), it reduces to the form

$$\begin{aligned} ds^2 = &- G_{tt}(r,\theta_0,\psi_{i0})dt^2 + g_{rr}(r,\theta_0,\psi_{i0})dr^2 + g_{\phi_i\phi_i}(r,\theta_0,\psi_{i0})d\phi_i'^{\,2} \\ &+ \sum_{i<j} 2g_{\phi_i\phi_j}(r,\theta_0,\psi_{i0})d\phi_i' d\phi_j' - 2g_{t\phi_i'}(r,\theta_0,\psi_{i0})dtd\phi_i'. \end{aligned} \tag{10}$$

Because in metric (10), g_{rr} is singular on the horizon, in order to study the motion of a tunneling particle, we need to eliminate such a coordinate singularity f rst. This can be realized through the following Painlevé coordinate transformation

$$dt = dT - \sqrt{\frac{g_{rr}(r,\theta_0,\psi_{i0}) - 1}{G_{tt}(r,\theta_0,\psi_{i0})}}dr, \tag{11}$$

where T is the Painlevé time coordinate [14–17]. In (11), because (θ_0, ψ_{i0}) are a set of constants, equation (11) is integrable. Under this coordinate transformation, we obtain

$$\begin{aligned} ds^2 = & - G_{tt}(r,\theta_0,\psi_{i0})dT^2 + 2\sqrt{G_{tt}(r,\theta_0,\psi_{i0})}\sqrt{g_{rr}(r,\theta_0,\psi_{i0})-1}\,drdT + dr^2 \\ & + g_{\phi_i\phi_i}(r,\theta_0,\psi_{i0})d\phi_i'^{\,2} + \sum_{i<j} 2g_{\phi_i\phi_j}(r,\theta_0,\psi_{i0})d\phi_i' d\phi_j' \\ & - 2g_{t\phi_i'}(r,\theta_0,\psi_{i0})d\phi_i'\left(dT - \sqrt{\frac{g_{rr}(r,\theta_0,\psi_{i0})-1}{G_{tt}(r,\theta_0,\psi_{i0})}}dr\right). \end{aligned} \tag{12}$$

As mentioned above, in the co-rotating frame, the tunneling particles have no motion on the ϕ_i' degrees of freedom, they satisfy $d\phi_i' = 0$; therefore, the line element of a tunneling particle is determined by

$$ds^2 = -G_{tt}(r,\theta_0,\psi_{i0})dT^2 + 2\sqrt{G_{tt}(r,\theta_0,\psi_{i0})}\sqrt{g_{rr}(r,\theta_0,\psi_{i0})-1}\,drdT + dr^2. \tag{13}$$

For the radiation of a zero-mass particle, its motion is determined by the null-geodesic equation $ds^2 = 0$. To solve this equation, we obtain

$$\dot{r} = \sqrt{G_{tt}(r,\theta_0,\psi_{i0}) \cdot g_{rr}(r,\theta_0,\psi_{i0})}\left(\pm 1 - \sqrt{1 - \frac{1}{g_{rr}(r,\theta_0,\psi_{i0})}}\right). \tag{14}$$

The plus and minus signs in (14) correspond to outgoing and ingoing radial null geodesics respectively. For an outgoing tunneling particle, $\dot{r}$ is positive, we have

$$\dot{r} = \sqrt{G_{tt}(r,\theta_0,\psi_{i0}) \cdot g_{rr}(r,\theta_0,\psi_{i0})}\left(1 - \sqrt{1 - \frac{1}{g_{rr}(r,\theta_0,\psi_{i0})}}\right). \tag{15}$$

Because $G_{tt}(r_h,\theta,\psi_i) = 0$, $g^{rr}(r_h,\theta,\psi_i) = 0$, near the horizon, $G_{tt}(r,\theta_0,\psi_{i0})$ and $g^{rr}(r,\theta_0,\psi_{i0})$ can be expanded by

$$G_{tt}(r,\theta_0,\psi_{i0}) = G_{tt}'(r_h,\theta_0,\psi_{i0})(r-r_h) + O((r-r_h)^2), \tag{16}$$

$$g^{rr}(r,\theta_0,\psi_{i0}) = g^{rr\prime}(r_h,\theta_0,\psi_{i0})(r-r_h) + O((r-r_h)^2). \tag{17}$$

Making use of (16) and (17) and expanding (15) near r_h, we obtain

$$\dot{r} = \frac{\sqrt{G_{tt}'(r_h,\theta_0,\psi_{i0})g^{rr\prime}(r_h,\theta_0,\psi_{i0})}}{2}(r-r_h) + O((r-r_h)^2). \tag{18}$$

In the Appendix, from the equation

$$\nabla^\mu(\xi^\mu\xi_\mu) = -2\kappa\xi^\mu \tag{19}$$

satisf ed by the Killing f eld $\xi^\mu = \frac{\partial}{\partial t} + \sum \Omega_i \frac{\partial}{\partial \phi_i}$, we have derived that the surface gravity of a higher-dimensional rotating black hole can be expressed by

$$\kappa(r_h) = \frac{\sqrt{G_{tt}'(r_h,\theta_0,\psi_{i0})g^{rr\prime}(r_h,\theta_0,\psi_{i0})}}{2}. \tag{20}$$

To substitute (20) into (18) we obtain

$$\dot{r} = \kappa(r_h)(r - r_h) + O((r - r_h)^2). \tag{21}$$

To consider the change of the entropy when a particle is emitted, we have

$$\Delta S = \int_{M_i}^{M_f} \frac{dS}{dM} dM. \tag{22}$$

Because in the co-rotating frame, the angular velocities of the horizon are zero, the angular momenta of a rotating black hole are zero in such a frame, and thus we have $\delta J_i = 0$ in the co-rotating frame. To consider the emission of a zero charge particle, we have $\delta Q_j = 0$. Therefore, from (3), we can write

$$\frac{dS}{dM} = \frac{1}{T_H} = \frac{2\pi}{\kappa(r_h)}. \tag{23}$$

To combine (21) and ignore the higher-order terms, we have

$$\frac{dS}{dM} = \frac{2\pi(r - r_h)}{\dot{r}}. \tag{24}$$

To substitute (24) into (22), we have

$$\Delta S = \int_{M_i}^{M_f} \frac{2\pi(r - r_h)}{\dot{r}} dM. \tag{25}$$

Because r_h is a singular point in the spacetime of a black hole, we can introduce a complex path from r_i to r_f to contain r_h and write

$$\pi = -\mathrm{Im} \int_{r_i}^{r_f} \frac{1}{r - r_h} dr. \tag{26}$$

Because $r_i > r_f$, the integral path is requested to round the pole r_h in a semicircle below the real axis. To substitute (26) into (25), we have

$$\Delta S = -2\mathrm{Im} \int_{M_i}^{M_f} \int_{r_i}^{r_f} \frac{dr}{\dot{r}} dM. \tag{27}$$

As we have argued that in the co-rotating frame, a tunneling particle only has its motion in radial direction, to make use of Hamilton's equation $\dot{r} = \frac{dH}{dp_r} = \frac{dM}{dp_r}$, we can write

$$\int_{M_i}^{M_f} \frac{1}{\dot{r}} dM = \int_{M_i}^{M_f} \frac{1}{\frac{dM}{dp_r}} dM = \int_0^{p_r} dp'_r. \tag{28}$$

To substitute (28) into (27), we obtain

$$\Delta S = -2\mathrm{Im} \int_{r_i}^{r_f} \int_{0}^{p_r} dp'_r dr = -2\mathrm{Im} \int_{r_i}^{r_f} p_r dr. \tag{29}$$

For a tunneling particle only having motion in radial direction, its classical action can just be written as $\mathcal{I} = \int_{r_i}^{r_f} p_r dr$. Thus, we have the relation

$$\Delta S = -2\mathrm{Im}\mathcal{I}. \tag{30}$$

To substitute (30) into (2), we obtain

$$\Gamma = \Gamma_0 \exp(-2\mathrm{Im}\mathcal{I}), \tag{31}$$

which is the tunneling rate given by Parikh and Wilczek. Therefore, we have derived the tunneling rate (1) for a higher-dimensional rotating black hole from the emission rate (2) and the f rst law of black hole thermodynamics.

Equation (31) for a higher-dimensional rotating black hole is derived from the emission rate (2) and the f rst law of black hole thermodynamics in the co-rotating frame. To transform back to the original frame of metric (4), it is reasonable to presume that the tunneling rate given by (31) is still correct; however, the action of a tunneling particle should contain the motion of the particle on the other coordinate degrees of freedom. A possible form of the action may be given by

$$\mathcal{I} = \int_{r_i}^{r_f} p_r dr + \int_{\phi_i(i)}^{\phi_i(f)} p_{\phi_i} d\phi_i + \int_{\psi_j(i)}^{\psi_j(f)} p_{\psi_j} d\psi_j + \int_{\theta_i}^{\theta_f} p_\theta d\theta, \tag{32}$$

which is the complete integral of the canonical momenta on their canonical coordinates of a tunneling particle moving in the spacetime of metric (4). However, to calculate the action in terms of (32) for a tunneling particle is not convenient. Therefore, in this paper, we have considered the effective motion of a tunneling particle from a rotating black hole which is equal to a dragging motion on the rotating degrees of freedom on the horizon and a radial motion from horizon to inf nity; and we point out that the dragging motion on the rotating degrees of freedom of a tunneling particle can be eliminated in a reference frame co-rotating with the event horizon.

3. Corrected Tunneling Rate of a Kerr Black Hole

There are many different methods for the calculation of the entropy correction of a black hole [61–64]. In [47–51], it was shown that the tunneling method had also offered an approach for the calculation of entropy correction of a black hole. In the tunneling method, if we know the correction of the Bekenstein–Hawking entropy of a black hole, then we can derive its corrected tunneling rate from formula (1) and (2). In this section, we will study the corrected tunneling rate of a four-dimensional Kerr black hole based on the corrected Bekenstein–Hawking entropy of Kerr black hole obtained in [50].

The corrected Bekenstein–Hawking entropy of a Kerr black hole has been obtained in [50] from the tunneling approach which is given by

$$\begin{aligned} S_{\text{Kerr}} &= S_{\text{BH}} + \frac{1}{90}\log S_{\text{BH}} + \text{higher order terms}, \\ &= \frac{A}{4} + \frac{1}{90}\log A + \text{constant} + \text{higher order terms}, \end{aligned} \tag{33}$$

with the horizon area of the Kerr black hole

$$A = 8\pi\left[M^2 + M\sqrt{M^2 - a^2}\right], \tag{34}$$

where $a = J/M$, J is the angular momentum of the Kerr black hole.

We consider the emission of a scalar particle carrying energy ω and orbital angular momentum j. Before the particle is emitted, to ignore a constant and higher order terms, the entropy of the black hole is

$$S_{\text{BH}}(M, J) = 2\pi\left[M^2 + \sqrt{M^4 - J^2}\right] + \frac{1}{90}\log\left[M^2 + \sqrt{M^4 - J^2}\right]. \tag{35}$$

After the particle is emitted, to ignore a constant and higher order terms, the entropy of the black hole is

$$\begin{aligned} S_{\text{BH}}(M-\omega, J-j) &= 2\pi\left[(M-\omega)^2 + \sqrt{(M-\omega)^4 - (J-j)^2}\right] \\ &\quad + \frac{1}{90}\log\left[(M-\omega)^2 + \sqrt{(M-\omega)^4 - (J-j)^2}\right]. \end{aligned} \tag{36}$$

For an emitted particle, we can suppose $\omega << M$ and $j << J$. From (35) and (36), to omit the higher-order terms of ω and j, we obtain

$$\begin{aligned} \Delta S &= S_{\text{BH}}(M-\omega, J-j) - S_{\text{BH}}(M, J) \\ &= -2\pi\left[2M + \frac{2M^3}{\sqrt{M^4 - J^2}}\right]\omega - \frac{2M + \frac{2M^3}{\sqrt{M^4-J^2}}}{90(M^2 + \sqrt{M^4 - J^2})}\omega \\ &\quad + \frac{2\pi J}{\sqrt{M^4 - J^2}}j + \frac{\frac{J}{\sqrt{M^4-J^2}}}{90(M^2 + \sqrt{M^4 - J^2})}j. \end{aligned} \tag{37}$$

The inverse temperature and the angular velocity of the Kerr black hole can be written in the form

$$\beta = \frac{1}{T_H} = \frac{4\pi(M^3 + M\sqrt{M^4 - J^2})}{\sqrt{M^4 - J^2}}, \tag{38}$$

$$\Omega_H = \frac{J}{2(M^3 + M\sqrt{M^4 - J^2})} = \frac{2\pi J T_H}{\sqrt{M^4 - J^2}}. \tag{39}$$

Making use of (38) and (39), (37) can be written in the form

$$\Delta S = -\frac{\omega - j\Omega_H}{T_H} - \frac{1}{90 S_{\text{BH}}} \cdot \frac{\omega - j\Omega_H}{T_H}. \tag{40}$$

Substituting (40) into (2), we obtain that the corrected tunneling rate of the Kerr black hole is given by

$$\Gamma = \Gamma_0 e^{-\beta(\omega - j\Omega_H) - \frac{1}{90S_{\mathrm{BH}}}\beta(\omega - j\Omega_H)} \approx \Gamma_0 \left[1 - \frac{\beta(\omega - j\Omega_H)}{90S_{\mathrm{BH}}}\right] e^{-\beta(\omega - j\Omega_H)}. \quad (41)$$

The dominant term of (41) is just the emission rate of the Kerr black hole without quantum corrections. The minus term in the brackets of (41) comes from quantum corrections, indicating a deviation of the radiation from pure thermal distribution. Because β/S_{BH} decreases generally when the mass of a black hole increases, the quantum corrections to the tunneling rate will decrease when the mass of the Kerr black hole increases. In (41), we need to add the condition $\omega - j\Omega_H > 0$, i.e. $j < \omega/\Omega_H$ to the tunneling particles. Such a condition makes the Boltzmann distribution be signif cative for the tunneling particles, which also means that we have excluded the super-radiance of the tunneling particles that is outside the validity of the semi-classical method and needs to be investigated from a full quantum f eld approach.

We can write (41) in the form

$$\Gamma = \Gamma_0 e^{-\beta_h(\omega - j\Omega_H)}, \quad (42)$$

where

$$\beta_h = \Big(1 + \frac{1}{90S_{\mathrm{BH}}}\Big)\beta \quad (43)$$

is the quantum corrected inverse temperature. Therefore, the corrected Hawking temperature of the Kerr black hole is given by

$$T_h = \frac{T_H}{1 + \frac{1}{90S_{\mathrm{BH}}}} \approx \Big(1 - \frac{1}{90S_{\mathrm{BH}}}\Big)T_H. \quad (44)$$

Thus, we can see that, for a Kerr black hole, to consider quantum corrections, its Bekenstein–Hawking entropy has a little increase, its tunneling rate has a little decrease, and its Hawking temperature also has a little decrease.

Conclusion

A semi-classical and direct-viewing description of black hole radiation has been proposed by Parikh and Wilczek in [14–17] which is called tunneling. Using this method, one can derive the Boltzmann distribution of the radiated particles, the Hawking temperature, the entropy correction of a black hole [18–51]. Some more recent studies and applications on this topic have been carried out in [52–56].

Because in [14–17], the tunneling rate of formula (1) was set up for a spherically symmetric black hole in fact, while many authors have used such a formula directly to the tunnelings of rotating black holes, it is necessary for us to look for a basis that formula (1) can be used to the tunneling of a rotating black hole. In this paper, from the relationship between the tunneling rate of a rotating black hole and the f rst law of black hole thermodynamics, we show that the tunneling rate $e^{-2\mathrm{Im}\mathcal{I}}$ can be set up from the emission rate $e^{\Delta S}$ and the f rst law of black hole thermodynamics for a general higher-dimensional rotating black hole. The study of the relationship between the tunneling rate of a spherically symmetric black hole and the f rst law of black hole thermodynamics has been carried out in [39]. In the derivation of this paper, we have used the method of rotating coordinate transformation that was f rst introduced in [24, 27].

In the tunneling approach, if we know the correction of the Bekenstein–Hawking entropy of a black hole, then we can derive its corrected tunneling rate from formula (1) and (2). In this paper, we have studied the corrected tunneling rate of the Kerr black hole from the corrected Bekenstein–Hawking entropy of Kerr black hole obtained in [50]. Then, from its corrected tunneling rate, we have obtained its corrected Hawking temperature.

Appendix. Surface Gravity of a Higher-Dimensional Rotating Black Hole from Black Hole Thermodynamics

In this appendix, we derive the expression for the surface gravity of a higher-dimensional rotating black hole from black hole thermodynamics. The metric of a stationary higher-dimensional rotating black hole is given by (4) generally. For metric (4), there exists the Killing f eld

$$\xi^\mu = \frac{\partial}{\partial t} + \sum \Omega_i \frac{\partial}{\partial \phi_i}, \tag{A.1}$$

where Ω_i are the angular velocities of the horizon corresponding to the coordinates ϕ_i. Because the horizon is a null surface and ξ^μ is normal to the horizon, we have

$$\xi^\mu \xi_\mu = 0 \tag{A.2}$$

on the horizon. For metric (4), we obtain

$$\xi^\mu \xi_\mu = g_{tt} - g_{\phi_i\phi_i}\Omega_i^2 - \sum_{i<j} 2g_{\phi_i\phi_j}\Omega_i\Omega_j + 2g_{t\phi_i}\Omega_i. \tag{A.3}$$

Here, we have def ned that the square of the norm of the Killing f eld is positive outside the horizon which can be verif ed for the case $\Omega_i = 0$. As in (7), we def ne

$$G_{tt} = g_{tt} - g_{\phi_i\phi_i}\Omega_i^2 - \sum_{i<j} 2g_{\phi_i\phi_j}\Omega_i\Omega_j + 2g_{t\phi_i}\Omega_i; \tag{A.4}$$

therefore, G_{tt} is a function depending on the variables (r, θ, ψ_i). Because of (A.2), we have

$$G_{tt}\Big|_{r=r_h} = 0. \tag{A.5}$$

We can introduce a scalar function λ and write [65]

$$\xi^\mu \xi_\mu = -\lambda^2, \tag{A.6}$$

λ is a constant on the horizon. According to (A.3), we have $\lambda^2 = -G_{tt}$ for metric (4). Let ∇^μ represent the covariant derivative operator; thus $\nabla^\mu(\xi^\nu\xi_\nu)$ is also normal to the horizon. Then, according to [6, 65], there exists a function κ satisfying the equation

$$\nabla^\mu(-\lambda^2) = -2\kappa\xi^\mu, \tag{A.7}$$

where on the horizon, $\kappa(r_h)$ is a constant and is just the horizon's surface gravity.

A lower index equation

$$\nabla_\mu(-\lambda^2) = -2\kappa\xi_\mu \tag{A.8}$$

can also be set up via the same approach. Thus, from (A.7) and (A.8) we have

$$\nabla^\mu(-\lambda^2)\nabla_\mu(-\lambda^2) = -4\kappa^2\lambda^2. \tag{A.9}$$

Because λ^2 is a scalar function, κ^2 is also a scalar function. Therefore the surface gravity of a higher-dimensional rotating black hole is invariant under general coordinate transformations, including the rotating coordinate transformation (5). From (A.3), (A.4), (A.6), (A.9), and the higher-dimensional axial symmetry of the metric, for metric (4), we obtain

$$4\kappa^2 G_{tt} = g^{rr}(\partial_r G_{tt})^2 + g^{\theta\theta}(\partial_\theta G_{tt})^2 + g^{\psi_i\psi_i}(\partial_{\psi_i} G_{tt})^2 + 2g^{\theta\psi_i}\partial_\theta G_{tt}\partial_{\psi_i} G_{tt} + \sum_{i<j} 2g^{\psi_i\psi_j}\partial_{\psi_i} G_{tt}\partial_{\psi_j} G_{tt}. \tag{A.10}$$

We need to take the limit $r \rightarrow r_h$ in both sides of (A.10). Because G_{tt} is a function depending on the variables (r, θ, ψ_i), we can write

$$\lim_{r\rightarrow r_h} \partial_\theta G_{tt}(r,\theta,\psi_i) = \left[\frac{\partial G_{tt}(r,\theta,\psi_i)}{\partial\theta}\right]\Bigg|_{r=r_h}. \tag{A.11}$$

The right hand side of (A.11) is the correct meaning of $\lim_{r\rightarrow r_h} \partial_\theta G_{tt}$. Because in (A.11), the partial derivative is taken with respect to the variable θ, while not related with the variable r, we have

$$\left[\frac{\partial G_{tt}(r,\theta,\psi_i)}{\partial\theta}\right]\Bigg|_{r=r_h} = \frac{\partial G_{tt}(r_h,\theta,\psi_i)}{\partial\theta}. \tag{A.12}$$

Because of (A.5), $G_{tt}(r_h, \theta, \psi_i)$ is a constant which is zero, we have

$$\frac{\partial G_{tt}(r_h,\theta,\psi_i)}{\partial\theta} = 0. \tag{A.13}$$

Therefore we have

$$\lim_{r\rightarrow r_h} \partial_\theta G_{tt}(r,\theta,\psi_i) = 0. \tag{A.14}$$

The same reason as above, we have

$$\lim_{r\rightarrow r_h} \partial_{\psi_i} G_{tt}(r,\theta,\psi_i) = 0. \tag{A.15}$$

Thus, taking the limit $r \rightarrow r_h$ in both sides of (A.10) yields

$$\kappa(r_h) = \lim_{r\rightarrow r_h} \frac{\partial_r\sqrt{G_{tt}}}{\sqrt{g_{rr}}}. \tag{A.16}$$

In (A.16), like that of (A.11), the partial derivative is taken before the limit. For a four-dimensional rotating black hole, $\kappa(r_h)$ is a constant on the horizon; for a higher-dimensional rotating black hole, $\kappa(r_h)$ is also a constant on the horizon. Therefore, $\kappa(r_h)$

can be evaluated at an arbitrary point on the horizon. Thus, for the azimuthal angles (θ, ψ_i), we can take them to be a set of constant values (θ_0, ψ_{i0}) in (A.16). To substitute (16) and (17) into (A.16), the surface gravity of a stationary higher-dimensional rotating black hole can be expressed by

$$\kappa(r_h) = \frac{\sqrt{G'_{tt}(r_h, \theta_0, \psi_{i0}) g^{rr\prime}(r_h, \theta_0, \psi_{i0})}}{2}. \tag{A.17}$$

Equation (A.17) is needed in Sec. 2 in obtaining expression (21) for a stationary higher-dimensional rotating black hole.

References

[1] Hawking, S. W. *Nature* (London) 1974, 248, 30.

[2] Hawking, S. W. *Commun. Math. Phys.* 1975, 43, 199.

[3] Bekenstein, J. D. *Lett. Nuo. Cim.* 1972, 4, 737.

[4] Bekenstein, J. D. *Phys. Rev. D* 1973, 7, 2333.

[5] Bekenstein, J. D. *Phys. Rev. D* 1974, 9, 3292.

[6] Bardeen, J. M.; Carter, B.; Hawking, S. W. *Commun. Math. Phys.* 1973, 31, 161.

[7] Wald, R. M. *Commun. Math. Phys.* 1975, 45, 9.

[8] DeWitt, B. S. *Phys. Rep.* 1975, 19, 295.

[9] Unruh, W. G. *Phys. Rev. D* 1976, 14, 870.

[10] Page, D. N. *Phys. Rev. D* 1976, 13, 198.

[11] Kraus, P.; Wilczek, F. *Nucl. Phys. B* 1995, 433, 403.

[12] Kraus, P.; Wilczek, F. *Nucl. Phys. B* 1995, 437, 231.

[13] Keski-Vakkuri, E.; Kraus, P. *Nucl. Phys. B* 1997, 491, 249.

[14] Parikh, M. K.; Wilczek, F. *Phys. Rev. Lett.* 2000, 85, 5042.

[15] Parikh, M. K. *Phys. Lett. B* 2002, 546, 189.

[16] Parikh, M. K. *Int. J. Mod. Phys. D* 2004, 13, 2351.

[17] Parikh, M. K. (2004). Energy Conservation and Hawking Radiation. arXiv:hep-th/0402166.

[18] Srinivasan, K.; Padmanabhan, T. *Phys. Rev. D* 1999, 60, 024007.

[19] Shankaranarayanan, S.; Srinivasan, K.; Padmanabhan, T. *Class. Quantum Gravit.* 2002, 19, 2671.

[20] Padmanabhan, T. *Mod. Phys. Lett. A* 2004, 19, 2637.

[21] Hemming, S.; Keski-Vakkuri, E. *Phys. Rev. D* 2001, 64, 044006.

[22] Medved, A. J. M. *Phys. Rev. D* 2002, 66, 124009.

[23] Vagenas, E. C. *Phys. Lett. B* 2003, 559, 65.

[24] Angheben, M.; Nadalini, M.; Vanzo, L.; Zerbini, S. *JHEP* 2005, 0505, 014.

[25] Nadalini, M.; Vanzo, L.; Zerbini, S. *J. Phys. A* 2006, 39, 6601.

[26] Arzano, M.; Medved, A. J. M.; Vagenas, E. C. *JHEP* 2005, 0509, 037.

[27] Kerner, R.; Mann, R. B. *Phys. Rev. D* 2006, 73, 104010.

[28] Kerner, R.; Mann, R. B. *Phys. Rev. D* 2007, 75, 084022.

[29] Zhang, J. Y.; Zhao, Z. *JHEP* 2005, 0510, 055.

[30] Zhang, J. Y.; Zhao, Z. *Phys. Lett. B* 2005, 618, 14.

[31] Zhang, J. Y.; Zhao, Z. *Phys. Lett. B* 2006 638, 110.

[32] Jiang, Q. Q.; Wu, S. Q.; Cai, X. *Phys. Rev. D* 2006, 73, 064003.

[33] Liu, W. B. *Phys. Lett. B* 2006, 634, 541.

[34] Wu, S. Q.; Jiang, Q. Q. *JHEP* 2006 0603, 079.

[35] Fang, H. Z. *Nucl. Phys. B* 2007, 767, 130.

[36] Di Criscienzo, R.; Nadalini, M.; Vanzo, L.; Zerbini, S.; Zoccatelli, G. *Phys. Lett. B* 2007, 657, 107.

[37] Chatterjee, B.; Ghosh, A.; Mitra, P. *Phys. Lett. B* 2008, 661, 307.

[38] Pilling, T. *Phys. Lett. B* 2008, 660, 402.

[39] Zhang, B. C.; Cai, Q. Y.; Zhan, M. S. *Phys. Lett. B* 2008, 665, 260.

[40] Ma, Z. Z. *Phys. Lett. B* 2008, 666, 376.

[41] Kerner, R.; Mann, R. B. *Class. Quantum Gravit.* 2008, 25, 095014.

[42] Kerner, R.; Mann, R. B. *Phys. Lett. B* 2008, 665, 277.

[43] Li, R.; Ren, J. R. *Phys. Lett. B* 2008, 661, 370.

[44] Chen, D. Y.; Jiang, Q. Q.; Zu, X. T. *Phys. Lett. B* 2008, 665, 106.

[45] Jiang, Q. Q. *Phys. Rev. D* 2008, 78, 044009.

[46] Yale, A.; Mann, R. B. *Phys. Lett. B* 2009, 673, 168.

[47] Zhang, J. Y. *Phys. Lett. B* 2008, 668, 353.

[48] Banerjee, R.; Majhi, B. R. *Phys. Lett. B* 2008, 662, 62.

[49] Banerjee, R.; Majhi, B. R. *JHEP* 2008, 0806, 095.

[50] Banerjee, R.; Modak, S. K. *JHEP* 2009, 0905, 063.

[51] Modak, S. K. *Phys. Lett. B* 2009, 671, 167.

[52] Jiang, Q. Q.; Cai, X. *JHEP* 2009, 0911, 110.

[53] Jiang, Q. Q.; Han, Y.; Cai, X. *JHEP* 2010, 1008, 049.

[54] Majhi, B. R.; Samanta, S. *Ann. Phys.* 2010, 325, 2410.

[55] Ghosh, T.; Sengupta, S. *Phys. Lett. B* 2011, 696, 167.

[56] Yale, A. *Phys. Lett. B* 2011, 697, 398.

[57] Massar, S.; Parentani, R. *Nucl. Phys. B* 2000, 575, 333.

[58] Myers, R. C.; Perry, M. J. *Ann. Phys.* 1986, 172, 304.

[59] Cvetic, M.; Youm, D. *Nucl. Phys. B* 1996, 477, 449.

[60] Gibbons, G. W.; Perry, M. J.; Pope, C. N. *Class. Quantum Gravit.* 2005, 22, 1503.

[61] Mann, R. B.; Solodukhin, S. N. *Nucl. Phys. B* 1998, 523, 293.

[62] Kaul, R. K.; Majumdar, P. *Phys. Rev. Lett.* 2000, 84, 5255.

[63] Carlip, S. *Class. Quantum Gravit.* 2000, 17, 4175.

[64] Das, S.; Majumdar, P.; Bhaduri, R. K. *Class. Quantum Gravit.* 2002, 19, 2355.

[65] Wald, R. M. (1984). General Relativity. Chicago, IL: University of Chicago Press.

In: Black Holes: Evolution, Theory and Thermodynamics ISBN: 978-1-61942-929-1
Editors: A. J. Bauer and D. G. Eiffel, pp. 117-137

Chapter 5

BLACK HOLES OR FROZEN STARS? A VIABLE THEORY OF GRAVITY WITHOUT BLACK HOLES

*I. Schmelzer**
Berlin, Germany

Abstract

Do observations of black hole candidates rule out alternative theories of gravity without horizon formation? This depends on the existence, viability and reasonableness of alternative theories of gravity without black holes.

Here a theory of gravity without black hole horizon formation is presented. The gravitational collapse stops shortly before horizon formation and leaves a stable frozen star. In the limit $\Xi, \Upsilon \rightarrow 0$ the Einstein equations of GR are recovered, and the frozen stars become observationally indistinguishable from GR black holes. The theory therefore provides a counterexample to recent claims that observational evidence from black hole candidates "all but requires the existence of a horizon".

The theory presented here shares its equations with RTG. Nonetheless, as is shown, there remain important conceptual and physical differences. In particular, some serious problems of RTG are not present in the theory proposed here. So it can be argued that the theory is a physically viable and conceptually sound alternative to GR.

Keywords: gravity, alternative theories, black holes, frozen stars

1. Motivation

Broderick, Loeb and Narayan [4] have claimed that "recent millimeter and infrared observations of Sagittarius A* (Sgr A*), the supermassive black hole at the center of the Milky Way, all but requires the existence of a horizon". Moreover, they specify that they "are able to frame this argument entirely in terms of observable quantities", and that "our results apply to all geometric theories of gravity that admit stationary solutions".

Is this claim correct? The aim of this text is to present a counterexample – a metric theory of gravity which admits, if one of the free paramters of the theory Υ is greater zero,

*E-mail address: ilja.schmelzer@gmail.com

stationary solutions. But in the limit $\Upsilon \to 0$ these stationary solutions become indistinguishable from a GR black hole solution if observation is restricted to outside observers.

The theory presented here as a counterexample has been defined in [7], [8] and will be named “general Lorentz ether theory” (GLET). The Lagrangian, and, therefore, the equations of motion are, for special conditions on the free parameters of the theory $\Xi, \Upsilon > 0$ and for Einstein’s cosmological constant $\Lambda < 0$, equivalent to those of the relativistic theory of gravity (RTG) proposed by Logunov and coworkers [24]. The results presented here for black holes in GLET for this parameter constellation are, therefore, applicable to RTG too, and one may think that GLET is only an irrelevant metaphysical interpretation of RTG. But the difference in the seemingly only metaphysical interpretation has an important influence on the physics, in form of different “causality conditions” in above theories. Now, the RTG causality condition is a quite problematic thing, because it is incompatible with the equations of motion: Even if it is fulfilled for some initial conditions, it may be violated later, thus, the whole theory seems conceptually inconsistent.

The metaphysics of GLET does not lead to such conceptual problems. While, for the different causality condition of GLET, a similar situation is possible too, it obtains a metaphysically meaningful interpretation: The equations of any continuous condensed matter theory becomes physically invalid at places where the density becomes zero and the material tears into parts. With this much better justification of the causality condition in GLET, another problem of RTG – the ghost instability – can be solved too. Last but not least, RTG requires a seemingly wrong sign $\Lambda < 0$ for Einstein’s cosmological constant, thus, needs some dark energy or quintessence to describe an accelerating expansion of the universe, while there is no such restriction in GLET.

Thus, the most powerful known arguments against RTG are not applicable against GLET. So it becomes a viable alternative to GR, devoid of the problems of RTG as well as the problems of GR with causality protection and singularities.

But is it only an unexpected exception that GLET leads to a modification of GR only in a region extremely close to the GR horizon? We argue that it is not, that it is, instead, very likely that many reasonable alternative theories of gravity will show similar behaviour. The argument we propose is metaphysical in its nature: What we experience as the flow of time is not adequately described by actual GR spacetime metaphysics. If one corrects this fault, one easily ends up with theories where GR is corrected by small non-covariant terms, which allow to define a preferred time. Such theories are, then, likely to differ from GR only very close to the horizon, because here a very natural candidate for a preferred time coordinate, harmonic time, becomes singular.

So, Broderick et al deserve praise for looking carefully at one of the domains where it is quite likely that GR may be empirically falsified, where a large class of reasonable alternatives to GR will make different empirical predictions.

2. Definition of the Theory

The theory of gravity defined in [7], [8], which is considered here, is defined in the following way: Similar to GR and other metric theories of gravity, we have the gravitational field $g_{\mu\nu}(x)$ together with some unspecified number of matter fields $\varphi_m(x)$. Different from GR, these fields are not defined on an arbitrary four-dimensional manifold, but on a classical Newtonian spacetime $\mathbb{R}^3 \times \mathbb{R}$, with a set of preferred background coordinates $\mathfrak{x}^\alpha$, in particular a preferred time $\mathfrak{x}^0 = \mathfrak{t}$. Latin indices $i, j, \ldots$ refer to the preferred spatial coordinates $\mathfrak{x}^i$, $1 \le i \le 3$, lower greek indices $\alpha, \beta, \ldots$ refer to preferred four-coordinates $\mathfrak{x}^\alpha$, $0 \le \alpha \le 3$, while upper greek indices $\mu, \nu, \ldots$ refer to general spacetime coordinates x^μ, $0 \le \mu \le 3$.

In the preferred coordinates $\mathfrak{x}^\alpha$, the action of the theory is

$$S = -\frac{1}{16\pi G}\int \left(R + 2\Lambda + \Upsilon g^{00} - \Xi g^{ii}\right)\sqrt{-g}\, d^4\mathfrak{x} + \int L_{matter}\sqrt{-g} d^4\mathfrak{x}, \qquad (1)$$

where L_{matter} a covariant Lagrangian for the matter f elds. Introducing the diagonal matrix $(\gamma_{\alpha\beta}) = \mathrm{diag}(\Upsilon, -\Xi, -\Xi, -\Xi)$, one can rewrite this as

$$S = -\frac{1}{16\pi G}\int \left(R + 2\Lambda + \gamma_{\alpha\beta}g^{\alpha\beta}\right)\sqrt{-g}\, d^4\mathfrak{x} + \int L_{matter}\sqrt{-g} d^4\mathfrak{x} \qquad (2)$$

or, in a covariant form, where the preferred coordinates appear as functions $\mathfrak{x}^\alpha(x)$ of general coordinates x, as

$$S = -\frac{1}{16\pi G}\int \left(R + 2\Lambda + \gamma_{\alpha\beta}g^{\mu\nu}\mathfrak{x}^\alpha_{,\mu}\mathfrak{x}^\beta_{,\nu}\right)\sqrt{-g}\, d^4x + \int L_{matter}\sqrt{-g} d^4x. \qquad (3)$$

We also need – for reasons which become clear below – the additional "causality condition" that the preferred time $\mathfrak{t} = \mathfrak{x}^0$ has to be time-like everywhere. With this additional condition, the def nition of the theory is already complete.

The Euler-Lagrange equations for the $g_{\alpha\beta}$ give

$$G^\alpha_\beta = 8\pi G(T_m)^\alpha_\beta + \left(\Lambda + \frac{1}{2}\gamma_{\gamma\delta}g^{\gamma\delta}\right)\delta^\alpha_\beta - g^{\alpha\gamma}\gamma_{\gamma\beta} \qquad (4)$$

while variation over the $\mathfrak{x}^\alpha$ in (3) gives the harmonic equations

$$\partial_\alpha(g^{\alpha\beta}\sqrt{-g}) = \Box_g \mathfrak{x}^\beta = 0, \qquad (5)$$

which are also a consequence of the equations (4). The harmonic equation has (modulo constants) the form of a conservation law

$$\partial_\alpha \mathcal{T}^\alpha_\beta = 0 \qquad (6)$$

with an (densitized) energy-momentum tensor def ned by

$$\mathcal{T}^\alpha_\beta = \frac{1}{8\pi G}\gamma_{\beta\gamma}g^{\alpha\gamma}\sqrt{-g}\,. \qquad (7)$$

The equations of motion (4) allow to split this energy-momentum tensor into a "matter part"

$$(\mathcal{T}_m)^\alpha_\beta = (T_m)^\alpha_\beta\sqrt{-g} \qquad (8)$$

and a "gravitational part"

$$(\mathcal{T}_g)^\alpha_\beta = \frac{1}{8\pi G}\left(\delta^\alpha_\beta\left(\Lambda + \frac{1}{2}\gamma_{\gamma\delta}g^{\gamma\delta}\right) - G^\alpha_\beta\right)\sqrt{-g} \qquad (9)$$

so that

$$\mathcal{T}^\alpha_\beta = (\mathcal{T}_m)^\alpha_\beta + (\mathcal{T}_g)^\alpha_\beta. \qquad (10)$$

3. The Condensed Matter Interpretation

The theory allows a condensed matter interpretation. In a variant of the ADM decomposition [1] we can use the preferred coordinates $\mathfrak{x}^i$, $\mathfrak{t}$, to split the gravitational f eld $g_{\mu\nu}$ into a scalar ρ, a three-vector v^i and a three-metric σ^{ij}:

$$g^{00}\sqrt{-g} = \rho, \tag{11a}$$
$$g^{0i}\sqrt{-g} = \rho v^i, \tag{11b}$$
$$g^{ij}\sqrt{-g} = \rho v^i v^j - \sigma^{ij}. \tag{11c}$$

Their condensed matter interpretation becomes obvious if we look what happens with the harmonic equations (5). They become

$$\partial_t \rho + \partial_i(\rho v^i) = 0, \tag{12a}$$
$$\partial_t(\rho v^j) + \partial_i(\rho v^i v^j - \sigma^{ij}) = 0. \tag{12b}$$

These are the continuity and Euler equations well-known from classical condensed matter theory if we identify ρ with the density of some fundamental medium, v^i with its velocity, and σ^{ij} with its stress tensor.

What about the matter f elds? If there would be other media interacting with our medium, this would require some momentum exchange, thus, some additional interaction terms in the Euler equation (12b). Now the equations (12) hold also in the case of interaction with matter f elds. But there is a simple solution of this problem: The matter f elds have to be interpreted as f elds which describe some other *material properties of the same medium* instead of properties of other media.

Of course, to give a condensed matter interpretation of a metric theory of gravity raises the problem of explanation of relativistic symmetry. Indeed, situations which are equivalent from point of view of the space-time interpretation are described, depending on the particular choice of the time coordinate $\mathfrak{t}$, in a physically different way, in particular as conf gurations with different ρ, v^i, and σ^{ij}. But such an explanation of relativistic symmetry is possible – the general Lagrangian can be derived from independent f rst principles. This derivation is given in appendix A.

With this interpretation, the theory proposed here appears as a natural generalization of the classical Lorentz ether to gravity. Let's note that many classical arguments against the Lorentz ether are no longer applicable to this generalization: It has a Lagrange formalism, thus, has the resulting "action equals reaction" symmetry, it is, in particular, no longer incompressible, and it provides an explanation of relativistic symmetry. Then it no longer describes only electromagnetism, but all f elds including gravity.

Thus, let's name it "general Lorentz ether theory" (GLET).

Of course, the theory describes, essentially, only the gravitational f eld. All other f elds are only characterized as other, unspecif ed "material properties" of the ether. In this sense, the theory is quite incomplete. It is only a general theory, for a whole class of more specif c models, where the material properties have to be specif ed in detail.

Such a proposal for a particular model of the material properties, which gives the fermions and gauge f elds of the standard model of particle physics, and which is at least conceptually compatible with this theory, has been proposed in [9].

4. Comparison with the Relativistic Theory of Gravity (RTG)

The Lagrangian (2) of the theory itself is not new. For the particular choice of the signs of the free parameters as $\Xi, \Upsilon > 0$ and $\Lambda < 0$, it is the Lagrangian of the "relativistic theory of gravity" (RTG). This theory has been f rst proposed by Freund, Maheshwari and Schonberg 1969 [35] and appears in a two-parameter family of Lagrangians constructed by Ogievetsky and Polubarinov 1965 [34] as the case $q = 0$, $p = -1$. Later, the Lagrangian has been rediscovered by Logunov and coworkers [22]– [25], who have named the theory "relativistic theory of gravity" (RTG). Because this name is widely used (see [20]– [32]) and neutral we will use it too (instead of, say, FMS theory). For an introduction into RTG see [24].

RTG is a bimetric theory. It contains, together with an effective physical metric $g_{\mu\nu}(x)$, also a f xed Minkowski background metric $\eta^{\mu\nu}$, which is, at the same time, the unique vacuum solution of the theory. The gravitational f eld is a spin 2 f eld $h_{\mu\nu}$ which moves on this Minkowski background. It uniquely def nes and is uniquely def ned by the effective metric $g_{\mu\nu}(x)$ by the formula

$$h^{\mu\nu} = g^{\mu\nu}\sqrt{-g} - \eta^{\mu\nu}\sqrt{-\eta}. \tag{13}$$

Conceptually all f elds, the gravitational f eld $h_{\mu\nu}$ as well as the matter f elds, move on the Minkowski background $\eta_{\mu\nu}(x)$. Gravitational interaction leads to an effective replacement of $\eta_{\mu\nu}(x)$ by the effective metric $g_{\mu\nu}(x)$. But this gravitational interaction cannot violate the Einstein causality of the background metric $\eta_{\mu\nu}(x)$. As a consequence, the light cone of $g_{\mu\nu}(x)$ should be inside the light cone of the background $\eta_{\mu\nu}(x)$. This is an additional "causality condition", which does not follow from the equations of motion of RTG.

Gravity in RTG is massive. The connection between the graviton mass $m > 0$, the background metric $\eta^{\mu\nu}$, and the parameters Υ, Ξ, Λ of the theory presented here is def ned by

$$\Lambda = -\frac{m^2}{2}, \quad \Xi = -\eta^{11}\frac{m^2}{2}, \quad \Upsilon = \eta^{00}\frac{m^2}{2}. \tag{14}$$

This implies $\Upsilon > 0, \Xi > 0, \Lambda < 0$. For this choice of the parameters of GLET, we obtain an exact correspondence between the Lagrangian, the equations of motion and the solutions of above theories. This exact correspondence can and will be used to take over mathematical results about RTG solutions to our theory.

In particular, we obtain for these choices of the parameters a constant (vacuum) solution

$$g^{\mu\nu}_{vac} = \eta^{\mu\nu} = -\Lambda^{-1}\gamma^{\mu\nu}. \tag{15}$$

It also follows that the equation (4) reduces in the Newtonian limit around $g^{\mu\nu}_{vac}$ to

$$(\Delta - m^2)V = \kappa\rho_{matter}, \tag{16}$$

so that the action (2) def nes a bimetric theory on a Minkowski background with massive graviton. This massive gravitational f eld "has spins 2 and 0 and represents a physical f eld in the Faraday–Maxwell spirit" [24].

The relation between GLET and RTG is therefore similar to the relation between the Lorentz ether interpretation and the Minkowski spacetime interpretation of special relativity. As far as we consider only the Lagrangian and the equation, and restrict ourself to the choice of signs $\Upsilon > 0$, $\Xi > 0$, $\Lambda < 0$, there is indeed a complete equivalence between the equations of the two theories.

Now, RTG is not a popular theory of gravity. There are some serious arguments against this theory. Given that the Lorentz ether interpretation of special relativity is also quite unpopular in comparison with the Minkowski spacetime interpretation, one may ask why one should consider an unpopular metaphysical interpretation of an already unpopular theory of gravity.

The reason is that the equivalence between the theories is not complete, and that these remaining differences become important if we evaluate the popular arguments against RTG – they become invalid as arguments against GLET.

4.1. Different Causality Conditions

So let's see where the similarity is not complete. The two different interpretations of the effective metric $g_{\mu\nu}$ – in our theory in terms of ρ, v^i, σ^{ij}, in RTG in terms of $\eta_{\mu\nu}, h_{\mu\nu}$ – have a different domain of applicability, which leads to different "causality conditions" for above interpretations. In GLET, the condition is that the density $\rho(\mathfrak{x})$ has to be non-negative. In RTG, the light cone of the effective metric $g_{\mu\nu}$ has to be inside the light cone of the fundamental metric $\eta_{\mu\nu}$ of RTG.

Above causality conditions are important for the meaningfulness of the metaphysical interpretations. And, in particular, above conditions enforce something which is missed in general relativity, namely chronology protection: A metric which fulf lls the causality conditions is a global hyperbolic metric, thus, it does not have any closed time-like curves (CTCs). The conceptual possibility of such CTCs is probably the most important metaphysical problem of the spacetime interpretation of GR. Thus, the causality conditions are conceptually important restrictions, which serve a purpose.

Unfortunately, above causality conditions appear to be in conf ict with the evolution equations of the theory. The additional terms in the equations do not def ne a mechanism for chronology protection. In particular, it is easy to f nd solutions of RTG where the causality condition is fulf lled for the initial conditions but becomes violated later. Such solutions exist even arbitrary close to the vacuum solution. And every RTG solution of this type also allows to def ne GLET solutions which violate the GLET causality condition.

But there exists an important metaphysical difference between violations of the RTG causality condition, interpreted from point of view of RTG metaphysics, and violations of the GLET causality condition interpreted from point of view of the condensed matter interpretation.

Namely, the RTG metaphysics leaves this situation completely unexplained and unexplainable. Something has to be wrong with these solutions, they cannot be solutions of RTG. But what is wrong with them? What is, in particular, wrong with the initial conditions, which do not violate the causality condition? There should be something wrong with them – once they do not lead to valid solutions of the theory. Or there should be something wrong with the equations of RTG – they should be modif ed in the critical region where the violation of the RTG causality condition begins.

Surprisingly, the metaphysical situation is much better in GLET. Here, the causality condition is violated if the density of the Lorentz ether becomes zero. Of course, this means that the condensed matter theory can no longer be applied. In particular, regions of the solution where the density is below zero are clearly physically meaningless.

But this is only a problem of the particular equations for the Lorentz ether. A condensed matter interpretation of some set of equations presupposes that the equations become invalid if the density becomes too small. So, the mathematics of the theory has to be changed in the regions with too small or zero density of the ether. This is natural from the very point of view of the interpretation, which assumes from the start that the condensed matter equations are only approximations of some different, more fundamental theory.

Nothing comparable can be said about RTG. Nothing in the theory and its interpretation suggests that the equations are only approximations of some other, different theory, and no idea is given about the nature of such a more fundamental theory.

4.2. Restrictions for Global Cosmology

Together with the Einstein-Hilbert Lagrangian of general relativity, Einstein's cosmological term is also part of above theories. But RTG imposes an additional restriction to this term: Einstein's cosmological constant has to be negative, $\Lambda < 0$.

As a consequence, RTG does not predict an acceleration of the expansion of the universe, as it seems to follow from observation. Thus, RTG needs some strange form of matter, called quintessence, to explain accelerated expansion.

This restriction is not present in GLET. The interpretation, as well as the derivation of the Lagrangian, do not contain anything which would fx the sign of Λ.

Therefore the observation of Λ as being positive def nes a strong empirical problem for RTG, but not for GLET.

Another, more subtle difference, which may nonetheless be important for cosmology, is the possible value of Ξ. Large enough values of Ξ do not allow for an accelerating expansion, but can make expansion close to constant: $a(\tau) \approx a_0\tau$ (also called "coasting"). Without it, with usual (even dark) matter only, we would have a deceleration of the expansion. Now, alternative explanations of the empirical evidence for accelerated expansion seem at least possible (see, for example, the explanation proposed by Wiltshire [11]). But, even if they fail, it may be that a "freely coasting" universe may remain viable (see [36]), while a decelerating universe without any additional terms fails.

In this case, the difference between the causality conditions of RTG and GLET may become important too. In RTG, the value for Ξ has to be chosen in such a way that the background metric always contains the physical metric, which is, in RTG, oscillating. As a consequence, Ξ has to be so small that it does not lead to any observable effects.

The situation is different in GLET, where the causality condition does not impose any conditions on Ξ. So it can be chosen large enough to have an inf uence on global cosmology, to enforce a freely coasting universe instead of one with decelerating expansion.

4.3. The Problem with the Ghost Instability

But there is also another popular argument against RTG, which is part of a general argument against massive theories of gravity. In the linear approximation, RTG, and, therefore, also GLET (for $\Xi, \Upsilon > 0, \Lambda < 0$), is such a massive theory of gravity.

At a f rst, naive look, there seems to be no problem to obtain GR in the natural limit $\Xi, \Upsilon \to 0$. Unfortunately it requires a more detailed consideration, because in the case of massive gravity with the Fierz-Pauli mass term it seems equally obvious but is wrong – it leads to the van Dam–Veltman–Zakharov discontinuity [13]–[16]. In [19], the discontinuity is understood as a consequence of an inf nite mass for the spin-0 component, which does not disappear if the mass for the spin-2 component becomes zero. The RTG mass term is not of Fierz–Pauli type, so that with $m \to 0$ the masses of above spins go to zero and no discontinuity appears.

But there was a reason for considering the Fierz–Pauli mass term. Before the discontinuity has been found, it was believed to be the only viable mass term: All other mass terms lead, in the linear approximation, to tachyons or ghosts. And, indeed, the linear approximation of RTG contains a ghost because of the spin-0 component. Ghosts are supposed to lead to negative energies and instabilities, therefore there is a strong prejudice against them.

Fortunately, it has been shown by Loskutov [21] that the f ow of gravitational energy from an isolated sourse is positive-def nite (see also [20]). Grishchuk, who has heavily criticized RTG in [17], [18] and mentioned these problems as part of his critique in [18], has later demonstrated (together with Babak) "the incorrectness of the widely held belief that the non-Fierz–Pauli theories possess negative energies and instabilities" [19].

But is the situation, therefore, completely satisfactory? There may be some doubt, because [21] considers only a particular situation, and, moreover, uses approximations. On the one hand, the problem has been formulated based on a simple, linearized approximation. So if the solution is presented in a similar form, using an approximation too, there seems not much to object. Nonetheless, some feeling of uncertainty remains.

Fortunately, the possibility of an instability may be considered also from a global point of view, and independent of the particular situation. So, let's consider a particular solution, and let's denote the preferred coordinates of this solution as x, so that the solution is $g_{\alpha\beta}(x)$. Now, assume there is an instability, so that a small difference in the initial values of the preferred coordinate leads to some other solution $\mathfrak{x}^\alpha$. Now, if we look at the Lagrangian:

$$S = -\frac{1}{16\pi G}\int \left(R + 2\Lambda + \Upsilon g^{\mu\nu}\mathfrak{t}_{,\mu}\mathfrak{t}_{,\nu} - \Xi g^{\mu\nu}\mathfrak{x}^i_{,\mu}\mathfrak{x}^i_{,\nu}\right)\sqrt{-g}\, d^4\mathfrak{x} + \int L_{matter}\sqrt{-g}\, d^4\mathfrak{x} \qquad (17)$$

it looks locally like that of GR together with scalar f elds $\mathfrak{t}(x)$, $\mathfrak{x}^i(x)$. And these scalar f elds would have the usual properties of scalar f elds for $\Xi > 0$, $\Upsilon < 0$. But the choice which corresponds to RTG is $\Upsilon > 0$. So, the sign of the $\mathfrak{t}(x)$ f eld seems to be the wrong one, with negative energy corresponding to large values of the $\mathfrak{t}(x)$ f eld.

So, it seems, if we consider quantum f uctuations, we have to be afraid of instabilities where the $\mathfrak{t}(x)$ f eld becomes inf nite.

But let's look how dangerous it is. Let's consider only f uctuations of the "dangerous" f eld – the time coordinate, assuming $\mathfrak{x}^i(x) = x^i$. Let's also ignore the back-reaction of the coordinates on the metric. Is it possible that $\mathfrak{t}(x)$, initially close to x^0, increases to inf nity for some f nite $t = x^0$? Not. Near the inf nity, the value of $g^{\alpha\beta}\mathfrak{t}_{,\alpha}\mathfrak{t}_{,\beta} = g^{00}\frac{\partial \mathfrak{t}}{\partial x^0}^2$ becomes large in comparison with g^{00}, and it would have to become inf nite if $\mathfrak{t}$ becomes inf nite. But that means that the conserved energy density $\rho = g^{00}\sqrt{g}$ would have to become inf nite at t.

Other types of instabilities also seem to be prevented by the theory itself. Heavy oscillations, for example, do not def ne valid systems of coordinates. In particular, $\mathrm{t}(t)$ cannot be a time-like coordinate similar to t initially, always remain time-like together with t, and later decrease in t. This also prevents $\mathrm{t} \to -\infty$ in some future.

What is the key of this consideration? Usual energy conservation, together with the positivity of energy for usual matter, as well as for the spatial coordinates considered as scalar f elds, prevent usual instabilities. Here, the only exception is the time t, which formally looks like a f eld with negative energy. But this f eld appears restricted in itself, f rst by its geometric character – it has to be a global time-like coordinate – and then by the remarkable other form of conservation law: The conservation of the ether itself, as described by the continuity equation (12a), together with the positivity $\rho > 0$ of the ether density.

So there is no reason to be afraid of ghost instabilities. Instead, the situation is much more satisfactory than in general relativity, where local energy and momentum conservation is undef ned.

4.4. Summary

So we have found that, in agreement with results of Loskutov [21] and Grishchuk [19], that we should not be afraid of negative energies and instabilities in RTG as well as GLET.

We have also found some serious weaknesses of RTG – the conceptual inconsistency between the RTG equations and the RTG causality condition, and the restrictions of the RTG parameters which may appear problematic for global cosmology. Above problems do not appear in GLET.

The initial presentations of RTG have been criticized also for other reasons, which have no relevance for the questions considered here. In particular, the original version of RTG has been motivated with incorrect claims about general relativity, like an inability of GR to give def nite predictions for gravitational phenomena [22]. This has been correctly criticized, for example, by Seldovitch and Grishchuk [17]. This has, clearly, not increased the popularity of RTG, but is irrelevant for the problems considered here.

So we can conclude that the comparison with RTG, in particular the evaluation if the known arguments against RTG can be applied against GLET too, has not given any reasons to doubt the viability of GLET.

5. Frozen Stars Instead of Black Holes

On the other hand, given that the equations of above theories are identical, we can reuse now the results of RTG about black holes.

The gravitational collapse and its f nal state – a black hole in GR, but something completely different in GLET – is a domain where the additional terms in the GLET Lagrangian becomes important. Moreover, it remains important even for arbitrary small values of Ξ, Υ. All we need is that $\Upsilon > 0$.

This is a surprising fact, which deserves a detailed consideration, and it is surprising for several reasons.

First of all, the additional terms in the GLET Lagrangian do not depend on derivatives of the metric. This makes them similar to Einstein's cosmological term. But, in this case, why not apply the same reasoning, which is applicable to Einstein's cosmological term, to these terms too? This would suggest that the additional terms become important only on cosmological distances: Terms which depend on derivatives of the metric become more important locally, where they may be large for small local variations, but less important on large distances.

And, indeed, the additional terms are important for global cosmology. For a homogeneous metric, which is

$$ds^2 = a^6(t)dt^2 - \beta^4 a^2(t)(dx^2 + dy^2 + dz^2). \tag{18}$$

in harmonic time, the use of proper time $d\tau = a^3 dt$ gives the usual FRW-ansatz for the f at universe with some scaling factor $\beta > 0$:

$$ds^2 = d\tau^2 - \beta^4 a^2(\tau)(dx^2 + dy^2 + dz^2). \tag{19}$$

This leads to ($p = k\varepsilon$, $8\pi G = c = 1$):

$$3\left(\frac{\dot a}{a}\right)^2 = -\frac{1}{2}\frac{\Upsilon}{a^6} + \frac{3}{2}\frac{\Xi}{\beta^4 a^2} + \Lambda + \varepsilon, \tag{20a}$$

$$2\frac{\ddot a}{a} + \left(\frac{\dot a}{a}\right)^2 = +\frac{1}{2}\frac{\Upsilon}{a^6} + \frac{1}{2}\frac{\Xi}{\beta^4 a^2} + \Lambda - k\varepsilon. \tag{20b}$$

This leads to interesting consequences, like a big bounce instead of a big bang for any value $\Upsilon > 0$, as considered in [7], which are outside the domain of interest here. But, whatever the values we have to use for Ξ, Υ from cosmology, it seems clear that they become completely irrelevant for Solar system observations, for the same reason that the value of the cosmological constant Λ is irrelevant here: The dominant terms for Solar system size effects will be those depending on derivatives of the metric.

For the parameter Υ, it is not only the consideration of cosmological vs. local effects which suggests that it is too small to have any physical consequences. It should be small on cosmological scale today too, because its inf uence on cosmology decreases with $a(\tau)$ as Υa^{-6}. But $a(\tau)$ has increased a lot during the expansion of the universe. So, while we clearly have no lower bound, we can be sure that Υ has to be extremely small, far too small to have any inf uence on stellar dynamics today.

So what is wrong with this qualitative argument? The point is that, different from Einstein's cosmological term, the additional terms of GLET also depend on the preferred coordinates, moreover, on derivatives of them. So, the argument works only if it is unproblematic to use local harmonic coordinates. This is correct in usual stellar systems. But it becomes very problematic near the horizon of a black hole. Near the horizon, the harmonic coordinates become singular. And, because of this singularity, even if multiplied with an extremely small value of Υ, the additional term becomes important.

And, indeed, for $\Upsilon > 0$, however small, the gravitational collapse, which in GR leads to a black hole, looks completely different. As already mentioned, we can reuse here the results of RTG. So we can refer the reader to sec. 11 of [23] (see also [24], [26]– [29]). The gravitational collapse stops shortly before horizon formation, and, as the result of the collapse, a "frozen star" with size slightly greater than its Schwarzschild horizon remains.

The difference exists only in a very small environment of the Schwarzschild radius. Outside this environment, the GR solution is a good approximation. Therefore in the GR limit $\Upsilon \to +0$ recovers only the outer part of the Schwarzschild solution.

This effect appears also in other theories of gravity with non-zero graviton mass [19]. Their common property is that the mass term depends on the f at background and therefore is able to feel locally that time dilation relative to the background becomes extremal.

Here, [32] gives (for RTG) a lower bound $z \gtrsim 10^{23} M_{\odot}/M$.

5.1. The Observational Argument of Broderick and Narayan

In [2]– [4], Broderick and Narayan claim that observations of black hole candidates, in particular "of Sagittarius A^* (Sgr A^*), the supermassive black hole at the center of the Milky Way, all but require the existence of a horizon" [4]. In fact, if there is a surface, in principle it should be possible to observe surface radiation. In [4] Broderick et al claim to "show that for such surface emission to remain undetected would require unphysically large radiative eff ciencies; greater than 99.6%, as compared to the typical eff ciencies in AGN of 10% ... Most importantly, we do this using only observable quantities, ensuring that these limits are largely independent of the gravitational theory employed. ... Our results hold for any geometric gravitational theory admitting stationary solutions."

Now, GLET is a theory of this type. If these claims would be correct, GLET would be falsif ed by observation. So, there is a reason to take a close look at these arguments. Of course, they also have to rely on some physical assumptions: Their "... argument depends critically upon three underlying, physically well motivated assumptions: (1) Sgr A^* is accretion powered, (2) has reached steady state and (3) compact surfaces are approximate blackbodies." They "wish to emphasize that relaxing any of these would require fundamental alterations to presently well understood physics (such as microscopic physics at pedestrian, and laboratory accessible, densities, temperatures and magnetic f eld strengths)." Nevertheless, they discuss each in detail, describing how these are justif ed in the context of Sgr A^*.

It appears that we do not have to question assumptions (1) and (2), because the assumption which is the critical one is the steady state assumption.

Now, the steady state assumption is well justif ed for usual compact objects, in particular also such objects like neutron stars. Infalling matter usually has some mass as well as kinetic energy. If we ignore possible transformations of mass into energy, the infalling mass increases the mass of the star, and the kinetic energy the temperature of the star. On the other hand, temperature decreases because of thermal radiation. The steady state assumption is, roughly speaking, that these effects are in equilibrium, thus, the average kinetic energy of infalling matter is radiated away as thermal radiation.

In this case, it would be, indeed, quite clear that one could distinguish a frozen star, with a surface in steady state, from a black hole. All one has to do is to estimate the amount of infalling matter. For the infalling matter, one can apply reasonable models based on known and well-understood physics. For the type of outcoming radiation, there is also a quite reasonable model – it has to be a blackbody radiation. If the infalling matter hits the surface, it is also quite clear that most of its energy is gravitational binding energy transformed into kinetic energy of the infalling body, thus, has to be radiated away. So one can compute the amount of blackbody radiation which would have to come out, if the black hole candidate is in a steady state. Once we do not see this – and, following [4], we don't – the case seems closed.

But it is not. What can be, and has to be, questioned is the steady state assumption.

To see why it may be reasonable to expect that the steady state has not yet been reached, it seems useful to have a look at the surface temperature, as measured by an observer on the surface, which would be necessary to create the steady state outgoing radiation.

Here we have to take into account that the outgoing radiation is highly redshifted, by a redshift factor z def ned (in GR approximation) by

$$(1+z) = \left(1 - \frac{R_S}{R}\right)^{-\frac{1}{2}}, \tag{21}$$

Second, only a small part of the surface radiation reaches inf nity. Most of it simply falls back on the surface because of gravitational lensing. The part which is able to reach inf nity is (in GR approximation) def ned by

$$\frac{d\Omega}{2\pi} = 1 - \left[1 - \frac{27}{4}\frac{1 - R_S/R}{(R_S/R)^2}\right]^{\frac{1}{2}} \tag{22}$$

which gives z^{-2} for large z [33]. The observable f ux at distance D of a black hole with surface temperature T for frequency ν' in the surface frame will be

$$F_{\nu'} = \frac{2h\nu'^3}{c^2}\frac{e^{-h\nu'/kT}}{1 - e^{-h\nu'/kT}}\frac{\pi R^2}{D^2}\frac{d\Omega}{2\pi}. \tag{23}$$

The f ux visible in inf nity decreases by a factor $(1+z)^{-1}$ because of time dilation. The frequency in the surface frame is $\nu' = (1+z)\nu$. This gives for large z

$$F_\nu = \left(\frac{kT}{1+z}\right)^3 F\left(\frac{h(1+z)\nu}{kT}\right)\frac{\pi R_S^2}{D^2}\frac{27}{8}^{-3} \tag{24}$$

with

$$F(\lambda) = \frac{2\lambda^3}{h^2c^2}\frac{e^{-\lambda}}{1 - e^{-\lambda}} \leq F_{max} \tag{25}$$

being a bounded function. Thus, for suff ciently small $\frac{kT}{1+z}$ surface radiation will become invisible. Thus, we would need extremely large surface temperatures $T_{surf} \sim z$ to be able to see the surface radiation at large distances.

In principle the heating by infalling matter could provide suff cient energy to reach such temperatures. To decide if this happens for a given z, we would need a hypothesis about the thermal capacity of the material of the frozen star at extremely large temperatures $T \sim z$. To exclude a surface completely, we would need such a hypothesis even for $T \to \infty$.[1] We obviously cannot have reliable data in this limit.

To justify the steady state assumption, in [4] two characteristic time scales are compared: "The dynamical timescale of Sgr A^* depends upon the nature of the object. Nevertheless, we may expect that it is comparable to the dynamical timescale of the corresponding

[1]Moreover, we would also have to assume the validity of the EEP even for such extremal redshifts.

black hole, $GM/c^3 \simeq 20$ s. This is supported by the ~ 10 min variability observed in the NIR and X-rays, presumed to be associated with material orbiting nearby. Both of these are much, much shorter than the estimated age of Sgr A^*, 10 Gyr, making it natural to assume that it has suff cient time to reach steady state." So they conclude that "even surfaces with extraordinary redshifts ($z \lesssim 10^{15}$) will have reached steady state".

But even for Sgr A^* with mass $M = 4.5 \pm 0.4 \cdot 10^6 M_\odot$ this argument does not give a steady state even for the minimal surface redshift $z \gtrsim 10^{23} M_\odot/M$ given in [32]. Thus, the minimal surface redshift which is, because of cosmological upper limits for Υ, required, is already large enough to make it impossible to apply the steady state assumption.

In a similar way it seems extremely unlikely that the impact of infalling matter leads to observable effects. In principle, an ideal point-like elastic infalling body would bounce back from the surface of a frozen star [23], [29]. But this idealization fails already for comets falling on the Earth. For the frozen stars, it follows from (22) that even a minor scattering away from the vertical direction is suff cient to prevent even an ideal elastic body to reach inf nity. Thus, even if, because of energy conservation, an infalling ideally elastic body has enough energy to go back to inf nity, it seems extremely unlikely that this leads to observable effects different from those predicted by GR.

Thus, the existence of a surface for black hole candidates does not endanger the viability of our theory even if the surface remains invisible. Instead, the central black hole singularity, which is unavoidable in GR, provides a strong conceptual argument against GR.

6. General Character of the Argument

What can be seen here, with the example of GLET, is not only a counterexample of a viable theory with stable stars instead of black holes. It seems worth to be noted that similar effects appear also in other theories of gravity. In particular, it seems to be a common property of theories of massive gravity.

For such a theory to follow the same scheme, one does not need that much. To see this, we only have to consider the substantial points necessary to obtain the effect. All one needs is some small non-covariant term in the equations which prefers a certain (for simplicity f at) background. Then, all we need is that the preferred coordinates of this background become invalid near the black hole horizon, similar to the harmonic time t in GLET, which becomes inf nite at the horizon. Then, however small the background-dependent term, it may become inf nite where the background coordinate becomes inf nite. So one can assume that this leads to a regularization, so that the preferred coordinate does not become inf nite. So, if the background coordinates are suff ciently close to harmonic coordinates, which seems not unreasonable, so that they degenerate similarly near horizon formation, one can expect that in this theory no horizon appears. But even if there remains a horizon, the situation has to change in an essential way, to prevent the inf nity of the preferred coordinates.

Then, the smaller the additional, non-covariant term, the larger the domain of applicability of the GR approximation. One can expect that, multiplying the additional term with a small enough number, we can obtain a theory where everything remains like in GR up to an arbitrary large surface time dilation, and only then it starts to differ from GR – but then it will differ essentially.

One should not think that it is the horizon itself which causes the problem. The problem is caused by the singularity of the preferred coordinates of the background. we can expect that close to the horizon formation something chances in a very essential way, probably in such a radical way that no horizon appears.

6.1. The Metaphysics of General Relativity

In this context it seem important to consider the problems of the metaphysics of the space-time interpretation of general relativity.

The f rst objection against the very notion of GR metaphysics is the claim that there is no such animal. All GR cares about are observables. But really? There is agreement about the fact that "proper time" def nes the results of clock measurements. Or, in neutral words, proper time is simply clock time. But is clock time everything?

What about the f ow of time we experience? As the result of relativistic time dilation, one thing is clear: We cannot measure it with clocks. I have a clock, you have a clock, at our f rst meeting, we start them showing the same time. At our second meeting, they show different time. The difference is minimal, but conceptually this does not matter – the result is the same: The clocks do not measure the f ow of time.

But, once there is nothing in GR which measures the f ow of time, maybe we should completely forget about this notion? Then there simply is no such thing as a f ow of time. Unfortunately, this is not a very reasonable choice. First of all, because we simply experience the f ow of time. Everything changes, that's fact. Everybody observes it. And, even more, there is a connection between the f ow of time and clock time, a qualitative one as well as a quantitative: The qualitative one is that time f ows in the direction of increasing clock time. And the quantitative one is that clock time works nicely as an approximate description of the f ow of time.

So, a theory which bans the f ow of time from physics is simply not complete. I'm not a worldline in some four-dimensional thing named spacetime. There is more, point. At the current moment, I may be located at some point on this worldline. Without this location on my worldline, something important is missed.

What can one do to meet, on the one hand, the positivistic ideal of a physical theory which describes only objective, exactly measurable things, but, on the other hand, the ideal of a complete theory, which does not exclude completely such an obviously experienced things as the f ow of time? One solution is to fall back to solipsism. Its only me who experiences the f ow of time, other people can tell me something about their experiences, but they may be simply robots with no real experiences. So, all one needs is to def ne the f ow of time on a single worldline, the own worldline. And, for this purpose, clock time works nicely.

One may hide the solipsistic character of this interpretation. Other people also experience a f ow of time, no problem. Their f ow of time is described by the f ow of proper time along their worldline. Let's name this the multisolipsistic interpretation.

This is the point where the simple twin experiment becomes a twin paradox. There is, last but not least, nothing strange in the fact that clocks are not as accurate as we would prefer them. There is always some inf uence of the environment. So if I leave one clock at home, and take one with me, there is nothing paradoxical if they show, later, different time – they have been in different environments. So if I meet, in the twin experiment, my twin later, after his journey, there is nothing problematic with the fact that our clocks show different numbers.

But, according to the multisolipsistic interpretation, I do not really meet the same twin, but something different – a sort of empty place on the worldline of my twin, while my twin is really on a different place on his worldline. It may be, possibly, occupied by another clon of my twin, who follows the same worldline. So, or we have to multiply clones without necessity, (violating Ockham's razor), or we have to accept that we sometimes communicate with empty places. But in the last case, it seems more reasonable to go back to pure solipsism.

Or we have to accept that the f ow of time on my worldline is not accurately described by my own clock time. So, if I meet my twin later, we have experienced the same objective f ow of time. But, once our clocks show different clock time, this same f ow of time is not described accurately, but only approximately, by clock time.

In the last case, there is also a quite simple object which can be used to describe the f ow of time, namely a global time-like coordinate. The condition that it is time-like meets the observation that clock time increases with the f ow of time. What is missed is that proper time is, at least locally and for small velocities, an accurate approximation for the f ow of time.

So one can conclude that there is no consistent metaphysics of the f ow of time in general relativity. There are four variants of it – no f ow of time at all, solipsistic f ow of time, the multi-solipsistic variant of it, and the f ow of time as an otherwise unspecif ed time-like coordinate. All of them are more or less unsatisfatory.

The positivistic prejudice against metaphysics allows to hide this unsatisfactory situation. It is simply not appropriate for a physicist to discuss such things at all. And, whenever somebody starts to discuss such things, one can fallback to the "no f ow of time" position.

A satisfactory interpretation of the f ow of time in general relativity would be, instead, the following: There is an objective f ow of time, and this f ow of time is def ned by a global time-like coordinate t. There should be an additional equation, which describes the f ow of time relative to clock time, but this additional equation is unknown. Thus, general relativity is simply an incomplete theory.

Given that, from point of view of simplicity, there is a nice candidate for this unknown equation for t, namely the harmonic condition: In harmonic coordinates, the Einstein equations themself are much simpler than in general coordinates.

But the particular choice of such an equation is, of course, already speculation beyond the scope of general relativity. The point of this section is a different one: Already the consideration of the GR metaphysics themself – if one seriously considers them, instead of following the positivistic refusal to discuss them at all – strongly suggests the incompleteness of GR, and the necessity to complete it with an equation for a preferred notion of time.

All we need for this argument are the following points:

- A complete physical theory should describe everything we observe;
- We observe a f ow of time, which is connected with clock time, and fulf lls the following properties:

- Clock time increases with the f ow of time;
- Clock time locally, and for small velocities, serves as an approximation of the f ow of time;

Everything else follows from the evaluation of various proposals to meet these conditions.

Conclusion

So, the claim made by Broderick at al that observations of black holes all but require horizon formation have to be rejected.

We have presented here a particular example of a metric theory of gravity, GLET, which is compatible with the observations of black hole candidates considered, but makes the same predictions about them as general relativity.

We have found that, even in principle, observations of black hole candidates can give only observational upper bounds for one of the parameters of the theory, $\Upsilon > 0$. At the current moment, it does not even give stronger upper bounds than those following from cosmological observations.

The mathematical result itself – the very existence of a metric theory of gravity with such properties – has not been the main point of this paper. In fact, given that the equations of GLET are identical with those of RTG, and that it is already known that in RTG the gravitational collapse does not lead to horizon formation, this would have been not that important.

But one thing is the existence of some theory with equations which have such properties, and another one the existence of a reasonable physical theory. So, it was also necessary to consider the very justif cation of the theory, as well as known counterarguments. Given that some arguments against RTG seem to be quite serious – in particular the cosmological observations like accelerated expansion of the universe, but also the questionable nature of the RTG causality condition – it is important that there is a theory which does not have these problems. The other popular argument against RTG – the fear of ghost-like instabilities – has been rejected as unjustif ed. This has been done in agreement with similar considerations of this problem by other researchers, but from a slightly different point of view, which seems useful to understand why the "ghost f eld" is not as dangerous as one may think looking at the linear approximation.

So we conclude that there is not only some theory with such properties, but a physically reasonable and viable one, which gives a counterexample.

Last but not least, let's note that Broderick at al deserve praise for looking at the region near horizon formation: This is the very region where it is likely that alternative theories will make predictions different from GR. Indeed, this is what we have argued about in the last section: The spacetime metaphysics of GR is far away from being consistent, from being able to describe elementary observations about the f ow of time in a consistent, complete, and reasonable way. Many problems make it reasonable to guess that GR is only an approximation of more fundamental theory, and that this more fundamental theory does not have full relativistic symmetry. And, in this case, the horizon formation of black holes seems likely to be a place where non-covariant properties of such a more fundamental theory may become observable.

A. Derivation of the Lagrangian

Given the condensed matter interpretation, it becomes reasonable to ask for an explanation of the relativistic symmetry of the Lagrangian. Surprisingly, there is a nice and simple derivation of the general Lagrangian from simple f rst principles. In essence, the relativistic symmetry is obtained as a consequence of the "action equals reaction" symmetry of the Lagrange formalism.

We have already used the so-called "parametrized formalism" where we have four explicit scalar functions $\mathfrak{x}^\alpha(x)$ for the "preferred coordinates" $\mathfrak{x}^\alpha$ in terms of general coordinates x^μ. In particular, the Lagrangian presented in this form is given by (3) and looks covariant if one forgets about the geometric meaning of the $\mathfrak{x}^\alpha(x)$ as preferred coordinates and considers them as usual scalar f elds.

The key of the derivation is the observation that in the parametrized formalism Euler-Lagrange equations for the "f elds" $\mathfrak{x}^\alpha(x)$, which can be def ned as usual by

$$\frac{\delta S}{\delta \mathfrak{x}^\alpha} = \frac{\partial L}{\partial \mathfrak{x}^\alpha} - \partial_\mu \frac{\partial L}{\partial \mathfrak{x}^\alpha_{,\mu}} + \partial_\mu \partial_\nu \frac{\partial L}{\partial \mathfrak{x}^\alpha_{,\mu\nu}}, - \cdots \tag{26}$$

give valid equations of motion.[2] Even more, for the translational symmetry $\mathfrak{x}^\alpha \to \mathfrak{x}^\alpha + c^\alpha$ of the Lagrangian they automatically give Noether conservation laws

$$\frac{\delta S}{\delta \mathfrak{x}^\alpha} = \partial_\mu \mathcal{T}^\mu_\alpha \tag{27}$$

with

$$\mathcal{T}^\mu_\alpha = -\frac{\partial L}{\partial \mathfrak{x}^\alpha_{,\mu}} + \partial_\nu \frac{\partial L}{\partial \mathfrak{x}^\alpha_{,\mu\nu}} \cdot - \cdots . \tag{28}$$

Assuming that we have a four-dimensional parametrized f eld theory with preferred coordinates $\mathfrak{x}^\alpha(x)$, the following postulate is suff cient for the derivation of the general Lagrangian:

Postulate 1. *The energy-momentum tensor* $\mathcal{T}^\alpha_\beta$ *defined by* (28) *has the following properties:*

1. *For some invertible constant matrix* $c^{\alpha\beta}$ *the tensor* $\mathfrak{g}^{\alpha\beta} = c^{\alpha\gamma}\mathcal{T}^\alpha_\gamma$ *is symmetric.*

2. *The* $\mathfrak{g}^{\alpha\beta}$ *are among the independent field variables.*

Indeed, this postulate gives together with (27)

$$\frac{\delta S}{\delta \mathfrak{x}^\alpha} = c_{\alpha\beta}\partial_\gamma \mathfrak{g}^{\beta\gamma}. \tag{29}$$

In the four-dimensional case, one can identify the variables $\mathfrak{g}^{\alpha\beta}$ with the usual metric variables $g_{\alpha\beta}$ by

$$\mathfrak{g}^{\alpha\beta} = g^{\alpha\beta}\sqrt{\det(g_{\alpha\beta})}, \quad g_{\alpha\beta} = \mathfrak{g}_{\alpha\beta}\sqrt{\det(\mathfrak{g}^{\alpha\beta})}, \tag{30}$$

[2]The only difference from the standard proof for f elds is that variations $\mathfrak{x}^\alpha + \delta\mathfrak{x}^\alpha$ may not def ne a valid system of coordinates. But for small enough $\delta\mathfrak{x}^\alpha$ with f nite support, which are suff cient for the proof, this problem does not appear.

so that (29) becomes

$$\frac{\delta S}{\delta \mathfrak{x}^\alpha} = c_{\alpha\beta}\partial_\gamma(g^{\beta\gamma}\sqrt{g}), \tag{31}$$

but the variables $\mathfrak{g}^{\alpha\beta}$ may be used as well as the independent variables. We can def ne now constants $\gamma_{\alpha\beta}$ by

$$c_{\alpha\beta} = -(8\pi G)^{-1}\gamma_{\alpha\beta}. \tag{32}$$

With these $\gamma_{\alpha\beta}$ the action

$$S_0 = -(16\pi G)^{-1}\int \gamma_{\alpha\beta} g^{\mu\nu}\mathfrak{x}^\alpha_{,\mu}\mathfrak{x}^\beta_{,\nu}\sqrt{-g}\, d^4x \tag{33}$$

def nes a particular solution of (29). As usual, the general solution S of the inhomogeneous problem (29) is the sum of this particular solution S_0 and the general solution of the corresponding homogeneous problem, which is in our case

$$\frac{\delta(S - S_0)}{\delta \mathfrak{x}^\alpha} = 0. \tag{34}$$

Thus, $S - S_0$ should not depend on the preferred coordinates $\mathfrak{x}^\alpha$. But this is simply a well-known way to def ne the most general Lagrangian of GR. So we obtain

$$S - S_0(g_{\alpha\beta}, \mathfrak{x}^a_{,\mu}) = S_{GR}(g_{\alpha\beta}) + S_{matter}(g_{\alpha\beta}, \varphi_m), \tag{35}$$

where only the term S_0 depends on the preferred coordinates $\mathfrak{x}^a$, while S_{GR} as well as S_{matter} have to be covariant, and only S_{matter} depends on the matter f elds φ_m. This action may differ from (2) because it can contain non-minimal interactions with matter f elds in S_{matter} and higher order terms like R^2, $R^{\mu\nu}R_{\mu\nu}$ and so on in S_{GR}. But our context is that of an effective f eld theory, so there is no reason to restrict the action to the lowest order terms (2). All higher order terms may be present, they will be simply suppressed in the large distance limit.

Note that this derivation not only derives the gravity part $S_0 + S_{GR}$ of the action, but also the main property – covariance – of the matter part S_{matter}, or, in other words, the Einstein equivalence principle. The main ingredients of this derivation of relativistic symmetry are the action-equals-reaction symmetry of the Lagrange formalism and the choice of the $g_{\alpha\beta}$ (or, equivalently, of the $\mathfrak{g}^{\alpha\beta}$) as independent variables so that the conservation laws do not depend on the matter f elds φ_m:

$$\frac{\delta}{\delta \mathfrak{x}^\alpha}\frac{\delta}{\delta\varphi_m}S = \frac{\delta}{\delta\varphi_m}\frac{\delta}{\delta \mathfrak{x}^\alpha}S = \frac{\delta}{\delta\varphi_m}(c_{\alpha\mu}\partial_\nu g^{\mu\nu}\sqrt{g}) = 0. \tag{36}$$

How restrictive is our postulate 1? This is an interesting question which we have to leave to future research.

References

[1] Arnowitt, R., Deser, S., Misner, C., Quantum theory of gravitation: General formalism and linearized theory, Phys. Rev. 113, 745–750, 1959.

[2] Broderick, A.E., Narayan, R., On The Nature of the Compact Dark Mass at the Galactic Center, Astrophys. J. 638, L21–L24, 2006, arXiv:astro-ph/0512211.

[3] Broderick, A.E., Narayan, R., Where are all the gravastars? Limits upon the gravastar model from accreting black holes, Class.Quant.Grav. 24: 659–666, 2007, arXiv:gr-qc/070115.

[4] Broderick, A.E., Loeb, A., Narayan, R., The event horizon of Sagittarius A^*, 2009, arXiv:0903.1105.

[5] Choquet-Bruhat, Y., Theoreme d'existence pour certain systemes d'equations aux derivees partielles non lineaires, Acta Math., 88, 141, 1952.

[6] Deser, S., Gen. Rel. and Grav 1, 9, 1970.

[7] Schmelzer, I., General ether theory – a metric theory of gravity with condensed matter interpretation, Proc. of the XXII international workshop on high energy physics and f eld theory, Protvino, June 23-25, 1999.

[8] Schmelzer, I., A generalization of the Lorentz ether to gravity with general-relativistic limit, Advances in Applied Clifford Algebras, 2011, DOI 10.1007/s00006-011-0303-7, arXiv:gr-qc/0205035.

[9] Schmelzer, I., A condensed matter interpretation of SM Fermions and Gauge f elds, Foundations of Physics, vol. 39, 1, p. 73, 2009, arXiv:0908.0591.

[10] Will, C.M., The confrontation between general relativity and experiment, arXiv:gr-qc/9811036.

[11] Wiltshire, D.L., Dark energy without dark energy, 2007, arXiv:0712.3984.

[12] Fierz, M., Pauli, W., Proceedings of the Royal Society of London, Vol. A 173, p. 211, 1939.

[13] Van Dam, H., Veltman, M., Massive and massless Yang–Mills and gravitational f elds, Nuclear Phys. B, 22, 397–411, 1970.

[14] Van Dam, H., Veltman, M., On the mass of the graviton, Gen. Rel. Grav., 3, 215–220, 1972.

[15] Zakharov, V.I., JETP Lett., 12, p. 312, 1970.

[16] Boulware, D., Deser, S., Can gravitation have a f nite range, Phys. Rev. D 6, 3368–3382, 1972.

[17] Zeldovich, Ya.B., Grishchuk, L.P., Sov. Phys. Usp., 31, p. 661, Russian version: vol. 155, p. 517, 1988.

[18] Grishchuk, L.P., Sov. Phys. Usp., 33, p. 669, Russian version: vol. 160, 8, p. 147, 1990.

[19] Babak, S.V., Grishchuk, L.P., Finite-range gravity and its role in gravitational waves, black holes and cosmology, Int. J. Mod. Phys. D 12, 1905, 2003, arXiv:gr-qc/0209006v2.

[20] Gershtein, S.S., Logunov, A.A., Mestvirishvili, M.A., On gravitational ow in the relativistic theory of gravitation, arXiv:gr-qc/0608015v1.

[21] Loskutov, Yu.M., Positive def niteness of intensity of gravitational radiation in the theory of gravitation with nonzero graviton mass, Teor. Mat. Fiz., 107, 329–343, 1996.

[22] Logunov A.A. et.al., Theor. Math. Phys., 69, 3, 328–340, 1986.

[23] Logunov, A.A., Theor. Math. Phys., 70, 1, 1987.

[24] Logunov A.A., The theory of gravity, Moscow: Nauka, 2001, arXiv:gr-qc/0210005.

[25] Logunov, A.A., Relativistic theory of gravity, Nauka, 2006 (in Russian).

[26] Gershtein, S.S., Logunov, A.A., Mestvirishvili, M.A., Impossibility of unlimited gravitational collapse, arXiv:gr-qc/0612177v1.

[27] Gershtein, S.S., Logunov, A.A., Mestvirishvili, M.A., On one fundamental property of gravitational f eld in the f eld theory, 2004, arXiv:gr-qc/0412122v5.

[28] Gershtein, S.S., Logunov, A.A., Mestvirishvili, M.A., Repulsive force in the f eld theory of gravitation, Theor. Math. Phys., 145, 1604–1618, Teor. Mat. Fiz., 145, 272–288, 2005, arXiv:gr-qc/0505070v3.

[29] Vlasov, A.A., Logunov, A.A., Bouncing from the Schwarzschild sphere in the relativistic theory of gravity with nonzero graviton mass, Theor. Math. Phys., 78, 229–233, 1989.

[30] Logunov, A.A., Mestvirishvili, M.A., Chugreev, Yu.V., Graviton mass and evolution of a Friedman universe, Theor. Math. Phys. 74, 1–20, 1988.

[31] Gershtein, S.S., Logunov, A.A., Mestvirishvili, M.A. Tkachenko, N.P., Graviton mass, quintessence and oscillatory character of the universe evolution, 2003, arXiv:astro-ph/0305125v1.

[32] Nieuwenhuizen, T.M., Exact solution for the interior of a black hole, arXiv:0805.4169v1.

[33] Abramowicz, M.A., Klu'niak, W., Lasota, J.-P., No observational proof of the black-hole event-horizon, arXiv:astro-ph/0207270v1.

[34] Ogievetsky, V.I., Polubarinov, I.V., Interacting f eld of spin 2 and the Einstein equations. Ann. Phys., 35, 167, 1965.

[35] Freund, P.G.O., Maheshwari, A., Schonberg, E., Finite-range gravitation. Astrophys. J., 157, 857, 1969.

[36] Gehlaut, S., Geetanjali, P.K., Lohiya, D., A concordant freely coasting cosmology, 2003, arXiv:astro-ph/0306448v1.

[37] Primack, J.R., astro-ph/9707285.

[38] Zakharov, A.D., Sov. Phys. Dokl. 12, 1040, 1968.

In: Black Holes: Evolution, Theory and Thermodynamics ISBN: 978-1-61942-929-1
Editors: A. J. Bauer and D. G. Eiffel, pp. 139-155

Chapter 6

SMOOTHLY EXOTIC BLACK HOLES

Torsten Asselmeyer-Maluga[1,*] ***and Carl Brans***[2,†]
[1] German Aerospace Center, Berlin, Germany
[2] Loyola University, New Orleans, Louisiana

Abstract

We review some implications of the existence of exotic smoothness (differentiable) structures on otherwise topologically trivial spaces, such as $\mathbb{R}^4$ or $\mathbb{R}^2 \times S^2$, for the application of standard Einstein equations to black hole structures. These results in differential topology have shown that in considering the Einstein or any other equations globally we have always implicitly made a choice of one from a possible infinity of smoothness structures in continuing solutions from one coordinate patch to another. Since there are no known physical reasons for restricting studies to this one choice it is important to consider implications of choosing others. Although there do not yet exist finite presentations of non-trivially smooth coordinate atlases, we should consider possible implications of globally extending what we obtain locally in one coordinate patch to solutions of the Einstein equations. This may be physically important since in the Kruskal presentation of black hole solutions we only have direct physical information in our local coordinates **outside** the horizons. We discuss implications of going beyond the standard smooth extension globally to the topological $\mathbb{R}^2 \times S^2$, Kruskal space, and point out that predictions of what happens beyond our local, physically observed, coordinate patch to the full topological Kruskal space may not be justified by any known physics. In particular, we have no *a priori* physical reason for using only the standard smoothness on topological $\mathbb{R}^2 \times S^2$ when predicting the nature of singularities within the horizon for solutions to the vacuum Einstein equations.

PACS 04.20.Gz, 04.70.Bw, 04.20.Dw

Keywords: Smooth exotic black hole, exotic smooth Kruskal space, Smoothing of singularities

AMS Subject Classification: 57R55, 57R65, 57R57

*E-mail address: Torsten.Asselmeyer-Maluga@dlr.de
†E-mail address: brans@loyno.edu

1. Introduction

From the very early days, say with the proposals of the Einstein-Rosen bridge, general relativity has opened the door to non trivial, "exotic," topologies. But in this chapter we will deal with the physically required properties associated with "differential topology." Informally, we can say that differential topology [1], [2], [3] is the branch of mathematics concerned with questions of how to do calculus globally on a topological space, M. In this chapter we will deal only with the class of spaces which are locally isomorphic to $\mathbb{R}^n$, with fixed dimension, n. Mathematicians have placed such spaces into various categories, TOP, PL, DIFF, ANALYTIC. This classification is provided by dividing M into a union of open subsets (balls), B, together with isomorphisms, ϕ_B, of each ball onto $\mathbb{R}^n$ which is given the standard Euclidean topology. Each ball is a *coordinate patch* and the corresponding identification of it with $\mathbb{R}^n$ gives the local *coordinates*. This collection of patches and coordinates is called an *atlas* for M. The category classification depends on how the local coordinates in one patch are related to those in each intersecting one, $\psi_{BB'} = \phi_B \cdot \phi_{B'}^{-1}$, $B \cap B' \neq$ null, which maps one open set of $\mathbb{R}^n$ isomorphically onto another. In particular, the manifold with a given atlas is said to be C, (*topological*) if each ϕ_B is a homeomorphism. Other categories are PL, (*piecewise linear*), C^∞, (*smooth* or *differentiable*), C^ω, (*analytic*), depending on whether each $\psi_{BB'}$ is piecewise linear, smooth, or analytic in the range space, $\mathbb{R}^n$, given the corresponding standard structures of real number analysis. In this chapter we will concentrate on the physically important case of smooth manifolds and their corresponding equivalence classes. Two smooth M_1 and M_2 are *diffeomorphic*, i.e. equivalent, if and only if there is a homeomorphism, $F : M_1$ onto M_2, which (together with F^{-1}) is smooth in the standard real variable sense when expressed in local coordinate patches of the two atlases on M_1 and M_2. It is very important to note that diffeomorphisms are **not** only different choices of coordinates, but rather have *global* significance. Two smooth manifolds are **physically inequivalent** spaces, even though they are homeomorphic, i.e., topologically equivalent, if they are **not** diffeomorphic. We will try to clarify this issue below.

To do physics we must do calculus on our manifolds, so we must use local coordinates provided by an atlas on a smooth manifolds. Physically we can think of the coordinates in a patch as a mathematical representation of the results of measurements of events in a (local) physical *reference frame*. The choice of different atlases **within a diffeomorphism class** does not lead to different physics. Rather it is just an expression of the *principle of general relativity* proclaiming the physical equivalence of different choices of (smooth) reference frames. The central point of these studies will be to ask the question:

- **Do there exist physically interesting manifolds with the same topology but which are not diffeomorphic, and thus not physically equivalent?**

Surprisingly in the 1980's Donaldson, Freedman et al. discovered that this is a non-trivial question in dimension four for topologically uncomplicated spaces such as $\mathbb{R}^4$ and $\mathbb{R} \times S^3$. This is a profoundly surprising and counter-intuitive result, and a very important one too since it had previously been proven that the standard n-dimensional (topological) coordinates on $\mathbb{R}^n$ provides the **only** smooth structure on the space for all n **except for the physically distinctive case** $n = 4$! Why is the $n = 4$ case special? Perhaps this can be explained by saying that this dimension is too big to apply the direct approaches for the first three dimensions, but not big enough to move the structures called "handles" around to eliminate critical points in Morse theory co-bordism available for dimensions five and up. Freedman[4] has provided a heuristic description of this problem.

Of course the physical principle of general relativity is realized in the process of forming the equivalence class of all smooth manifolds under diffeomorphisms, i.e., topological homeomorphisms which are smooth and smoothly invertible in terms of the local coordinate patches. Such diffeomorphisms are the mathematical expression of the physical notion of coordinate transformations.

To consider simple examples of the notions of homeomorphism and diffeomorphism let $\mathbb{R}_P$ be the usual real line as a set of points, $p \in \mathbb{R}_P$ each of which can be uniquely identified by a real number, its "value," $v(p)$. We take care at this point to distinguish the the points of the point set $\mathbb{R}_P$ from coordinates which we can and will introduce. To turn $\mathbb{R}_P$ into a topological space we can choose to define the metric distance between to points p, q as the absolute value $|v(p) - v(q)|$ and define the "standard" Euclidean topology, converting $\mathbb{R}_P \to \mathbb{R}_T$, a (metric) topological space. Now for a fine point: To display this topological space, $\mathbb{R}_T$, as a topological manifold we need to define an atlas with coordinate patches (neighborhoods) and coordinates for the points, p, say $x(p)$, which have unique inverses, $p(x)$, at least locally (over the coordinate patch). Functions on $\mathbb{R}_T$ will then be defined to be continuous if and only they are continuous in the usual sense when explicitly displayed in terms of coordinates. Thus, if $\bar{f}$ is a real valued function, $\bar{f} : \mathbb{R}_T \to \mathbb{R}$, then define $f(x) : \mathbb{R} \to \mathbb{R}$ by $f(x) = \bar{f}(p(x))$. $\bar{f}$ is then said to be continuous if and only if $f(x)$ is in terms of real functions of one real variable. The standard choice of course is to let the entire space be the one coordinate patch with coordinate $x(p) = v(p)$. But, is this the only one? A simple counter example is provided by the global coordinate, $y(p) = (v(p))^3$ with inverse $p(y) = y^{1/3}$. [1]. It is easy to see that using y instead of $x = y^{1/3}$ leads to an homeomorphic topological manifold with an identical class of continuous functions since $x \to x^3$ is one-to-one, onto, and continuous in the usual sense of real numbers. Thus the two coordinations (that is, choice of topological coordinates) provided by x and y produce homeomorphic manifolds. If we speak only of topology then these produce the same equivalence class of topological manifolds on the real line. Call these two atlases A_x, A_y respectively. In this simple example each atlas has only a single coordinate patch, the entire space $\mathbb{R}_P$.

However, in order to do physics it is necessary to do calculus, i.e., differentiate, etc., so we must decide how we can use the standard calculus of real variables (one in this case) to do differentiation. So we define a smooth manifold from $\mathbb{R}_P$ by choosing either the function $x(p)$ or $y(p)$ to provide the smooth coordinate. For atlas A_x, and resulting smooth manifold, $\mathbb{R}_{Sx}$, a physical function $\bar{f} : \mathbb{R}_T \to \mathbb{R}$ will be *smooth* or *differentiable* in A_x if and only if the real valued $\bar{f}_x : \mathbb{R}_T \to \mathbb{R}$ defined, as above, by $f(x) = \bar{f}(p(x))$ is smooth, i.e., differentiable, in the usual real value sense. However, for A_y amd $\mathbb{R}_{Sy}$, $\bar{f}$ is smooth if and only if $f(y) = \bar{f}(p(y))$ is smooth. Hence we see that A_x and A_y are *different* smoothness structures, since the coordinate function, $x(p) = v(p)$ is by trivial definition smooth in A_x, but produces the real function, $y(p) = (v(p))^{1/3}$, for which the derivative does *not* exist at the origin and thus is not smooth in A_y. So these are different smoothness structures. But a simple homeomorphic rearrangements of the points of $\mathbb{R}_P$, $p \to p' = p^3$

[1] In these discussion we will always use the unique real cube root.

clearly provides a *diffeomorphism,* i. e., a physical equivalence, for $\mathbb{R}_{Sx} \leftrightarrow \mathbb{R}_{Sy}$. In fact, for dimension $n = 1$, $\mathbb{R}^n_T$ has one and only one smooth structure as we will see below.

It may seem that this discussion is unnecessarily complicated but as we will see the investigation of each of these steps leads to surprisingly non-trivial results when applied to the Euclidean 4-space.

In keeping with the main topic of this book we note that black holes, or at least their exteriors, are usually represented by Kruskal coordinates, as shown in Figure 1 below. Kruskal space is topologically $\sim \mathbb{R}^2 \times S^2$ which in fact has **different, non-diffeomorphic smooth structures**. Thus the physics present in Kruskal space is not necessarily globally that given by the standard smooth version. A heuristic proof would be to excise a line, $\mathbb{R}^1$, from a **large** $\mathbb{R}_\Theta$. Simply choose a line that does not intersect a compact subset which is not contained in the interior of any smoothly embedded S^3. Then the resulting $\sim \mathbb{R}^2 \times S^2$ contains the same compact set which cannot be in the interior any smoothly embedded S^3. For more rigor and generalized topologies see the article by Ding[5]. However for our purposes it is sufficient to note that these differential topological results show that the representation of physics in an exterior *local* coordinate patch, or reference frame, such as that represented by the white region in Figure 1, **may not be continued smoothly outside this region** if we choose some exotic smoothness. In particular, the $u = v+$ constant path of the light ray reaching an event, P, in the local coordinate patch (white), may not be continued outside this patch as a smooth function of u, since u (and v) may not be smooth globally, so **inferences about the physics near the distant source from such an assumption may not be justified.** There is no *a priori* reason for choosing standard smoothness in which (u, v) are globally smooth, so it is our purpose to point out that we are making a possibly unjustified assumption with this choice. In other words, it is entirely possible to do physics on an exotic Kruskal space in which our extrapolation of path easily understand in some spacetime neighborhood outside of this neighborhood smoothly, since our directly measured physical coordinates my not be continued smoothly elsewhere!

Either the full Kruskal (u, v, S^2) or the exterior of the origin, the two hyperboloids, $u^2 - v^2 = 1$, is clearly TOP $\sim \mathbb{R}^2 \times S^2$. The Swarzschild metric on this space is "maximal analytic continuation" of Schwarschild. However, there is no *a priori* reason to believe that the usual Kruskal coordinates are globally smooth and thus physically significant beyond some local coordinate patch. We now proceed to a brief discussion of some of the more technical details.

2. Construction of Exotic $\mathbb{R}^4$ and Exotic $S^2 \times \mathbb{R}^2$

Einstein's insight that gravity is the manifestation of geometry leads to a new view of the structure of spacetime. From the mathematical point of view, spacetime is a smooth 4-manifold endowed with a (smooth) metric as basic variable for general relativity. Later on, the existence question for Lorentz structure and causality problems (see Hawking and Ellis [6]) gave further restrictions on the 4-manifold: causality implies non-compactness, Lorentz structure needs a codimension-1 foliation. Usually, one starts with a globally foliated, non-compact 4-manifold $\Sigma \times \mathbb{R}$ fulfilling all restrictions where Σ is a smooth 3-manifold representing the spatial part. But other non-compact 4-manifolds are also possible, i.e. it is

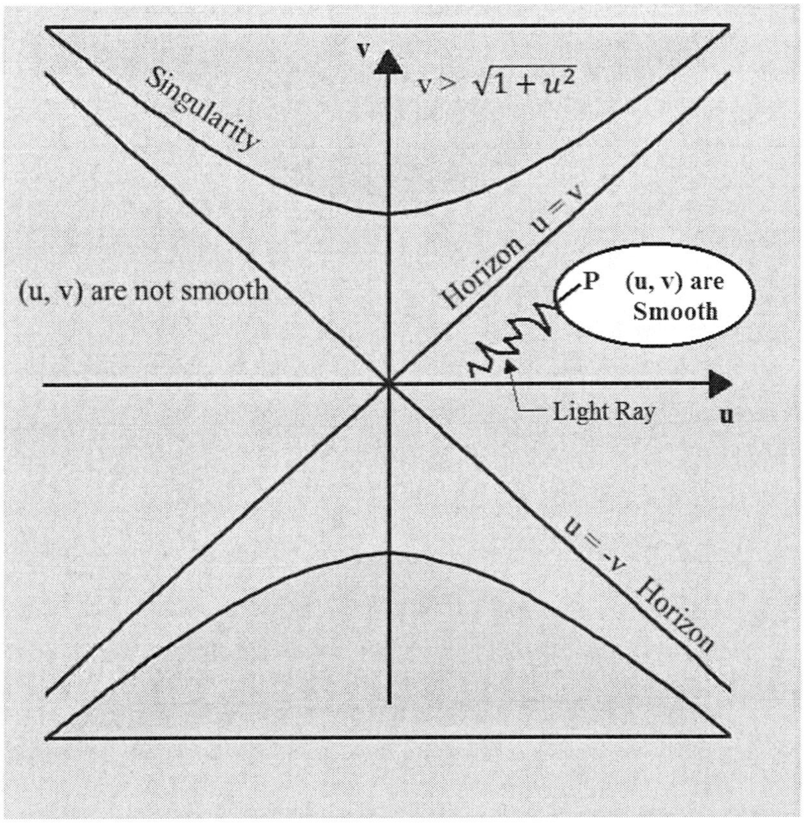

Figure 1. Topological Kruskal space:(u, v) need not be globally smooth, and cannot be for exotic $\mathbb{R}^2 \times S^2$.

enough to assume a non-compact, smooth 4-manifold endowed with a codimension-1 foliation.

All these restrictions on the representation of spacetime by the manifold concept are clearly motivated by physical questions. Among the properties there is one distinguished element: the smoothness. Usually one assumes a smooth, unique atlas of charts (i.e. a smooth or differential structure) covering the manifold where the smoothness is induced by the simplest, standard, smooth structure on $\mathbb{R}^4$. But that is not the full story. Especially in dimension 4, there are an infinity of possible other smoothness structures (i.e. a smooth atlas) non-diffeomorphic to each other. For a deeper insight we refer to the book [3].

2.1. Exotic Smoothness in Short

We use the term **exotic** to describe the choice of a smooth structure on a manifold M if this smoothness is not diffeomorphic (physically equivalent) to the "standard" definition of smoothness on M. In particular, such standard choices occur naturally for manifolds such as $\mathbb{R}^4, \mathbb{R}^1 \times S^3, \mathbb{R}^2 \times S^2$, etc.

The implications for physics are tremendous because we rely on the smooth calculus to formulate field theories. Thus non-diffeomorphic smoothness structures have to represent different physical situations leading to different measurable results[2]. But it should be stressed that *exotic smoothness is not exotic physics.* Exotic smoothness is part of topology (differential topology), i.e. a finer determination of the topology (the smooth atlas of the manifold) fulfilling the condition of smoothness. Therefore we obtain another property besides geometry to vary a manifold with fixed topology. Usually one starts with a topological manifold M and introduces structures on them. Then one has the following ladder of possible structures:

$$\begin{array}{ccc} \text{Topology} \rightarrow & \text{piecewise-linear(PL)} \rightarrow & \text{Smoothness} \\ \rightarrow & \text{bundles, Lorentz, Spin etc.} \rightarrow & \text{metric,...} \end{array}$$

We do not want to discuss the first transition, i.e. the existence of a triangulation on a topological manifold. But we remark that the existence of a PL structure implies uniquely a smoothness structure in all dimensions smaller than 7 [7].

Given two homeomorphic manifolds M, M', how can we compare both manifolds to decide whether both are diffeomorphic? A mapping between two manifolds M, M' can be described by a $(n+1)-$dimensional manifold W with $\partial W = M \sqcup M'$, called a cobordism W. In the following, the two homeomorphic manifolds M, M' are simply-connected. The celebrated h-cobordism theorem of Smale [8, 9] gives a simple criteria when a diffeomorphism between M and M' for dimension greater than 4 exists: there is a cobordism W between M and M' where the inclusions $M, M' \hookrightarrow W$ induce homotopy-equivalences between M, M' and W. W is then called an h-cobordism. Therefore the classification problem of smoothness structures in higher dimensions (> 4) is a homotopy-theoretic problem. The set of smooth structures (up to h-cobordisms) of the $n-$sphere S^n has the structure of an abelian group Θ_n (under connected sum $\#$). Via the h-cobordism theorem this group is identical to the group of homotopy spheres which was analyzed by Kervaire and Milnor [10] to be a finite group. Later on the result about the exotic spheres can be extended to any smoothable manifold [7], i.e. there is only a finite number of non-diffeomorphic smoothness structures. In all dimensions smaller than four, there is an unique smoothness structure, the standard structure. But all methods failed for the special case of dimension four.

2.2. Exotic $\mathbb{R}^4$

One of the most surprising aspects of exotic smoothness is the existence of exotic $\mathbb{R}^4$'s. In all other dimensions [11], the Euclidean space $\mathbb{R}^n$ with $n \neq 4$ admits a unique smoothness structure, up to diffeomorphism. Beginning with the first examples [12], Taubes [13] and Freedman/DeMichelis [14] constructed countably many large and small exotic $\mathbb{R}^4$'s, respectively. A small exotic $\mathbb{R}^4$ embeds smoothly in the 4-sphere whereas a large exotic $\mathbb{R}^4$ cannot be so embedded. For the following we need some simple definitions: the connected sum $\#$ and the boundary connected sum $\natural$ of manifolds. Let M, N be two n-manifolds with boundaries $\partial M, \partial N$. The *connected sum* $M \# N$ is the procedure of cutting out a disk D^n from the interior $int(M) \setminus D^n$ and $int(N) \setminus D^n$ with the boundaries $S^{n-1} \sqcup \partial M$ and $S^{n-1} \sqcup \partial N$, respectively, and glueing them together along the common boundary component S^{n-1}.

[2]Recall that a diffeomorphism is the mathematical generalization of coordinate transformation, or change of physical reference frame.

The boundary $\partial(M\#N) = \partial M \sqcup \partial N$ is the disjoint sum of the boundaries $\partial M, \partial N$. The *boundary connected sum* $M\natural N$ is the procedure of cutting out a disk D^{n-1} from the boundary $\partial M \setminus D^{n-1}$ and $\partial N \setminus D^{n-1}$ and gluing them together along S^{n-2} of the boundary. Then the boundary of this sum $M\natural N$ is the connected sum $\partial(M\natural N) = \partial M\#\partial N$ of the boundaries $\partial M, \partial N$.

2.2.1. Large Exotic $\mathbb{R}^4$

Large exotic $\mathbb{R}^4$ can be constructed using the failure to arbitrarily split a compact, simply-connected 4-manifold. For every topological 4-manifold one knows how to split this manifold *topologically* into simpler pieces using the work of Freedman [15]. But as shown by Donaldson [16], some of these 4-manifolds do not exist as smooth 4-manifolds. This contradiction between the continuous and the smooth case produces the first examples of exotic $\mathbb{R}^4$. Below we discuss one of these examples.

One starts with a compact, simply-connected 4-manifold X classified by the intersection form [15]

$$Q_X : H_2(X,\mathbb{Z}) \times H_2(X,\mathbb{Z}) \to \mathbb{Z}$$

a quadratic form over the second integer homology group. In the first construction of a large exotic $\mathbb{R}^4$, one starts with the K3 surface as 4-manifold having the intersection form

$$Q_{K3} = E_8 \oplus E_8 \oplus (\oplus_3 \begin{pmatrix} 0 & 1 \\ 1 & 0 \end{pmatrix}) := 2E_8 \oplus 3H \tag{1}$$

with the the matrix E_8:

$$E_8 = \begin{bmatrix} 2 & 1 & 0 & 0 & 0 & 0 & 0 & 0 \\ 1 & 2 & 1 & 0 & 0 & 0 & 0 & 0 \\ 0 & 1 & 2 & 1 & 0 & 0 & 0 & 0 \\ 0 & 0 & 1 & 2 & 1 & 0 & 0 & 0 \\ 0 & 0 & 0 & 1 & 2 & 1 & 0 & 1 \\ 0 & 0 & 0 & 0 & 1 & 2 & 1 & 0 \\ 0 & 0 & 0 & 0 & 0 & 1 & 2 & 0 \\ 0 & 0 & 0 & 0 & 1 & 0 & 0 & 2 \end{bmatrix}.$$

The work of Donaldson [16] shows that a closed, **smooth**, simply-connected, compact 4-manifold $X_{E_8\oplus E_8}$ with intersection form $E_8 \oplus E_8$ does not exist. But Freedman [15] showed there is a topological splitting

$$K3 = X_{E_8\oplus E_8}\#\left(\#_3(S^2 \times S^2)\right) \tag{2}$$

with the $m-$times connected sum $\#_m$ (see above) which fails to be smooth. This splitting means that we glue together the two manifolds $\#_3(S^2 \times S^2) \setminus D^4$ and $X_{E_8\oplus E_8} \setminus D^4$ along the common boundary $S^3 = \partial D^4$ (D^4 is the 4-disk or 4-ball). Now we define the interior $X = \#_3(S^2 \times S^2) \setminus Int\, D^4$. The splitting (2) gives a way to represent the $3H$ part of the intersection form (1) by using X but that fails smoothly. So, choosing a topological splitting

$$\begin{aligned} K3 &= X_{E_8\oplus E_8}\#\left(\#_3(S^2 \times S^2)\right) \\ &= \left(X_{E_8\oplus E_8} \setminus D^4\right) \cup \left(S^3 \times [0,1]\right) \cup \left(\#_3(S^2 \times S^2) \setminus D^4\right) \end{aligned}$$

gives a $S^3 \times [0,1]$ inside the $K3$. The interior of $S^3 \times [0,1]$ defines a manifold $S^3 \times [0,1)$ glued to a (topological) 4-disk $D^4 \subset \#_3(S^2 \times S^2) \setminus D^4$ along the common boundary, i.e. $W = D^4 \cup S^3 \times [0,1)$ topologically. W is homeomorphic to $\mathbb{R}^4$ but the non-existence of the smooth splitting implies that it is an exotic $\mathbb{R}^4$ and there is no smooth embedded S^3 (otherwise the topological splitting is smooth). This failure for a smooth embedding implies also that such exotic $\mathbb{R}^4$'s do not embed in the 4-sphere, i.e. it is a large exotic $\mathbb{R}^4$. The details of the construction can be found in our book [3] in section 8.4.

Gompf [17] introduced an important tool for finding new exotic $\mathbb{R}^4$ from others, the end-sum $\natural_e$. Let R, R' be two topological $\mathbb{R}^4$'s. The end-sum $R\natural_e R'$ is defined as follows: Let $\gamma : [0,\infty) \to R$ and $\gamma' : [0,\infty) \to R'$ be smooth properly embedded rays with tubular neighborhoods $\nu \subset R$ and $\nu' \subset R'$, respectively. For convenience, identify the two semi-infinite intervals with $[0,1/2)$, and $(1/2,1]$ leading to diffeomorphisms, $\phi : \nu \to [0,1/2) \times \mathbb{R}^3$ and $\phi' : \nu' \to (1/2,1] \times \mathbb{R}^3$. Then define

$$R\natural_e R' = R \cup_\phi I \times \mathbb{R}^3 \cup_{\phi'} R'$$

as the **end sum** of R and R'. With a little checking, it is easy to see that this construction leads to $R\natural_e R'$ as another topological $\mathbb{R}^4$. However, if R, R' are themselves exotic, then so will $R\natural_e R'$ and in fact, it will be a "new" exotic manifold, since it will not be diffeomorphic to either R or R'. Gompf used this technique to construct a class of exotic $\mathbb{R}^4$'s none of which can be embedded smoothly in the standard $\mathbb{R}^4$.

By an extension of Donaldson theory for a special class of open 4-manifolds, so-called end-periodic 4-manifolds, by Taubes [13] gives a continuous family of exotic $\mathbb{R}^4$ which was extended by Gompf to a continuous 2-parameter family $R_{s,t}$.

2.2.2. Small Exotic $\mathbb{R}^4$

Small exotic $\mathbb{R}^4$'s are again the result of anomalous smoothness in 4-dimensional topology but of a different kind than for large exotic $\mathbb{R}^4$'s. In 4-manifold topology [15], a homotopy-equivalence between two compact, closed, simply-connected 4-manifolds implies a homeomorphism between them (a so-called h cobordism). But Donaldson [18] provided the first smooth counterexample, i.e. both manifolds are generally not diffeomorphic to each other. The failure can be localized in some contractible submanifold (Akbulut cork) so that an open neighborhood of this submanifold is a small exotic $\mathbb{R}^4$. The whole procedure implies that this exotic $\mathbb{R}^4$ can be embedded in the 4-sphere S^4. Below we discuss the details for one of these examples.

In 1975 Casson (Lecture 3 in [19]) described a smooth 5-dimensional h-cobordism between compact 4-manifolds and showed that they "differed" by two proper homotopy $\mathbb{R}^4$'s (see below). Freedman knew, as an application of his proper h-cobordism theorem, that the proper homotopy $\mathbb{R}^4$'s were $\mathbb{R}^4$. After hearing of Donaldson's work in March 1983, Freedman realized that there should be exotic $\mathbb{R}^4$'s and, to find one, he produced the second part of the construction below involving the smooth embedding of the proper homotopy $\mathbb{R}^4$'s in S^4. Unfortunately, it was necessary to have a compact counterexample to the smooth h-cobordism conjecture, and Donaldson did not provide this until 1985 [18]. The idea of the construction is simply given by the fact that every such smooth h-cobordism between non-diffeomorphic 4-manifolds can be written as a product cobordism except for a compact contractible sub-h-cobordism V, the Akbulut cork. An open subset $U \subset V$ homeomorphic to $[0,1] \times \mathbb{R}^4$ is the corresponding sub-h-cobordism between two exotic $\mathbb{R}^4$'s. These exotic $\mathbb{R}^4$'s are called ribbon $\mathbb{R}^4$'s. They have the important property of being diffeomorphic to open subsets of the standard $\mathbb{R}^4$. That stands in contrast to the previous defined examples of Kirby, Gompf and Taubes.

To be more precise, consider a pair (X_+, X_-) of homeomorphic, smooth, closed, simply-connected 4-manifolds. The transformation from X_- to X_+ visualized by a h-cobordism can be described by the following construction.

Let W be a smooth h-cobordism between closed, simply connected 4-manifolds X_- and X_+. Then there is an open subset $U \subset W$ homeomorphic to $[0,1] \times \mathbb{R}^4$ with a compact subset $K \subset U$ such that the pair $(W \setminus K, U \setminus K)$ is diffeomorphic to a product $[0,1] \times (X_- \setminus K, U \cap X_- \setminus K)$. The subsets $R_\pm = U \cap X_\pm$ (homeomorphic to $\mathbb{R}^4$) are diffeomorphic to open subsets of $\mathbb{R}^4$. If X_- and X_+ are not diffeomorphic, then there is no smooth 4-ball in $R_\pm$ containing the compact set $Y_\pm = K \cap R_\pm$, so both $R_\pm$ are exotic $\mathbb{R}^4$'s.

Thus, remove a certain contractible, smooth, compact 4-manifold $Y_- \subset X_-$ (called an Akbulut cork) from X_-, and re-glue it by an involution of ∂Y_-, i.e. a diffeomorphism $\tau : \partial Y_- \to \partial Y_-$ with $\tau \circ \tau = Id$ and $\tau(p) \neq \pm p$ for all $p \in \partial Y_-$. This argument was modified above so that it works for a contractible *open* subset $R_- \subset X_-$ with similar properties, such that R_- will be an exotic $\mathbb{R}^4$ if X_+ is not diffeomorphic to X_-. Furthermore R_- lies in a compact set, i.e. a 4-sphere or R_- is a small exotic $\mathbb{R}^4$. In the next subsection we will see how this results in the construction of handlebodies of exotic $\mathbb{R}^4$. In [14] Freedman and DeMichelis constructed also a continuous family of small exotic $\mathbb{R}^4$.

2.2.3. Handle Decomposition of the Small Exotic $\mathbb{R}^4$ and Casson Handles

As of now, we only know of exotic $\mathbb{R}^4$'s represented by infinite number of coordinate patches. This naturally makes it difficult to provide an explicit description of a metric. However, in §3 below, a suggestion to overcome this limitation is provided by the consideration of periodic explicitly described coordinate patches making use of more complex pieces, so-called handles, and even more complex gluing maps. Then one also gets infinite structures of handles but with a clear picture: the coordinate patches have a periodic structure.

Handles Every 4-manifold can be decomposed using standard pieces such as $D^k \times D^{4-k}$, the so-called k-handle attached along $\partial D^k \times D^{4-k}$ to the $0-$handle $D^0 \times D^4 = D^4$. In the following we need two possible cases: the 1-handle $D^1 \times D^3$ and the 2-handle $D^2 \times D^2$. These handles are attached along their boundary components $S^0 \times D^3$ or $S^1 \times D^2$ to the boundary S^3 of the $0-$handle D^4 (see [2] for the details). The attachment of a 2-handle is defined by a map $S^1 \times D^2 \to S^3$, the embedding of a circle S^1 into the 3-sphere S^3, i.e. a knot. This knot into S^3 can be thickened (to get a knotted solid torus). The important fact for our purposes is the freedom to twist this knotted solid torus (so-called Dehn twist). The (integer) number of these twists (with respect to the orientation) is called the framing number or the framing. Thus the gluing of the 2-handle on D^4 can be represented by a knot or link together with an integer framing. The simplest example is the unknot with framing ± 1 representing the complex projective space $\mathbb{C}P^2$ or with reversed orientation

$\overline{\mathbb{C}P}^2$, respectively. The 1-handle will be glued by the map of $S^0 \times D^3 \to S^3$ represented by two disjoint solid 2-spheres D^3. Akbulut [20] introduced another description. He observed that a 1-handle is something like a surgered 2-handle with a fixed framing. We remark that all details can be found in [2]. Now we are ready to discuss the handle body decomposition of an exotic $\mathbb{R}^4$ by Bizaca and Gompf [21].

Handle decomposition of small exotic $\mathbb{R}^4$ Firstly it is very important to notice that the exotic $\mathbb{R}^4$ is the **interior** of the handle body described below (since the handle body has a non-null boundary and is compact). The construction of the handle body can be divided into two parts. The first part is known as the Akbulut cork, a contractable 4-manifold with boundary a homology 3-sphere (a 3-manifold with the same homology like the 3-sphere) used to construct exotic smoothness on compact 4-manifolds (see subsection 2.2.2). The Akbulut cork is given by a linking between a 1-handle and a 2-handle of framing 0. The second part is the Casson handle CH where we use the simplest non-trivial example explained below. In short the Casson handle is a periodic structure of disk neighborhoods (immersion of $D^2 \times D^2$) where one disk admits self-intersections (see below for the details) represented by by a 1-handle (knot with dot) linked to a 2-handle with framing 0.

Start with the well-known example of the Akbulut cork A as a contractible 4-manifold, the Mazur manifold [20], with boundary the homology 3-sphere $\Sigma(2,5,7)$. This homology 3-sphere is given by the set

$$\Sigma(2,5,7) = \left\{x, y, z \in \mathbb{C} \,|\, x^2 + y^5 + z^7 = 0\,,\ |x|^2 + |y|^2 + |z|^2 = 1\right\}$$

Now it is easy to define the interior $int(A)$ of the cork as the set

$$int(A) = \left\{x, y, z \in \mathbb{C} \,|\, x^2 + y^5 + z^7 = 0\,,\ |x|^2 + |y|^2 + |z|^2 < 1\right\}$$

This set is a smooth manifold which can be covered by a finite number of charts. But the smoothness structure of the exotic $\mathbb{R}^4$ depends on the infinite Casson handle.

Casson handle Now consider the Casson handle and its construction in more detail. Briefly, a Casson handle CH is the result of attempts to embed a disk D^2 into a 4-manifold. In most cases this attempt fails and Casson [19] looked for a substitute, which is now called a Casson handle. Freedman [15] showed that every Casson handle CH is homeomorphic to the open 2-handle $D^2 \times \mathbb{R}^2$ but in nearly all cases it is not diffeomorphic to the standard handle [22, 23]. The Casson handle is built by iteration, starting from an immersed disk in some 4-manifold M, i.e. an injective smooth map $D^2 \to M$. Every immersion $D^2 \to M$ is an embedding except on a countable set of points, the double points. One can kill one double point by immersing another disk into that point. These disks form the first stage of the Casson handle. By iteration one can produce the other stages. Finally consider not the immersed disk but rather a tubular neighborhood $D^2 \times D^2$ of the immersed disk including each stage. The union of all neighborhoods of all stages is the Casson handle CH. So, there are two input data involved with the construction of a CH: the number of double points in each stage and their orientation $\pm$. Thus we can visualize the Casson handle CH by a tree: the root is the immersion $D^2 \to M$ with k double points, the first stage forms the next level of the tree with k vertices connected with the root by edges etc. The edges are evaluated using the orientation $\pm$. Every Casson handle can be represented by such an infinite tree. The Casson handle $CH(R_+)$ having an immersed disk with one (positive oriented) self-intersection (or double point) is the simplest Casson handle represented by the simplest tree

R_+ having one vertex in each level connected by one edge with evaluation $+$. We will now go into more detail. The reader not interested in very technical terms can go directly to the next subsection.

Each building block of a Casson handle, sometimes called a "kinky" handle, is diffeomorphic to $\natural(S^1 \times D^3)$ with two attaching regions. Technically speaking, one region is a tubular neighborhood of band sums of Whitehead links connected with the previous block. The other region is a disjoint union of the standard open subsets $S^1 \times D^2$ in $\#S^1 \times S^2 = \partial(\natural S^1 \times D^3)$ (this is connected with the next block). The number of end-connected sums is exactly the number of self intersections of the immersed two handle. The simplest Casson handles have $S^1 \times D^3$ as their building blocks represented. Consider the simple Casson handle $CH(R_+)$, a periodic Casson handle with positive oriented self-intersections. As shown in [24], the Casson handle is a so-called end-periodic manifold, i.e. a manifold with a periodic structure of building blocks, discussed in the next subsection. As described above a Casson handle can be expressed by an infinite tree with one end point and with a sign $\pm$ on each edge. Casson handles will be also important for the construction of exotic $S^2 \times \mathbb{R}^2$, the exotic Kruskal space.

2.3. Exotic Kruskal Space $S^2 \times \mathbb{R}^2$

In [5] Ding constructed a continuous family of exotic $\mathbb{R}^2$-bundles M_S over a closed, connected surface S which includes also the case $S^2 \times \mathbb{R}^2$. The idea of the construction is very simple. The end-sum of Gompf is generalized to open 4-manifolds $M_1 \natural_e M_2$ in the obvious way. Then we construct the open 4-manifold

$$M_{S,\theta} = M_S \natural_e R_{s,t}$$

and Ding [5] is able to show that one gets a continuous family of exotic M_S. In particular, we obtain a continuous family

$$S^2 \times_\theta \mathbb{R}^4 = \left(S^2 \times \mathbb{R}^2\right) \natural_e R_{s,t}$$

of exotic $S^2 \times \mathbb{R}^2$. This family is again large in the sense that an emebdding of $S^2 \times_\theta \mathbb{R}^2$ fails to embed in S^4 or in the standard $\mathbb{R}^4$. But for the construction of the exotic Kruskal space we need an exotic $S^2 \times \mathbb{R}^2$ which embeds into S^4 or $\mathbb{R}^4$.

For this construction we start with a Casson handle $(CH, \partial CH)$ having an circle as attaching region. By using Freedmans work [15] we know that $(CH, \partial CH)$ is homeomorphic to the open 2-handle $(D^2 \times \mathbb{R}^2, \partial D^2 \times \mathbb{R}^2)$. The gluing of a disk D^2 along its boundary to $D^2 \times \mathbb{R}^2$ produces $S^2 \times \mathbb{R}^2$. Therefore this gluing produces $S^2 \times \mathbb{R}^2$ topologically. There are many examples of exotic Casson handles [22, 23] especially the simplest handle $CH(R_+)$ is also included [25]. All Casson handle have a circle[3] as attaching region. Therefore we glue a disk D^2 to the Casson handle CH along ∂CH to get

$$S^2 \times_\theta \mathbb{R}^2 = D^2 \cup_{\partial CH} CH$$

an exotic $S^2 \times \mathbb{R}^2$ which depends on the Casson handle CH. This exotic space embedds into S^4 and $\mathbb{R}^4$.

[3]But this circle is not the boundary of a disk.

But the Kruskal space is not only determined by the topology of $S^2 \times \mathbb{R}^2$ but also by a Lorentz metric on it. So we will discuss the existence of Lorentz structures on exotic (non-compact) 4-manifolds in the next subsection.

2.4. Lorentz Structure on Exotic Open 4-manifolds

A manifold admits a Lorentzian structure if (and only if) there is a line element field (i.e. a non-vanishing vector field of arbitrary sign, see Hawking and Ellis [6]). Then a compact manifold admits a Lorentzian metric if the Euler characteristics vanishes. For multiply-connected compact 4-manifolds there is the problem of causality (time loops etc., see also the topological censorship by Schleich and Witt [26]). In most cases causality can be implemented on 4-manifolds which are not compact. Both conditions, Lorentz structure and causality, can be trivially fulfilled on 4-manifolds $\Sigma\times\mathbb{R}$ including also the case $\Sigma = S^2\times\mathbb{R}$. But, is this choice unique?

Non-compactness of the 4-manifold $\Sigma \times \mathbb{R}$ as the result of the causality condition is a purely topological condition. The second condition, the existence of a Lorentz structure, is related to the structure of the tangent bundle $T(\Sigma \times \mathbb{R})$, i.e. to the smoothness structure. In case of the Lorentz structure, we have to deal with the tangent bundle $T(\Sigma \times \mathbb{R})$. We mentioned above that the existence of a Lorentz structure is connected with the existence of a line element. This line element exists only if the tangent bundle admits a splitting

$$T(\Sigma \times \mathbb{R}) = \xi \oplus \chi$$

where χ is a 3-dimensional subbundle and ξ is a 1-dimensional subbundle of the tangent bundle. A section of the bundle ξ is the line element. Now let $\Sigma \times_\Theta \mathbb{R}$ be an exotic 4-manifold, i.e. $\Sigma \times_\Theta \mathbb{R}$ is homeomorphic to $\Sigma \times \mathbb{R}$ but not diffeomorphic to it. Normally the cross product $\times$ denotes the topological formation of the product structure. Since we now know of the independence of smoothness from topology, we extend the notation to the smooth situation as well. We know that every 3-manifold Σ has a unique smoothness structure. Therefore $\Sigma\times\mathbb{R}$ represents smoothly the standard structure and we choose $\Sigma\times_\Theta\mathbb{R}$ to indicate the exotic smoothness structure. From this point of view we have a big difference between the bundle $T(\Sigma \times \mathbb{R})$ and $T(\Sigma \times_\Theta \mathbb{R})$. The first bundle admits a splitting

$$T(\Sigma \times \mathbb{R}) = T\Sigma \times T\mathbb{R}$$

whereas the second is not splittable in that manner. But it is known that the tangent bundle of every 3-manifold is trivial [27], i.e. $T\Sigma = \Sigma \times \mathbb{R}^3$. From this fact we obtain a splitting

$$T(\Sigma \times_\Theta \mathbb{R}) = \xi \oplus \chi$$

as needed for a Lorentz structure. This splitting is induced by the embedding $\Sigma \to \Sigma \times_\Theta \mathbb{R}$ and by the bundle splitting

$$T\Sigma = \Sigma \times \mathbb{R}^3 = (\Sigma \times \mathbb{R}^2) \oplus (\Sigma \times \mathbb{R}^1)$$

into a 1- and 2-dimensional subbundle. Therefore the exotic 4-manifold $\Sigma \times_\Theta \mathbb{R}$ admits also a Lorentz structure but of a different kind: a codimension-1 foliation of the 3-manifold Σ induces a Lorentz structure on $\Sigma \times_\Theta \mathbb{R}$.

3. Exotic Black Holes

In [28] the existence of an exotic black hole (as exotic Kruskal space) using an exotic $\mathbb{R}^4$ was suggested. The idea was simply to consider the complement $\mathbb{R}^4 \setminus (D^3 \times \mathbb{R}) = S^2 \times \mathbb{R}^2$ where $\times$ was only understood topologically. In case of the exotic small $\mathbb{R}^4$ given by a Casson handle, we can reproduce our construction of an exotic $S^2 \times \mathbb{R}^2$ by using a Casson handle.

Therefore we will here concentrate on the representation of the exotic $S^2 \times \mathbb{R}^2$ by using the Casson handle CH to get

$$S^2 \times_\Theta \mathbb{R}^2 = D^2 \cup_{\partial CH} CH .$$

In [24] the analytical properties of the Casson handle were discussed. The main idea is the usage of the theory of end-periodic manifolds, i.e. an infinite periodic structure generated by W glued along a compact set K to get

$$S^2 \times_\Theta \mathbb{R}^2 = K \cup_N W \cup_N W \cup_N \cdots$$

the end-periodic manifold. The definition of an end-periodic manifold is very formal (see [13]) and we omit it here. All Casson handles generated by a balanced tree have the structure of end-periodic manifolds as shown in [24]. By using the theory of Taubes [13] one can construct a metric on $\cdots \cup_N W \cup_N W \cup_N \cdots$ by using the metric on W. Then a metric g in $S^2 \times_\Theta \mathbb{R}^2$ transforms to a periodic function $\hat{g}$ on the infinite periodic manifold

$$\tilde{Y} = \cdots \cup_N W_{-1} \cup_N W_0 \cup_N W_1 \cup_N \cdots$$

where W_i is the building block W at the ith place. To reflect the number of the building block, we have to extend $\hat{g}$ to $Y \times \mathbb{C}^*$by using a metric $\hat{g}_z$holomorphic in $z \in \mathbb{C}^* = S^1$ with $Y = W/i$ where i identifies the two boundaries of W. From the formal point of view we have the *generalized Fourier-Laplace transform (or Fourier-Laplace transform for short)*

$$\hat{g}_z(.) = \sum_{n=0}^{\infty} a_n z^n \cdot \hat{g}(.) \tag{3}$$

where the coefficient a_n represents the building block W_n in $\tilde{Y}$ (see our last paper [29]). Without loss of generality we can choose the coordinates x in M so that the 0th component x_0 is related to the integer $n = [x_0]$ via its integer part $[\,]$. Using the inverse transformation we can construct a smoth metric g in $\tilde{Y}$ at the nth building block via

$$(\tilde{T}^n g)(x) = \frac{1}{2\pi i} \int\limits_{|z|=s} z^{-n} \hat{g}_z(\pi(x)) \frac{dz}{z} \tag{4}$$

for $x \in \tilde{Y}$, $s \in (0, \infty)$, $n = [x_0]$ with the projection $\pi : \tilde{Y} \to Y$ (mathematically: $\tilde{Y}$ is the universal cover of Y like $\mathbb{R}$ is the universal cover of S^1) [29].

In the case of the Kruskal space we have the metric

$$ds^2 = \left(\frac{2M^3}{r}\right) \exp\left(-\frac{r}{2M}\right) \left(-dv^2 + du^2\right) + r^2 \left(d\theta^2 + \sin^2\theta d\phi^2\right) \tag{5}$$

in the usual units with a singularity at $r = 0$ used for the whole space $S^2 \times \mathbb{R}^2$. The coordinates (u, v) together with the relation

$$u^2 - v^2 = \left(\frac{r}{2M} - 1\right) \exp\left(-\frac{r}{2M}\right) \tag{6}$$

represent $\mathbb{R}^2$ and the angles (θ, ϕ) the 2-sphere S^2 parametrized by the radius r. Clearly this metric can be also used for each building block W having the topological structure $D^3 \times S^1 = W$ with two attaching regions topologically given by $D^2 \times S^1 = N$ forming the boundary $S^2 \times S^1 = \partial(D^3 \times S^1)$ (see the description of a Casson handle above). Remember that the Casson handle is topologically the subset $D^2 \times \mathbb{R}^2 \subset S^2 \times \mathbb{R}^2$. Now we consider the decomposition $W = D^2 \times (D^1 \times S^1)$ and the part $D^1 \times S^1$ will be later the $\mathbb{R}^2$ part of the Casson handle. The size of the D^2 is parametrized by r as above. Then we obtain the metric (5) for the building block W.

Our model of the black hole based on the implicit dependence of the two coordinates (u, v) on the parameter r, the radius of the 2-sphere. Therefore we choose for the coordinate $z \in \mathbb{C}^*$ the relation $z = \exp(ir)$ and obtain a metric $\hat{g}$ on $Y \times \mathbb{C}^*$. So, we make the assumptions:

1. The coordinate z is related to the radius by $z = \exp(ir)$.
2. Only the (u, v) part of the metric is periodic and we do not change the other component $r^2 \left(d\theta^2 + \sin^2\theta d\phi^2\right)$ of the metric.
3. The integer part $n = [v]$ of the coordinate v gives the number of the building block W_n in the Casson handle (seen as end-periodic manifold).
4. The metric on $S^2 \times_\theta \mathbb{R}^2$ is given by a Fourier transformation (4) of the (u, v) part of the metric in the building block W.

Some more comments are in order. The number $n = [v]$ is related to the coordinate v as substitute of "time". The metric g in $\tilde{Y}$ is smooth with respect to v and we obtain the number of the building block by $n = [v]$. To express this property we have to identify (u, v) with the coordinates of $D^1 \times S^1$. Then we obtain the metric on $S^2 \times_\theta \mathbb{R}^2$ by the generalized Fourier-Laplace transformation of the metric on $Y = W/i$ using the metric of the building block W und the coordinate z similar to (4)

$$g(v, u, \theta, \phi) = \int \exp(irv)\, \hat{g}_r(v, u, \theta, \phi) dr$$

Especially the singular part of the metric (i.e. the (u, v) part) on the building block W

$$(\hat{g}_r)_{00} = \left(\frac{2M^3}{r}\right) \exp\left(-\frac{r}{2M}\right) = (\hat{g}_r)_{11}$$

transforms to the Heaviside jump function

$$g_{00} = g_{11} = 2M^3 \Theta(v^2 - u^2 - 1)$$

using the relation (6) having no singularity. But the metric vanishes for large values of v in the interior of the black hole. This sketch of some arguments gives a hint that the transformation of the smoothness to exotic smoothness could possibly smooth out some singularity in the black hole case.

Conclusion

Perhaps we can best summarize this rather mathematically technical paper by presenting past analogies. For many years space was thought to have some absolute properties, defining e.g., rest, motion, etc. However, through various well known stages, we have progressed through the principle of special relativity to regard the idea that space provides some absolute rest reference frame, to general relativity, in which arbitrary reference frames are allowed and moreover the resulting spacetime models are allowed, or even required, to have geometries other than the *standard, simplest* Euclidean form. At the time that Einstein made this last great geometric leap he was fortunate to have at hand the necessary mathematical tools of non-Euclidean, differential, geometry, to explore these non-standard, "*exotic*," geometric structures.

Unfortunately as of today, we do not have such explicit mathematical tools available to explicitly explore exotic smoothness. We have tried in this chapter to present a brief review of some of the fairly recent mathematical discoveries in differential topology which seem to point the way to a serious consideration of whether we can assume an indefinite continuation of our *local* smooth coordinates to all of spacetime, in the very small as well as in the large, cosmological, structures. We only see light arriving at earth with properties measured by earth, or near earth, observational tools as sketched in Figure 1. From these measurements we assume that this light has forever followed the null geodesic which we describe by continuing our local coordinates indefinitely into regions containing the sources of the light, such as near a black hole horizon. However, as we hope to have shown, this is an *explicit* assumption which mathematicians have pointed out to us may not be valid. So, we close by asking whether exotic smoothness is analogous to exotic (i.e. non-flat) differential geometry. Unfortunately we do not have the explicit coordinate presentations of any four-dimensional model to make explicit predictions concerning this question, but the infinite constructions such as those involving Casson handles may provide some guide in the not too distant future, perhaps as sketched in §3 above.

Acknowledgment

We are grateful to Patrick Garrity for providing the illustration in Figure 1.

References

[1] M. W. Hirsch. *Differential Topology*. Springer-Verlag, New York, 1976.

[2] R.E. Gompf and A.I. Stipsicz. *4-manifolds and Kirby Calculus*. American Mathematical Society, 1999.

[3] T. Asselmeyer-Maluga and C.H. Brans. *Exotic Smoothness and Physics*. WorldScientific Publ., Singapore, 2007.

[4] M.H. Freedman. There is no room to spare in four-dimensional space. *Noticess Amer. Math. Soc.*, **31**:3–6, 1984.

[5] F. Ding. Smooth structures on some open 4-manifolds. *Topology*, **36**:203–207, 1997.

[6] S.W Hawking and G.F.R. Ellis. *The Large Scale Structure of Space-Time*. Cambridge University Press, 1994.

[7] R. Kirby and L.C. Siebenmann. *Foundational essays on topological manifolds, smoothings, and triangulations*. Ann. Math. Studies. Princeton University Press, Princeton, 1977.

[8] S. Smale. Generalized Poincare's conjecture in dimensions greater than four. *Annals of Mathematics*, **74**:391, 1961.

[9] S. Smale. On the structure of manifolds. *Amer. J. of Math.*, **84**:387–399, 1962.

[10] M.A. Kervaire and J. Milnor. Groups of homotopy spheres: I. *Ann. Math.*, **77**:504 – 537, 1963.

[11] J. Stallings. Piecewise-linear structure of euclidean space. *Proc. Cambridge Phil. Soc.*, 58:481, 1962.

[12] R.E. Gompf. Three exotic R^4's and other anomalies. *J. Diff. Geom.*, **18**:317–328, 1983.

[13] C.H. Taubes. Gauge theory on asymptotically periodic 4-manifolds. *J. Diff. Geom.*, **25**:363–430, 1987.

[14] S. DeMichelis and M.H. Freedman. Uncountable many exotic $\mathbb{R}^4$'s in standard 4-space. *J. Diff. Geom.*, 35:219–254, 1992.

[15] M.H. Freedman. The topology of four-dimensional manifolds. *J. Diff. Geom.*, **17**:357 – 454, 1982.

[16] S. Donaldson. An application of gauge theory to the topology of 4-manifolds. *J. Diff. Geom.*, **18**:279–315, 1983.

[17] R Gompf. An infinite set of exotic $\mathbb{R}^4$'s. *J. Diff. Geom.*, **21**:283–300, 1985.

[18] S. Donaldson. Irrationality and the h-cobordism conjecture. *J. Diff. Geom.*, **26**:141–168, 1987.

[19] A. Casson. *Three lectures on new infinite constructions in 4-dimensional manifolds*, volume 62. Birkhäuser, progress in mathematics edition, 1986. Notes by Lucian Guillou, first published 1973.

[20] S. Akbulut and R. Kirby. Mazur manifolds. *Mich. Math. J.*, **26**:259–284, 1979.

[21] Ž. Bižaca and R. Gompf. Elliptic surfaces and some simple exotic $\mathbb{R}^4$'s. *J. Diff. Geom.*, **43**:458–504, 1996.

[22] R. Gompf. Infinite families of casson handles and topological disks. *Topology*, **23**:395–400, 1984.

[23] R. Gompf. Periodic ends and knot concordance. *Top. Appl.*, **32**:141–148, 1989.

[24] T. Kato. ASD moduli space over four-manifolds with tree-like ends. *Geom. Top.*, 8:779 – 830, 2004. arXiv:math.GT/0405443.

[25] Z. Bizaca. An explicit family of exotic Casson handles. *Proc. AMS*, **123**:1297 – 1302, 1995.

[26] J.L. Friedman, K. Schleich, and D.M. Witt. Topological censorship. *Phys. Rev. Lett.*, 71:1486–1489, 1993. Erratum-ibid. 75 (1995) 1872; arXiv:gr-qc/9305017.

[27] J. Milnor and J. Stasheff. *Characteristic Classes*. Ann. Math. Studies,76. Princeton Univ. Press, Princeton, N.J., 1974.

[28] C.H. Brans. Localized exotic smoothness. *Class. Quant. Grav.*, **11**:1785–1792, 1994.

[29] T. Asselmeyer-Maluga and C.H. Brans. Gravitational sources induced by exotic smoothness. arXiv: 1101.3168, subm. to Gen. Rel. Grav.

In: Black Holes: Evolution, Theory and Thermodynamics ISBN: 978-1-61942-929-1
Editors: A. J. Bauer and D. G. Eiffel, pp. 157-164

Chapter 7

Lower Dimensional Quantum Black Holes

***Xavier Calmet*[1,*] *and Greg Landsberg*[2,†]**
[1] Physics and Astronomy, University of Sussex, Falmer, Brighton, UK
[2] Department of Physics, Brown University, Providence, RI, US

Abstract

There are theoretical reasons to think that space-time is lower dimensional at very short distances. If this is the case, quantum black holes produced at the LHC or in cosmic rays scattering live in lower dimensions. We discuss production cross section and signatures for the corresponding quantum black holes at the LHC within several different models of low-dimensional space-time.

Models with low-scale quantum gravity (see, e.g. [1] for a recent review) have received a considerable amount of attention over the last decade. One of the most striking signatures of these models is the possibility of producing mini black holes at colliders [2–4].

The formation of a classical black hole in a collision of two particles with zero or non-zero impact parameter is remarkably well understood, thanks to the beautiful work by Penrose, unfortunately unpublished, which was followed by D'Eath and Payne [5] and more recently by Eardley and Giddings [6]. The construction of Penrose is valid in the limit where the collision energy, and thus the mass of the black hole, is much larger than the Planck mass. This seminal work establishes that the production cross section for black holes in the collision of two particles at very high energy is given by $\sigma \approx \pi r_s^2$ where r_s is the Schwarzschild radius, which depends on the mass of the black hole (and hence of the center-of-mass energy) and on the dimensionality of space.

An extension of this construction in the semi-classical regime was proposed by Hsu [7] using a path-integral approach. This construction holds if there is a hierarchy between the Planck mass and the center-of-mass energy. One typically estimates that the semi-classical black hole mass is some 5 to 20 times larger than the Planck mass. For this reason, it is

*E-mail address: x.calmet@sussex.ac.uk
†E-mail address: landsberg@hep.brown.edu

now well understood that the likelihood of producing semi-classical black holes at the LHC, which are thermal objects decaying into a multiparticle f nal state, is quite small.

Whereas it is hard to produce a semi-classical black hole at the LHC, quantum black holes (QBHs), which we def ne as non-thermal mini black holes with masses comparable to the Planck mass [8], could be produced copiously. Unfortunately, very little is known about QBHs since their description would require a complete understanding of quantum gravity. The physics of QBHs can be, to a certain extent, extrapolated from the understanding we have of the semi-classical black holes. If one considers the limit where the mass of a semi-classical black hole decreases, i.e., its temperature then increases, the black hole becomes less and less thermal, and the number of the f nal-state particles decreases as a result.

A truly quantum black hole is expected to decay into only a handful of particles, which would resemble strong gravitational rescattering. QBHs can be thought of as small gravitational bound states, which are short-lived. At a proton-proton collider, QBHs will be produced in the collisions of quarks and gluons. Therefore most of the time they will carry color and electric charges and can be classif ed according to the representations of the groups $SU(3)_{\text{color}}$ and $U(1)_{\text{QED}}$, since gauge symmetries are expected to be conserved by quantum gravity. Note that this does not conf ict with conf nement since the typical length scale of QCD, Λ_{QCD}^{-1}, is much larger than the Planck length $\sim M_P^{-1}$, where M_P is the Planck mass. This allows one to make rather precise predictions for the LHC. However, one crucial piece of information is diff cult to obtain: it is that of the cross section of the QBH formation. In the past, this cross section has been extrapolated from the semi-classical approximation as well. This could be too naïve.

We do not have much knowledge of the structure of space-time are very short distances, i.e. at the energies relevant to the formation of QBHs. However, it has recently been argued by Ambjorn, Jurkiewicz, and Loll [9] using a causal dynamical triangulations approach to quantum gravity, which is a modif cation of quantum Regge calculus where space-time is discretized, that space-time is 1+1-dimensional at very short distances. There is a clear connection of this idea with the models proposed by Horava [10] and more recently by one of us and collaborators [11].

If space-time is truly 1+1-dimensional at very short distances, then QBHs are expected to live in 1+1 dimensions. Before we turn our attention to black holes, let us outline how a model of reduced space-time dimensions with strong gravity at the scale of dimensionality reduction could be constructed. Further details will appear in [15]. The idea is to rely on the compactif cation of two spacial dimensions which we take perpendicular to the direction of motion of the particle. If one compactif es dimensions and the volume of the compactif ed dimensions is small, one expects the lower dimensionsal Planck mass to be large compared to the one in the theory with a larger number of dimensions. This is the inverse mechanism of that proposed by Arkani-Hamed and al. to explain the apparent weakness of gravity in four dimensions. From 4 dimensions to 2, one expects $M_2^2 = a^2 M_4^2$, where a is the compactif cation radius. If a is very small, one gets $M_2 \gg M_4$ in "appropriate" units. Unfortunately def ning gravity in 2 dimensions is not as straightforward. The literature on 2-dimensional gravity is rich, see e.g. [12–14]. Here we will follow the approach outlined by Grumiller and Jackiw [14]. Starting from General Relativity in 4-dimensions, we compactify two spatial dimensions:

$$d^2s = g_{\mu\nu}dx^\mu dx^\nu = g_{\alpha\beta}dx^\alpha dx^\beta + \frac{1}{\lambda}\phi d\Omega_2^2 \quad (1)$$

ϕ/λ f xes the compactif cation radius a. One obtains the following action for gravity in 2-dimensions

$$S = -\frac{1}{2\kappa}\int d^2x\sqrt{g}(\phi R - (\nabla\phi)^2 - \lambda e^{-2\phi} - \Lambda), \quad (2)$$

where we have introduced a cosmological constant Λ and κ is a dimensionless gravitational coupling constant. One can obtain a dynamical compactif cation if one assumes that the compactif cation radius a is renormalization scale dependent. We assume that for high energy collisions, i.e., in our case $\sqrt{s} \sim 1$ TeV, $a(\mu)$ becomes small (i.e. $a(1\text{ TeV}) \sim 0$ GeV^{-1}). and space-time becomes effectively two dimensional while on large distances $a(\mu < 1\text{ TeV}) \to \infty$ and space-time is 3+1 dimensional. Note that one expects the strength of gravity is modif ed as well as explained above and gravity will become strong at short distances. We point out that M_2, i.e. the two-dimensional Planck mass, is dimensionless and there are no graviton degree of freedom. Def ning the strength of gravity is slightly tricky, but can be done in terms of the interaction of the dilaton ϕ with the energy-momentum tensor.

Classical black hole solutions are known in two-dimensional gravity. However, here we are concerned with non-thermal quantum black holes. As in the case of small black holes with large extra-dimensions, we can extrapolate the cross section for lower dimensional small black hole production from the classical ones. If space-time is only two-dimensional at very short distances of the order of the inverse Planck mass, there are important implications for the production cross section of QBHs, namely the cross section has to be dimensionless since it corresponds to 1+1-dimensional physics. This two-dimensional cross section is either a constant ($\sigma_2 =$ constant) or a function of the ratio of the center-of-mass energy to the cosmological constant, which is the only other dimensionless parameter one can build from the variables in the action ($\sigma_2 = f(\sqrt{\hat{s}}/\Lambda)$). We need to convert the two-dimensional cross section to the one which is measured at macroscopic distances, in 3+1 dimensions. The important physical quantity is the f ux of particles contained in the cylinder of diameter $\sim b$, where b is the impact parameter of the scattering between the two particles. To achieve the correct dimensionality, the four-dimensional cross section is then $\sigma_4 \sim \sigma_2 b^2\theta(\sqrt{\hat{s}} - b^{-1})$.

Let us now discuss different models of space-time at very short distances. We consider the gravitational scattering of two particles in a collider. We shall f rst assume that gravity becomes strong at 1 TeV. Strong gravitational physics is felt by the colliding particles if $b^{-1} \sim M_P$ where M_P is the effective Planck mass of the model. This could be realized in models with large extra dimensions or a large hidden sector, see, e.g. [1] for a recent review. Two scenarios are possible. The f rst one is that quantum black holes never form because the parameters of the action (2) are such that gravity at very short distances is weak and does not have a minimal length. This theory is likely to be pathological as it would imply that there are real singularities in astrophysical black holes or during the Big Bang. If this was the case, semi-classical black holes would still be formed at high energy, but when particles collide at center-of-mass energies similar to the Planck mass, they would just fy by another without interacting gravitationally. We could call this model asymptotically-free gravity in 1+1 dimensions.

A second, more exciting possibility is that the parameters of the action (2) are such that gravity is strong as well in the 1+1-dimensional regime. In that case we expect quantum black holes to form in collisions at center-of-mass energies of the order of $\sqrt{\hat{s}} \sim b^{-1} \sim M_P$. We expect that the correct 4D cross section for quantum black holes should be a limit of the semi-classical black hole cross section:

$$\sigma = a_n \pi r_n^2, \tag{3}$$

where n is the number of extra dimensions in space at the macroscopic distances, a_n is a calculable numerical factor of order one and, r_n is the $4+n$ Schwarzschild radius given by:

$$r_s(\hat{s}, n, \bar{M}_P) = \left[\frac{\Gamma\left((3+n)/2\right)}{(2+n)} 2^n \sqrt{\pi}^{n-3} \frac{\sqrt{\hat{s}}}{\bar{M}_P} \right]^{\frac{1}{1+n}} \frac{1}{\bar{M}_P}, \tag{4}$$

where $\bar{M}_P$ is the reduced Planck mass. We propose the following cross section for QBHs from the four-dimensional point of view:

$$\hat{\sigma}_{QBH} = \frac{1}{16\pi \bar{M}_P^2} \theta(\sqrt{\hat{s}} - \bar{M}_P). \tag{5}$$

This cross section can be obtained from the one for classical black holes by taking the limit $\sqrt{\hat{s}} \to \bar{M}_P$. Note that the cross section is universal and does not grow with the quantum black hole mass. This ansatz is particularly interesting since it might provide a solution to the unitarity problem of General Relativity, since this cross section does not grow with energy. The step function implies that the two colliding partons must be within a distance of $\bar{M}_P^{-1}$ from one the other to form a quantum black hole.

We are assuming that the production cross section for QBHs is independent of the number of dimensions of space-time at intermediate energy scales $\sim M_P$. Indeed, there are different frameworks in which lower-dimensional quantum black holes could be produced. The most obvious one is that of a four-dimensional theory with a large hidden sector [1]. In the case of large extra dimensions at the macroscopic scale, this scenario is realizable if large extra dimensions open up at short distances, before collapsing to two dimensions in the full quantum-gravity regime. Since space and time are dynamical objects, this is easily imaginable. This is why we posit a universal cross section, which is independent of the number of dimensions.

Quantum black holes [8], in contrast to semi-classical ones, are very short-lived non-thermal objects and will thus decay immediately into a few particles, typically just two. As suggested by Meade and Randall [16], the corresponding physics would resemble strong gravitational scattering. As we have pointed out, this would take place in 1+1 space-time dimensions. Clearly this will affect the way one searches for QBHs at the LHC. The angular distribution of the two particles produced by the decaying black hole is model dependent. If there is no space-time foam, we can imagine that the particles colliding with energies close to the Planck mass will generate gravitational f elds, which will resemble that of the

Aichelburg-Sexl metric [17], i.e., they will look like shock waves. In that case, the angular distribution is very peculiar. A QBH will decay into two particles back-to-back, which will in most cases form two energetic jets. These jets will essentially be in the direction of the colliding beams and as such very tough to see. We still expect a suppression of the dijet cross section at the energies beyond the crossover energy b^{-1}, which may be the only detectable signature of quantum black hole formation.

An alternative model for space-time at short distances is that of space-time foam. In that case we can assume that the quantum black hole forms as previously described, but decays in a pair of jets emitted back-to-back in a random direction, according to the stochastic f uctuations of space-time at short distances. In this model we expect modif cation of dijet angular distributions above the formation threshold of QBHs, which would resemble that for a compositeness signal (i.e., an excess of centrally produced jets, unlike mostly forward jets from the t-channel exchange dominating pure QCD dijet production mechanism).

As in [8], we assume that processes involving QBHs conserve QCD and $U(1)$ charges since local gauge symmetries are not violated by gravity. Note that we make no similar assumption about global charges. For example, B-L, Lorentz invariance, or f avor number do not have to be preserved in quantum gravitational interaction. Note that the Lorentz invariance is expected to be violated if the dimensionality of space-time shrinks at short distances. The total cross section for black hole production in our model is quite large. We f nd $\sigma(p+p \rightarrow \text{QBH}) \approx 140$ pb for a 14 TeV LHC. For the start-up LHC operating energy of 7 TeV, the cross section is still sizable: ≈ 18 pb. The different possible f nal states allowed by gauge invariance have been discussed in [8]. Quantum black holes produced at a proton-proton collider can be classif ed according to representations of $SU(3)_{\text{color}}$ and $U(1)_{\text{QED}}$. This enables us to make predictions for production of certain exotic f nal states, shown in Table 1. Note that we are allowing transitions, which violate Lorentz symmetry, such as, e.g., those described by effective operators of the type $qq\bar{q}\gamma$. Since, in our scenario, the dimensionality of space-time shrinks at high energy, Lorentz invariance is not expected to hold above the dimensional crossover. In deriving the branching fractions listed in Table 1, we assume that all the transitions not explicitly protected by a gauge symmetry take place. For example, we include transitions described by effective operators of the type $qqql$ or $q\bar{q}e^+\mu^-$.

The branching fractions in Table 1 are model dependent. We assume that there are about hundred possible degrees of freedom accessible to QBHs in models with large extra dimensions. In the case of a large hidden sector, there are $\sim 10^{33}$ degrees of freedom. Neutral QBHs will thus decay predominantly into the hidden sector. Hence the branching fraction of QBHs in that scenario to two oppositely charged leptons is essentially zero. On the other hand, charged QBHs in this scenario, will still predominantly decay into about hundred possible degrees of freedom of the standard model, because of gauge invariance. Independently of the model, most QBHs produced at the LHC carry QCD and QED charges since there are produced in collisions of quarks and gluons. These QBHs will decay mainly to colored particles of the standard model and will thus lead to dijets events. Table 1 focuses on smocking-gun exotic signatures for QBHs, with potentially little background from standard model physics. The cross sections for different operating energies at the LHC are plotted in Fig. 1.

(5) at the 14 TeV LHC.

Table 1. Some possible exotic fina states in quantum black hole decays. The branching fraction $\mathbf{Br}(p+p \rightarrow \mathbf{QBH} \rightarrow Z+\mathbf{jet}) = 3/2 \times \mathbf{Br}(p+p \rightarrow \mathbf{QBH} \rightarrow \gamma+\mathbf{jet})$. Note that the fina state for neutral black holes depends on the model for strong gravity under consideration. In a large hidden sector model, neutral black holes decay predominantly into the hidden sector. In models with large extra dimensions, these black holes can have sizable decay into two charged leptons., e.g., $e^+\mu^-$

	Branching fractions for interesting decay modes
$\mathrm{Br}(p+p \rightarrow \mathrm{QBH}_{\bar{3}}^{4/3} \rightarrow l^+ + \bar{d})$	$\sim 0.14\%$
$\mathrm{Br}(p+p \rightarrow \mathrm{QBH}_{\bar{3}}^{-2/3} \rightarrow l^- + \bar{d})$	$\sim 0.04\%$
$\mathrm{Br}(p+p \rightarrow \mathrm{QBH}_{\bar{3}}^{1/3} \rightarrow \nu_i + \bar{d})$	$\sim 0.1\%$
$\mathrm{Br}(p+p \rightarrow \mathrm{QBH}_{\bar{3}}^{-2/3} \rightarrow \nu_i + \bar{u})$	$\sim 0.04\%$
$\mathrm{Br}(p+p \rightarrow \mathrm{QBH}_{\bar{3}}^{-2/3} \rightarrow \gamma + \bar{u})$	$\sim 0.04\%$
$\mathrm{Br}(p+p \rightarrow \mathrm{QBH}_{\bar{3}}^{1/3} \rightarrow \gamma + \bar{d})$	$\sim 0.1\%$
$\mathrm{Br}(p+p \rightarrow \mathrm{QBH}_1^0 \rightarrow e^+ + \mu^-)$	$\sim$ 0 to 0.006 %

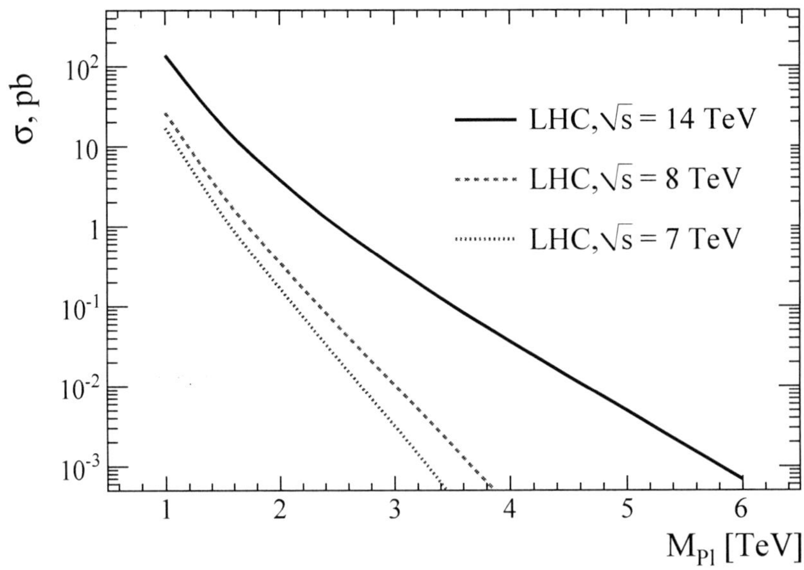

Figure 1. This f gure shows the cross section for different operating energies at the LHC using CTEQ parton distribution functions [18].

Another potential model is that without large extra dimensions or a large hidden sector, but with space-time becoming two-dimensional at short distances as discussed in [11]. In this model semi-classical black holes will not form in a collision of two partons at TeV energies; however, if the parameters of the action (2) are such that 1+1 gravity is strong, quantum black holes could still form. In that case, the impact parameter is of the order

$b \sim (1\ \text{TeV})^{-1}$ and the cross section could depend on the parameters of the fundamental action (2). We thus would expect $\sigma_4 = \sigma_2 b^{-2} = f(\sqrt{\hat{s}}/\Lambda))b^{-2}$. The function f is unknown and it is thus diff cult to make numerical predictions. However, if as assumed in [11], lowering the dimensionality of space-time solves the hierarchy problem, the cross sections for quantum black holes would be still of the order of $b^2 \sim 1\ \text{TeV}^{-2}$ and thus comparable in magnitude to those obtained above.

Note that in all of the above scenarios, which allow formation of semi-classical and classical black holes, their physics is not expected to be affected since their Schwarzschild radii are much larger than the typical scale b where the dimensionality of space-time changes.

Conclusions: We have considered quantum black holes at the LHC in the context of models of quantum gravity, which are lower-dimensional at short distances. Different scenarios are discussed. The f rst ones are within the context of well-studied models with a large extra-dimensional volume or a large hidden sector. In these scenarios, it is possible to either suppress quantum black hole production completely, or on the contrary produce them copiously at the LHC. We f nd that the cross sections for small non-thermal black holes can be quite different from those considered previously in the literature. The cross section can be quite large and does not depend on the black hole mass. The angular distribution of the f nal state particles would give important insight into the nature of space-time at short distance and will help to differentiate different models of space-time foam. The f nal state in the black hole decay will allow to determine the model at the origin of strong gravitational physics in the TeV range and help to probe the symmetries of quantum gravity. Clearly, the search strategies for mini black holes at the LHC have to be adapted to be able to detect the QBHs described in the paper. Finally, if space-time becomes 1+1-dimensional, as recently proposed to address the hierarchy problem of the standard model, gravity could become strongly coupled when space-time becomes 1+1-dimensional. In that case, only quantum black holes would be produced in particle collisions.

Acknowledgments

The work of XC is supported in part by the European Cooperation in Science and Technology (COST) action MP0905 “Black Holes in a Violent Universe”. The work of GL is partially supported by the US Department of Energy, under Grant No. DE-FG02-91ER40688. We thank the organizers of the Experimental Searches for Quantum Gravity Workshop ESQG 2010, where this work has originated.

References

[1] X. Calmet, *Mod. Phys. Lett. A* **25**, 1553 (2010) [arXiv:1005.1805 [hep-ph]].

[2] S. Dimopoulos and G. L. Landsberg, *Phys. Rev. Lett.* **87**, 161602 (2001) [arXiv:hep-ph/0106295].

[3] T. Banks and W. Fischler, arXiv:hep-th/9906038.

[4] S. B. Giddings and S. D. Thomas, *Phys. Rev. D* **65**, 056010 (2002) [arXiv:hep-ph/0106219].

[5] P. D. D'Eath and P. N. Payne, *Phys. Rev. D* **46**, 658 (1992); *Phys. Rev. D* **46**, 675 (1992); *Phys. Rev. D* **46**, 694 (1992).

[6] D. M. Eardley and S. B. Giddings, *Phys. Rev. D* **66**, 044011 (2002) [arXiv:gr-qc/0201034];

[7] S. D. H. Hsu, *Phys. Lett. B* **555**, 92 (2003) [arXiv:hep-ph/0203154].

[8] X. Calmet, W. Gong and S. D. H. Hsu, *Phys. Lett. B* **668**, 20 (2008) [arXiv:0806.4605 [hep-ph]].

[9] J. Ambjorn, J. Jurkiewicz and R. Loll, *Phys. Rev. Lett.* **93**, 131301 (2004) [arXiv:hep-th/0404156]; *Phys. Lett. B* **607**, 205 (2005) [arXiv:hep-th/0411152]; *Phys. Rev. D* **72**, 064014 (2005) [arXiv:hep-th/0505154].

[10] P. Horava, *Phys. Rev. D* **79**, 084008 (2009) [arXiv:0901.3775 [hep-th]]; *Phys. Rev. Lett.* **102**, 161301 (2009) [arXiv:0902.3657 [hep-th]]; P. Horava and C. M. Melby-Thompson, arXiv:1007.2410 [hep-th].

[11] L. Anchordoqui, D. C. Dai, M. Fairbairn, G. Landsberg and D. Stojkovic, arXiv:1003.5914 [hep-ph].

[12] C. G. . Callan, S. B. Giddings, J. A. Harvey and A. Strominger, *Phys. Rev. D* **45**, 1005 (1992) [arXiv:hep-th/9111056].

[13] J. P. S. Lemos, *Class. Quant. Grav.* **12**, 1081 (1995) [arXiv:gr-qc/9407024].

[14] D. Grumiller and R. Jackiw, arXiv:0712.3775 [gr-qc].

[15] X. Calmet and D. Fragkakis, in preparation.

[16] P. Meade and L. Randall, *JHEP* **0805**, 003 (2008) [arXiv:0708.3017 [hep-ph]].

[17] P. C. Aichelburg and R. U. Sexl, *Gen. Rel. Grav.* **2**, 303 (1971).

[18] H. L. Lai *et al.* [CTEQ Collaboration], *Eur. Phys. J. C* **12**, 375 (2000) [arXiv:hep-ph/9903282].

In: Black Holes: Evolution, Theory and Thermodynamics ISBN: 978-1-61942-929-1
Editors: A. J. Bauer and D. G. Eiffel, pp. 165-170

Chapter 8

NON THERMAL SMALL BLACK HOLES

Xavier Calmet*, Dionysios Fragkakis† and Nina Gausmann‡
Physics and Astronomy, University of Sussex, Falmer, Brighton, UK

Abstract

In this chapter we review the current theoretical state of the art of small black holes at the LHC. We discuss the production mechanism for small non thermal black holes at the LHC and discuss new signatures due to a possible discrete mass spectrum of these black holes.

One of the major remaining challenges of theoretical physics is to understand gravity at the quantum level. While quantum gravity is most likely impossible to probe experimentally in conventional models of short distance space-time where quantum effects are expected to become relevant at energies of some 10^{19} GeV, there is a class of models with low scale quantum gravity which is being probed directly by measurements at the Large Hadron Collider (LHC) at CERN which predict that quantum gravitational effects could become important at a few TeVs. These models require the existence of a large extra-dimensional volume [1, 2] or a large number of particles which lead to a running of the Planck mass [3], see e.g. [4] for a recent review. The production of small black holes at the LHC would be one of the most intriguing signals of low scale quantum gravitational physics. We emphasize that recent studies have shown that brane world models as well as models with a large number of particles typically have unitarity problems well below the effective Planck mass and new physics effects, responsible for f xing the perturbative unitarity of the S-matrix, should f rst be observed before quantum gravitational effects are seen at the LHC [5–7].

Black holes are fascinating objects because they involve physics under extreme conditions. Indeed, if the scale of quantum gravity is truly as low as a few TeV, the most striking feature of these models is the prediction that colliders such as the LHC may be able to create small black holes [8–18] which would allow us to probe quantum gravity directly in an

*E-mail address: x.calmet@sussex.ac.uk
†E-mail address: d.fragkakis@sussex.ac.uk
‡E-mail address: n.gausmann@sussex.ac.uk

experiment. It is important to realize that if produced at the LHC, these black holes would be quite different from astrophysical black holes in the sense that their masses would be close to the Planck mass, i.e., in these scenarios, a few TeVs. Depending on the ratio of their masses to the Planck mass, we shall refer to these small black holes as semi-classical black holes or quantum black holes.

The last ten years have resulted in impressive progress in our understanding of the production of small black holes in the collision of two particles at very high energy. In the early days of black hole formation at colliders, the hoop conjecture [19] due to Kip Thorne was used as a criteria for gravitational collapse in the collision of two particles head to head. The hoop conjecture states that if an amount of energy E is conf ned to a spherical region of space-time with a radius R with $R < E$, then that region will eventually evolve into a black hole. Natural units were used: $\hbar, c$ and Newton's constant are set to unity. While the hoop conjecture is to a certain extent at the hand waving level, there is a convincing proof of gravitational collapse in the case of a collision of two particles with non zero impact parameter. It had been known since the works of Penrose (unpublished) and later on by D'Eath and Payne [20] that a small black hole will be formed in the head to head collision of two particles with zero impact parameter. However, the relevant case for the LHC is that of a non vanishing impact parameter. The resolution of the problem was given by Eardley and Giddings [21], see also [22], who were able to construct a closed trapped surface. Apparently Penrose had also derived the latter result, but it was never published.

The construction of Eardley and Giddings [21] is valid in the limit where the mass of the black holes and hence the center of mass energy is much larger than the effective reduced Planck mass. In other words, the construction applies to two colliding planets with a very high center of mass energy tending towards inf nity. The black hole formed in that limit is essentially classical. It was shown by Hsu [22], using a path integral formulation, that this construction could be extended to the semi-classical regime if the small black hole mass is somewhat larger than the Planck mass. The ratio between the f rst semi-classical black hole mass and that of the Planck mass can be estimated. In the case of ADD, it is typically taken to be of the order of 5, while it could easily be 20 for RS [15]. Semi-classical black holes are thermal objects that are expected to decay via Hawking radiation to many particles, typically of the order of 20, after a spin down phase. This f nal explosion would lead to a spectacular signature in a detector. It is however now well understood [12, 15, 16] that it is very unlikely that semi-classical black holes will be produced at the LHC because the center of mass energy is not high enough. The main reasons are that not all the energy of the partons is available for black hole formation and the parton distribution functions tend to fall off very fast.

It has been proposed in [16] to extrapolate the semi-classical black hole into the quantum regime of quantum gravity and to consider the production of quantum black holes (QBHs) at the LHC. QBHs are def ned as the quantum analogs of ordinary black holes as their mass and Schwarzschild radius approach the quantum gravity scale. QBHs do not have semi-classical space-times and are not necessarily well-described by the usual Hawking temperature or black hole thermodynamics. In other words, they are non thermal. In many respects they are perhaps more analogous to strongly coupled resonances or bound states than to large black holes. QBHs presumably decay only to a few particles, each with Compton wavelength of order the size of the QBH. It seems unlikely that they would decay to a much larger number of longer wavelength modes.

An important question is whether the mass of the quantum black holes is quantized or continuous as expected in the case of macroscopic black holes. Most studies so far are assuming that QBHs have a continuous mass spectrum despite some recent warnings that the quantum black hole masses ought to be quantized [23]. We shall now describe the production cross section of quantum black holes at the LHC. We shall f rst discuss the continuous mass spectrum and then the discrete mass spectrum. In both cases, we shall assume that the cross sections can be extrapolated from the cross section obtained for semiclassical black holes, i.e. the geometrical cross section πr_s^2 where r_s is the Schwarzschild radius.

Continuous mass spectrum

In that case, the LHC production cross section for QBHs with a continuous mass spectrum is assumed to be of the form

$$\sigma^{pp}(s,x_{min},n,M_D) = \int_0^1 2zdz \int_{\frac{(x_{min}M_D)^2}{y(z)^2 s}}^1 du \int_u^1 \frac{dv}{v} \times F(n)\pi r_s^2(us,n,M_D) \sum_{i,j} f_i(v,Q) f_j(u/v,Q) \tag{1}$$

where M_D is the n dimensional reduced Planck mass, $z = b/b_{max}$, $x_{min} = M_{BH,min}/M_D$, n is the number of extra-dimensions, $F(n)$ and $y(z)$ are the factors introduced by Eardley and Giddings and by Yoshino and Nambu [24]. The n dimensional Schwarzschild radius is given by

$$r_s(us,n,M_D) = k(n)M_D^{-1}[\sqrt{us}/M_D]^{1/(1+n)} \tag{2}$$

where

$$k(n) = \left[2^n \sqrt{\pi}^{n-3} \frac{\Gamma((3+n)/2)}{2+n}\right]^{1/(1+n)}. \tag{3}$$

The fact that these QBHs are non thermal is ref ected in the assumption that they decay only to a few particles immediately after their creation.

Discrete mass spectrum

We now consider the discrete mass spectrum case. The cross section is given by

$$\sigma_{QBH}^{pp}(s,M_{QBH},n,M_D) = \pi r_s^2(M_{QBH},n,M_D) \int_0^1 2zdz \int_{\frac{(M_{QBH})^2}{y(z)^2 s}}^1 du \int_u^1 \frac{dv}{v} \times F(n) \sum_{i,j} f_i(v,Q) f_j(u/v,Q) \tag{4}$$

with the constant Schwarzschild radius given by

$$r_s(M_{QBH}^2,n,M_D) = k(n)M_D^{-1}[\sqrt{M_{QBH}^2}/M_D]^{1/(1+n)} \tag{5}$$

where as previously

$$k(n) = \left[2^n \sqrt{\pi}^{n-3} \frac{\Gamma((3+n)/2)}{2+n}\right]^{1/(1+n)}. \tag{6}$$

Note that the parton level cross section is constant in that case. The physics of QBHs with a discrete mass spectrum is very different from the continuous case. They are expected to behave as heavy resonances that will decay to a few particles. The total QBH cross section is given by the sum of the individual QBH production cross sections:

$$\sigma_{tot}^{pp}(s,n,M_D) = \sum_i \sigma_{QBH}^{pp}(s,M_{QBH}^i,n,M_D). \qquad (7)$$

We expect that the mass spectrum is quantized in terms of the Planck mass because of the existence of a minimal length [25, 26] in models incorporating quantum mechanics and general relativity. For a Planck mass at 1 TeV, one might expect 5 QBH states between the Planck mass and the semi-classical regime.

Decay modes

It is assumed that QBHs are defned by three quantities: their mass, spin and gauge charges. Importantly, QBHs can have a QCD, or color, charge. This is not in contradiction with confnement since the typical length scale of QCD, i.e., a Fermi, is much larger than the size of a QBH, e.g., TeV^{-1}. The formation and decay of a QBH takes place over a small space-time region – from the QCD perspective it is a short distance process, and hadronization occurs only subsequently. Their decomposition modes depend on a few assumptions which are as follows

I) Processes involving QBHs conserve QCD and U(1) charges since local gauge symmetries are not violated by gravity. Note that no similar assumption is made about global charges.

II) QBH coupling to long wavelength and highly off-shell perturbative modes is suppressed.

Assumption (II) is necessary so that precision measurements, or, possibly, proton decay do not force the quantum gravity scale to be much larger than the TeV range. It is not implausible that a nonperturbative QBH state couples only weakly to long distance or highly off-shell modes, but strongly to modes of size and energy similar to that of the hole. This is analogous to results obtained for (B+L) violating processes in the standard model: (B+L) violation is exponentially small in low energy reactions, but of order one for energies above the sphaleron mass.

Generically speaking, QBHs form representations of $SU(3)_c$ and carry a QED charge. The process of two partons p_i, p_j forming a quantum black hole in the c representation of $SU(3)_c$ and charge q as: $p_i + p_j \to \text{QBH}_c^q$ is considered in [16]. The following different transitions are possible at a proton collider:

a) $\mathbf{3} \times \overline{\mathbf{3}} = \mathbf{8} + \mathbf{1}$

b) $\mathbf{3} \times \mathbf{3} = \mathbf{6} + \overline{\mathbf{3}}$

c) $\mathbf{3} \times \mathbf{8} = \mathbf{3} + \overline{\mathbf{6}} + \mathbf{15}$

d) $\mathbf{8} \times \mathbf{8} = \mathbf{1}_S + \mathbf{8}_S + \mathbf{8}_A + \mathbf{10} + \overline{\mathbf{10}}_A + \mathbf{27}_S$

Most of the time the black holes which are created carry a $SU(3)_c$ charge and come in different representations of $SU(3)_c$ as well as QED charges. This allows to predict how they will be produced or decay. For example the production cross-section of a QBH_1^0 is given by

$$\begin{aligned} \sigma^{pp}(s, x_{min}, n, M_D) &= \int_0^1 2zdz \int_{\frac{(x_{min}M_D)^2}{y(z)^2 s}}^1 du \int_u^1 \frac{dv}{v} \\ &\times F(n)\pi r_s^2(us, n, M_D) \\ &\left(\frac{1}{9}\sum_{i,j=q,\bar{q}} f_i(v,Q) f_{\bar{j}}(u/v,Q) + \frac{1}{64} f_g(v,Q) f_g(u/v,Q)\right) \end{aligned} \quad (8)$$

where i, j runs over all the quarks and anti-quarks subject to the constraint of QED charge neutrality, and f_q, f_g are the quark and gluon parton distribution functions. For the production of a specifc member (i.e., with specifed color) of the octet QBH_8^0, one fnds the same expression. In case of a discrete mass spectrum us in r_s should be replaced by M_{QBH}^2. The details of the fnal states have been considered elsewhere [16] and applies to the discrete mass spectrum case as well.

In conclusion, we have reviewed the production mechanism for small non thermal black holes at the LHC and discussed new signatures due to a possible discrete mass spectrum of these black holes.

Acknowledgments

This work is supported in part by the European Cooperation in Science and Technology (COST) action MP0905 "Black Holes in a Violent Universe". The work of NG is supported by a SEPnet PhD fellowship.

References

[1] N. Arkani-Hamed, S. Dimopoulos, G. R. Dvali, *Phys. Lett.* **B429**, 263-272 (1998) [hep-ph/9803315]; I. Antoniadis, N. Arkani-Hamed, S. Dimopoulos *et al.*, *Phys. Lett.* **B436**, 257-263 (1998) [hep-ph/9804398].

[2] L. Randall, R. Sundrum, *Phys. Rev. Lett.* **83**, 3370-3373 (1999) [hep-ph/9905221].

[3] X. Calmet, S. D. H. Hsu, D. Reeb, *Phys. Rev.* **D77**, 125015 (2008) [arXiv:0803.1836 [hep-th]].

[4] X. Calmet, *Mod. Phys. Lett. A* **25**, 1553 (2010) [arXiv:1005.1805 [hep-ph]].

[5] M. Atkins and X. Calmet, *Phys. Lett. B* **695**, 298 (2011) [arXiv:1002.0003 [hep-th]].

[6] M. Atkins and X. Calmet, *Eur. Phys. J. C* **70**, 381 (2010) [arXiv:1005.1075 [hep-ph]].

[7] I. Antoniadis, M. Atkins and X. Calmet, *JHEP* **1111**, 039 (2011) [arXiv:1109.1160 [hep-ph]].

[8] S. Dimopoulos and G. L. Landsberg, *Phys. Rev. Lett.* **87**, 161602 (2001) [arXiv:hep-ph/0106295].

[9] T. Banks and W. Fischler, arXiv:hep-th/9906038.

[10] S. B. Giddings and S. D. Thomas, *Phys. Rev. D* **65**, 056010 (2002) [arXiv:hep-ph/0106219].

[11] J. L. Feng and A. D. Shapere, *Phys. Rev. Lett.* **88**, 021303 (2002) [arXiv:hep-ph/0109106].

[12] L. A. Anchordoqui, J. L. Feng, H. Goldberg and A. D. Shapere, *Phys. Lett. B* **594**, 363 (2004) [arXiv:hep-ph/0311365].

[13] L. A. Anchordoqui, J. L. Feng, H. Goldberg and A. D. Shapere, *Phys. Rev. D* **65** (2002) 124027 [arXiv:hep-ph/0112247].

[14] L. A. Anchordoqui, J. L. Feng, H. Goldberg and A. D. Shapere, *Phys. Rev. D* **68**, 104025 (2003) [arXiv:hep-ph/0307228].

[15] P. Meade and L. Randall, *JHEP* **0805**, 003 (2008) [arXiv:0708.3017 [hep-ph]].

[16] X. Calmet, W. Gong and S. D. H. Hsu, *Phys. Lett. B* **668**, 20 (2008) [arXiv:0806.4605 [hep-ph]].

[17] X. Calmet, D. Fragkakis and N. Gausmann, *Eur. Phys. J. C* **71**, 1781 (2011) [arXiv:1105.1779 [hep-ph]].

[18] X. Calmet and G. Landsberg, arXiv:1008.3390 [hep-ph].

[19] K. S. Thorne, *Nonspherical gravitational collapse: A short review*, in J. R . Klauder, *Magic Without Magic*, San Francisco (1972), 231–258.

[20] P. D. D'Eath and P. N. Payne, *Phys. Rev. D* **46**, 658 (1992); *Phys. Rev. D* **46**, 675 (1992); *Phys. Rev. D* **46**, 694 (1992).

[21] D. M. Eardley and S. B. Giddings, *Phys. Rev. D* **66**, 044011 (2002) [arXiv:gr-qc/0201034];

[22] S. D. H. Hsu, *Phys. Lett. B* **555**, 92 (2003) [arXiv:hep-ph/0203154].

[23] G. Dvali, C. Gomez and S. Mukhanov, arXiv:1106.5894 [hep-ph].

[24] H. Yoshino and Y. Nambu, *Phys. Rev. D* **67**, 024009 (2003) [arXiv:gr-qc/0209003]. Improved results have been obtained in H. Yoshino and V. S. Rychkov, *Phys. Rev.* D **71**, 104028 (2005) [Erratum-ibid. D **77**, 089905 (2008)] [arXiv:hep-th/0503171].

[25] L. J. Garay, *Int. J. Mod. Phys. A* **10**, (1995) 145 [arXiv:gr-qc/9403008]; C. A. Mead, *Phys. Rev.* **135**, (1964) B849; T. Padmanabhan, *Class. Quant. Grav.* **4**, (1987) L107.

[26] X. Calmet, M. Graesser and S. D. H. Hsu, *Phys. Rev. Lett.* **93**, (2004) 211101 [arXiv:hep-th/0405033]; *Int. J. Mod. Phys. D* **14** (2005) 2195 [arXiv:hep-th/0505144]; X. Calmet, *Eur. Phys. J. C* **54**, 501 (2008) [hep-th/0701073].

In: Black Holes: Evolution, Theory and Thermodynamics ISBN: 978-1-61942-929-1
Editors: A. J. Bauer and D. G. Eiffel, pp. 171-223

Chapter 9

FOOTPRINTS OF EXTRA DIMENSIONS IN THE RADIATION SPECTRA OF BLACK HOLES

Panagiota Kanti*
Division of Theoretical Physics, Department of Physics
University of Ioannina,
Ioannina, Greece

Abstract

In the context of the scenario with Extra Dimensions (either Large or slightly Warped), we consider the prospect of the creation of mini black holes during high-energy particle collisions at ground-based accelerators. We then focus on the main observable signal of their creation, that is the emission of Hawking radiation in the form of either scalars, fermions, gauge bosons or gravitons during both the spherically-symmetric ("Schwarzschild") and axially-symmetric ("spin-down") phases. We look for footprints in the radiation spectra of brane-localised Standard Model particles that could help us determine fundamental parameters of the theory, such as the number of additional spacelike dimensions that may exist in nature and the angular momentum of the black hole. The angular distribution of the emitted particles on our brane is studied, too, as an additional source of information and a distinctive feature of the Hawking radiation emitted by a rotating black hole. We also attempt to shed light on the question of the energy balance between the brane and the bulk emitted by the black hole: we consider the emission of all particles allowed to live on the brane and in the bulk and we compute their relevant emissivities. Finally, we discuss the effect of the mass on the scalar radiation spectra of a spinning black hole both in the bulk and on the brane as an indicative example of the role of the mass of the emitted particles.

Keywords: Black Holes, Extra Dimensions, Hawking Radiation

1. Introduction

The emergence, more than a decade ago, of the new theories [4–6] postulating the existence of additional spacelike dimensions in nature virtually redef ned the theoretical framework

*E-mail address: pkanti@cc.uoi.gr

in which gravitational and particle physics was studied. The fact that the mathematical formulation of the General Theory of Relativity allowed its extension to more than 4 dimensions was known for more than a century – the f rst unif cation theory of gravity and electromagnetism, the Kaluza–Klein theory, was a 5-dimensional one and was formulated only a few years after Einstein completed his theory of gravity. Klein's idea of the existence of a f fth spacelike dimension, that was compact and extremely small, was used extensively in the formulation of the superstring theory. Nevertheless, certain aspects of the new theories with extra dimensions were ground-breaking, and changed for ever the way we thought about extra dimensions in nature [1].

The scenario with Large Extra Dimensions [4, 5] was the f rst one that appeared in the literature. It generalized the Kaluza–Klein idea by postulating the existence of a number n of additional, f at, spacelike dimensions, apart from the usual inf nite-sized three. For simplicity, these extra dimensions were assumed to have the same size $\mathcal{R}$ and to make up a compact space of volume $V \sim \mathcal{R}^n$. In the context of this scenario, our $(3+1)$-dimensional world is a hypersurface, a *3-brane*, embedded in the $(4+n)$-dimensional spacetime, the *bulk*. All usual matter made up of and interacting through Standard-Model particles, including ourselves, is restricted to live on the brane. On this hypersurface, the scale where gravity becomes comparable to the other forces of nature is the Planck scale $M_P = 10^{19}$ GeV. On the other hand, gravitons, being the degrees of freedom of the gravitational f eld itself, are allowed to propagate in the whole of spacetime and thus mediate a higher-dimensional gravitational force. When two test particles with masses m_1 and m_2 are separated by a distance $r \ll \mathcal{R}$, the expression of the latter will have the form

$$F_G = G_D \frac{m_1 m_2}{r^{n+2}}, \tag{1}$$

where G_D is the higher-dimensional Newton's constant. If we demand that, by gradually increasing the separation between the two particles, we should recover the expression for the gravitational force in four dimensions at $r \gg \mathcal{R}$, it is found that [4]

$$G_D \simeq G_4 \mathcal{R}^n. \tag{2}$$

If we further def ne $G_D \sim 1/M_*^{n+2}$, and use the relation $G_4 \sim 1/M_P^2$, the above formula takes the form

$$M_P^2 \simeq \mathcal{R}^n M_*^{2+n}, \tag{3}$$

where M_* is the energy scale where gravitational forces become strong in the context of the higher-dimensional theory. In the framework of the traditional string theory, it was assumed that the size of the extra dimensions were of the order of the Planck scale $\mathcal{R} = l_P = 10^{-33}$ cm, and as a result the fundamental scale M_* was the same as the "effective" one M_P. However, in the scenario of Large Extra Dimensions, the ground-breaking assumption that $\mathcal{R}$ could be as large as 1 mm was made. Then, depending on the value of the number of extra dimensions n, M_* could be considerably smaller than M_P resulting in a much stronger gravitational force in the context of the higher-dimensional theory.

The topological structure of the higher-dimensional spacetime in the case of the scenario with Warped Extra Dimensions [6] is radically different. In contrast to a number n of

[1]For some early attempts to construct higher-dimensional gravitational models, see Refs. [7–10].

extra dimensions, here there is only one extra spacelike dimension. The fve-dimensional bulk is flled with a negative cosmological constant and it is thus an Anti de-Sitter spacetime. What is remarkable is that the extra dimension is generically non-compact but it may be compactifed at will if a second brane is introduced in the model. Solving the hierarchy problem, i.e. eliminating the gap between the gravity scale and the electroweak scale, was the objective of this model, too: assuming that the *visible* brane, where all SM felds live, is placed at a fnite distance $y = L$ from a *hidden* brane located at $y = 0$, it may be shown that the electroweak symmetry breaking on the visible brane takes place at a scale [6]

$$M_{EW} = e^{-k_\Lambda L} M_* \,. \tag{4}$$

In the above, k_Λ is the curvature scale associated with the negative cosmological constant that flls the 5-dimensional spacetime of the model, and it has been assumed that all fundamental scales at the hidden brane are of the order of M_*. The effective Planck scale M_P is now related to the fundamental one M_* through the equation

$$M_P^2 = \frac{M_*^3}{k} \left(1 - e^{-2k_\Lambda L}\right). \tag{5}$$

In both of the above two models, a low fundamental scale for gravity can be realised. The most optimistic case would be the one in which M_* is close to the energy scale of present particle experiments, i.e. $M_* \sim 1$ TeV. Since the above two models appeared in the literature, a number of works have placed limits on the fundamental energy scale M_*. These followed from either particle physics experiments, Cavendish-type measurements, astrophysical and cosmological processes – given in increasing order of the severity of the constraint derived in each case but also of the magnitude of the systematic errors in the limits produced (for more details on this, see [11]). Focusing on the frst two types of measurements, that are considered to be the most accurate, they give $M_* \geq 1.45$ TeV [12] and $M_* \geq 3.2$ TeV [13], respectively, for the case $n = 2$. We note that the case $n = 1$ in the context of the model with Large Extra Dimensions is already excluded if we wish to have $M_* \sim 1$ TeV, and, as the above limits show, the same happens with the case $n = 2$. As all types of limits become more relaxed as the number of additional spacelike dimensions increases, we may either keep the value of M_* low but admit three or more extra dimensions, or keep the number n small but push M_* to higher values.

What is even more remarkable than the truly revolutionary aspects that these two models contained is the number of paths for research that they opened. The objective of both models was to solve the hierarchy problem, i.e. to bridge the gap in the strength appearing at the microscopic level between the gravitational force and the other three forces present in nature. Nevertheless, it was soon realised that these models had a great potential for advancing our understanding of some of the most important questions in present-day research, such as the structure and true dimensionality of our spacetime and the formulation of the long-pursued fundamental theory describing all forces. As a result, the new theories soon became the framework in which all problems and quests of theoretical and mathematical physics were re-formulated and re-examined.

2. Creation and Properties of Higher-Dimensional Black Holes

The new theories with extra dimensions led to the realisation that not only our spacetime is fundamentally higher-dimensional and we live on a slice of it, but also that the higher-dimensional regime could be much closer, in terms of energy, than we thought. If M_* can be as low as a few TeV, as the most optimistic scenario assumes, then current particle-physics collision experiments, that operate at exactly these energies, may probe the strong-gravity regime.

As was very soon suggested, one of these strong-gravity effects that we may witness during particle collisions with center-of-mass energy E larger than M_* is the creation of black holes [14–16]. This could take place during the collision of two ordinary brane-localised particles with impact parameter $b < r_H(E)$, where $r_H(E)$ is the Schwarzschild radius that corresponds to E. According to General Relativity and the Hoop Conjecture [17,18], if the two colliding particles pass within a region with a radius of roughly the order of $r_H(E)$, then an event horizon will be formed and the particles will never escape their mutual gravitational attraction. Several aspects of the problem of the creation of black holes, that is in fact a non-trivial boundary-value problem especially in the higher-dimensional spacetime, have been studied over the years and detailed limits on the impact parameter have been derived. Also the production cross-section of black holes was estimated, with the consensus currently being that this takes values that are considerably higher than the ones for Standard-Model processes at such a high energy (for more details on these, see [11,19]).

In order to study further the properties of the produced black hole, we will need a line-element that describes the spacetime around such a higher-dimensional object. It turns out that, we may write such a line-element only for a black hole living in a spacetime with Large Extra Dimensions or Slightly Warped ones. After more than ten years of intense scientif c research, no analytic form for a regular, localised-on-our-brane, higher-dimensional black hole has been found in the context of a model where the presence of the brane has been taken into account. The only such solutions available in the literature are numerical [20] and all have a horizon radius which is smaller than the curvature radius ℓ_{AdS} of the AdS spacetime. In the limit of large ℓ_{AdS}, both the cosmological constant in the bulk and the tension of the brane are negligible and may be ignored when compared to the mass of the black hole. The latter should in any case be larger than the fundamental scale for gravity in order to be considered a classical object, and thus we may naturally assume that $r_H \ll \ell_{AdS}$. In that case, even if the bulk is warped, the black hole cannot see its curvature, and, to a f rst approximation, lives in a f at spacetime. Therefore, in what follows, we assume that the warping of the spacetime, if present, is small, and we work in the context of the model with Large Extra Dimensions.

The simplest prototype that we may use to describe such a higher-dimensional black hole is the spherically-symmetric Schwarzschild-Tangherlini one [21,22]

$$ds^2 = -\left[1-\left(\frac{r_H}{r}\right)^{n+1}\right] dt^2 + \left[1-\left(\frac{r_H}{r}\right)^{n+1}\right]^{-1} dr^2 + r^2 d\Omega_{2+n}^2, \qquad (6)$$

where $d\Omega_{2+n}^2$ is the line-element of a $(2+n)$-dimensional unit sphere

$$d\Omega_{2+n}^2 = d\theta_{n+1}^2 + \sin^2\theta_{n+1}\left(d\theta_n^2 + \sin^2\theta_n\left(\cdots + \sin^2\theta_2\left(d\theta_1^2 + \sin^2\theta_1\, d\varphi^2\right)\cdots\right)\right). \qquad (7)$$

Table 1. Horizon radius and temperature of the Schwarzschild-Tangherlini black hole as a function of the number of extra dimensions, for $M_* = 1$ TeV and $M_{BH} = 5$ TeV [11]

n	1	2	3	4	5	6	7
r_H (10^{-4} fm)	4.06	2.63	2.22	2.07	2.00	1.99	1.99
T_H (GeV)	77	179	282	379	470	553	629

If we apply the Gauss law in $D = 4+n$ dimensions, we f nd the following relation between the black-hole horizon radius and its mass [22]

$$r_H = \frac{1}{\sqrt{\pi} M_*} \left(\frac{M_{BH}}{M_*}\right)^{\frac{1}{n+1}} \left(\frac{8\Gamma(\frac{n+3}{2})}{(n+2)}\right)^{1/(n+1)} . \tag{8}$$

The above formula is a generalized one holding for a higher-dimensional black hole – the well-known linear relation between the mass and the horizon radius holding in 4 dimensions arises if we set $n = 0$ in the above expression, and also replace the fundamental Planck scale M_* with the 4-dimensional one M_P. Some indicative values of the horizon radius of miniature black holes are given in Table 1, where we have considered the scenario with $M_* = 1$ TeV and $M_{BH} = 5$ TeV. Then, Eq. (8) gives that $r_H = (4.06 - 1.99)\,10^{-4}$ fm, as n varies from 1 to 7 [11] – we note that these values are more than 30 orders of magnitude larger than the four-dimensional one for the same black-hole mass.

An important property of the black hole, that may be associated with observable signatures of their creation, is its temperature. This is def ned in terms of the surface gravity κ of the black hole, and for the spherically-symmetric case of Eq. (6) is given by

$$T_H = \frac{\kappa}{2\pi} = \frac{(n+1)}{4\pi r_H} . \tag{9}$$

By using the above formula and the corresponding values of r_H, we evaluate the temperature of the black hole, for $M_* = 1$ TeV and $M_{BH} = 5$ TeV, and display the results in Table 1. The value of the temperature def nes the peak of the radiation spectra associated with the Hawking process, and, as we clearly see, this lies in a regime accessible by present-day detection techniques.

The natural framework, in which such high-energy particle collisions and their strong-gravity consequences should be studied, is the one of the Quantum Theory of gravity. However, as this theory has not been formulated yet, one may still draw valuable conclusions by using purely classical arguments. In [23], it was thus assumed that a black hole could be created if the Compton wavelength $\lambda_C = 4\pi/E$ of the colliding particle of energy $E/2$ lies within the corresponding Schwarzschild radius $r_H(E)$ (8). This translates to

$$\frac{4\pi}{E} < \frac{1}{\sqrt{\pi} M_*} \left(\frac{E}{M_*}\right)^{\frac{1}{n+1}} \left(\frac{8\Gamma(\frac{n+3}{2})}{(n+2)}\right)^{1/(n+1)} . \tag{10}$$

The above formula may be solved for the value of the center-of-mass energy of the experiment necessary for the creation of the black hole. We f nd that, for $n = 2-7$, the minimum energy of the collision should be $E = (8.0-11.2)M_*$, respectively [23]. So far, no black-hole creation effect has been observed in any of the experiments looking for beyond-the-Standard-Model signatures. The most recent limits are the ones announced by the CMS [24] and ATLAS [25] collaborations of the Large Hadron Collider from proton-proton collisions with center-of-mass energy of 7 TeV. No excess was found above the predicted QCD background, and therefore it was concluded that, at 95% CL, no black holes exist in models with M_* up to 3 TeV. Nevertheless, according to the above criterion, the window of energy for the creation of black holes might not have been opened yet, therefore, we should wait for LHC to reach the ultimate goal of 14 TeV before drawing our f nal conclusions.

If these black holes are f nally produced, we expect them to go through a 'balding' phase f rst [15]. During this phase, the black hole will shed all quantum numbers and multipole moments apart from its mass M, electromagnetic charge Q, and angular momentum J. At the end of this phase, the black hole will settle down to an axially-symmetric state, the 'spin-down' phase, during which it will emit Hawking radiation thus loosing gradually its angular-momentum. In principle, a 'Schwarzschild' phase follows next during which the emission of Hawking radiation continues resulting in the loss of the actual mass of the black hole. When its mass is reduced down to M_*, no classical methods can be used to describe this object: the black hole can either emit a few energetic quanta and disappear (with a lifetime of the order of $\tau = (1.7-0.5)\times 10^{-26}$ sec, for M_* and M_{BH} as above), or form some type of stable remnant.

The line-element (6) can therefore be used to describe the gravitational background around a higher-dimensional black hole during its spherically-symmetric Schwarzschild phase. For the spin-down phase, a different line-element should be used that describes a rotating black hole in a higher-dimensional spacetime. For this purpose, the class of solutions found by Myers and Perry [22] is the most appropriate one. These solutions describe black holes living in a D-dimensional spacetime and characterized by $N = [(D-1)/2]$ angular-momentum components. However, throughout this review we will focus on the case of the simply-rotating black hole with only one non-vanishing angular-momentum component which, in addition, lies on the brane. This can be justif ed by the fact that the black hole is created by particles localised on the brane and thus having a non-zero impact parameter only along a direction lying on the brane. The line-element describing such a black hole is then given by the expression [22]

$$\begin{aligned} ds^2 = & \left(1-\frac{\mu}{\Sigma r^{n-1}}\right)dt^2 + \frac{2a\mu\sin^2\theta}{\Sigma r^{n-1}}\,dt\,d\varphi - \frac{\Sigma}{\Delta}\,dr^2 - \Sigma\, d\theta^2 \\ - & \left(r^2+a^2+\frac{a^2\mu\sin^2\theta}{\Sigma r^{n-1}}\right)\sin^2\theta\, d\varphi^2 - r^2\cos^2\theta\, d\Omega_n^2, \end{aligned} \tag{11}$$

where

$$\Delta = r^2 + a^2 - \frac{\mu}{r^{n-1}}, \qquad \Sigma = r^2 + a^2\cos^2\theta\,, \tag{12}$$

and $d\Omega_n^2$ is the line-element on a unit n-sphere. The mass and angular momentum (transverse to the $r\varphi$-plane) of the black hole are given by

$$M_{BH} = \frac{(n+2)A_{n+2}}{16\pi G_D}\mu, \quad J = \frac{2}{n+2}M_{BH}\,a\,, \tag{13}$$

with G_D being again the $(4+n)$-dimensional Newton's constant, and A_{n+2} the area of a $(n+2)$-dimensional unit sphere given by

$$A_{n+2} = \frac{2\pi^{(n+3)/2}}{\Gamma[(n+3)/2]}\,. \tag{14}$$

The horizon of the black hole is determined through the equation $\Delta(r_H) = 0$ and it may be written as $r_H^{n+1} = \mu/(1+a_*^2)$, where $a_* \equiv a/r_H$. Two important properties of the simply-rotating black hole is the angular velocity Ω and the temperature T_H of the black hole that are given by the following expressions

$$\Omega = \frac{ma}{r_H^2 + a^2}, \quad T_H = \frac{(n+1)+(n-1)a_*^2}{4\pi(1+a_*^2)r_H}\,. \tag{15}$$

As was mentioned, it is during the spin-down and the Schwarzschild phase that the emission of Hawking radiation, i.e. the emission of elementary particles, takes place. According to General Relativity, no particle can escape through the horizon of the black hole. Nevertheless, when a virtual pair of particles is produced outside the horizon of the black hole and the antiparticle falls inside the black hole, then the particle is free to propagate towards inf nity. The particles that reach asymptotic inf nity make up the Hawking radiation which is characterized by a thermal spectrum [26,27]. The emission takes place both on the brane and in the bulk. In the former case, during the Schwarzschild phase, the differential energy emission rate per unit time and unit frequency is given by the expression

$$\frac{d^2E(\omega)}{dt\,d\omega} = \frac{1+\delta_{s,1}}{2\pi}\sum_{j=s}^{\infty}\sum_{m=-j}^{j}\frac{|\mathcal{A}_{sjm}(\omega)|^2\,\omega}{\exp(\omega/T_H)\mp 1}\,. \tag{16}$$

In the above, ω is the energy of the emitted particles and ± 1 a statistics factor for fermions and bosons, respectively. The factor $|\mathcal{A}_{sjm}(\omega)|^2$ is the Absorption Probability (or, *greybody factor*) and determines the number of particles that will escape the strong gravitational f eld of the black hole to reach inf nity. This quantity is determined by solving a scattering problem for a particle with spin s propagating on the brane in the vicinity of the black hole.

For emission on the brane during the spin-down phase, the energy emission rate is given by a similar expression

$$\frac{d^2E(\omega)}{dt\,d\omega} = \frac{1+\delta_{s,1}}{2\pi}\sum_{j=s}^{\infty}\sum_{m=-j}^{j}\frac{|\mathcal{A}_{sjm}(\omega)|^2\,\omega}{\exp(k/T_H)\mp 1}\,. \tag{17}$$

with the only difference being the substitution of the energy ω in the exponential by the quantity $k = \omega - m\Omega$. For the spin-down phase particularly, we may also def ne the differential energy emission rate per unit time, unit frequency and angle of emission through the formula

$$\frac{d^3E(\omega)}{d(\cos\theta)\,dt\,d\omega} = \frac{1+\delta_{s,1}}{4\pi}\sum_{j=s}^{\infty}\sum_{m=-j}^{j}\frac{|\mathcal{A}_{sjm}(\omega)|^2\,\omega}{\exp(k/T_H)\mp 1}\left[({}_{-h}S_{s,j}^m)^2 + ({}_hS_{s,j}^m)^2)\right], \tag{18}$$

where ${}_{\pm h}S^{m}_{s,j}$ is the angular part of the wavefunction of the spin-s field propagating on the brane. As we will see, the above formula gives the angular distribution of the emitted particles and is one of the most distinct features of the Hawking radiation from a rotating black hole. Let us finally note that the black hole emits radiation also in the bulk, the higher-dimensional space, in the form of gravitons and possibly scalar fields. The differential energy emission rate is given by an expression similar to the above, however its exact form depends on the structure and symmetries of the higher-dimensional line-element, and it will be given when particular cases are studied in section 4.

3. Decay of Higher-Dimensional Black Holes via Emission on the Brane

We now proceed to study the decay of miniature higher-dimensional black holes, that may be created during a particle collision, through the emission of Hawking radiation directly on our brane. We will study the cases of both rotating and spherically-symmetric black holes, thus covering the evolution during the two intermediate phases in the life of the black hole, the spin-down and the Schwarzschild phase, during which the emission of Hawking radiation can be better studied using semi-classical techniques.

3.1. The "Master" Equation for Particles Living on the Brane

In order to study the emission of Hawking radiation on the brane, we will need the equations of motion of all Standard-Model particles (scalars, fermions and gauge bosons) propagating in the gravitational background that the black hole induces locally on the brane. This 4-dimensional background follows after fixing the values of all the additional angular coordinates, that correspond to the n spacelike, compact dimensions. Then, the line-element $d\Omega_n^2$ of the unit n-dimensional sphere appearing in Eq. (11) is eliminated; the remaining part stays unaltered and describes, after an appropriate redefinition of the coordinates θ and φ, the brane-induced gravitational background which has the form

$$ds_4^2 = \left(1-\frac{\mu}{\Sigma r^{n-1}}\right)dt^2 + \frac{2a\mu\sin^2\theta}{\Sigma r^{n-1}}\,dt\,d\varphi - \frac{\Sigma}{\Delta}dr^2 - \Sigma d\theta^2$$
$$- \left(r^2+a^2+\frac{a^2\mu\sin^2\theta}{\Sigma r^{n-1}}\right)\sin^2\theta\, d\varphi^2 . \qquad (19)$$

In the above, the metric functions Δ and Σ are still given by Eqs. (12).

The equations of motion of particles with an arbitrary value of spin may be derived by using the well-known Newman-Penrose formalism [28, 29]; a similar analysis to the one followed by Teukolsky [30] in the pure 4-dimensional case will allow us to write the field equations for particles with spin 0, 1/2 and 1 in the form of a single "master" equation where the spin s appears as a parameter. Note that the brane-induced line-element (19), due to the explicit dependence of its metric components on n, is not identical to the one describing a 4-dimensional Kerr black hole. Therefore, despite the similarities, the derivation of the master equation needs to be repeated starting from first principles.

This derivation was f rst performed for all species of brane-localised particles in a previous review [11] written by the present author in the early days of the study of Hawking radiation from higher-dimensional brane black holes. The analysis employed the Newman-Penrose formalism but covered only the case of positive spin-weight components ($h = +s$) of gauge bosons and fermions. Soon after, the study of the emission of Hawking radiation from a higher-dimensional rotating black hole was initiated, and it became clear that, for the complete study of the effect, the full set of equations describing all radiative components of the given f eld should be derived. As this review focuses on higher-dimensional rotating black holes, it will be useful to present here the main steps of the analysis leading to the master equation on the brane, and improve it by deriving the f eld equations for all radiative components.

Following the Newman-Penrose formalism [28, 29], we choose a tetrad basis of null vectors $(\ell^\mu, n^\mu, m^\mu, \bar{m}^\mu)$: ℓ and n are real vectors, whereas m and $\bar{m}$ are a pair of complex conjugate vectors. These two pairs of vectors satisfy the relations: $\mathbf{l}\cdot\mathbf{n} = 1$, $\mathbf{m}\cdot\bar{\mathbf{m}} = -1$, while all other products are zero. We f nd that we can write such a tetrad basis in the following form:

$$\begin{aligned} \ell^\mu &= \left(\frac{r^2+a^2}{\Delta}, 1, 0, \frac{a}{\Delta}\right), \quad n^\mu = \left(\frac{r^2+a^2}{2\Sigma}, -\frac{\Delta}{2\Sigma}, 0, \frac{a}{2\Sigma}\right), \\ m^\mu &= \left(ia\sin\theta, 0, 1, \frac{i}{\sin\theta}\right)\frac{1}{\sqrt{2}\bar{\rho}}, \quad \bar{m}^\mu = (m^\mu)^*, \end{aligned} \tag{20}$$

where $\bar{\rho} = r + ia\cos\theta$. By using the metric (19), we may also derive the expressions of the null vectors with their index down:

$$\begin{aligned} \ell_\mu &= \left(1, -\frac{\Sigma}{\Delta}, 0, -a\sin^2\theta\right), \quad n_\mu = \left(\frac{\Delta}{2\Sigma}, \frac{1}{2}, 0, -\frac{a\sin^2\theta\Delta}{2\Sigma}\right), \\ m_\mu &= \left[ia\sin\theta, 0, -\Sigma, -i\sin\theta(r^2+a^2)\right]\frac{1}{\sqrt{2}\bar{\rho}}, \quad \bar{m}_\mu = (m_\mu)^*. \end{aligned} \tag{21}$$

The above two sets of vectors allow us then to construct the λ_{abc} coeff cients, which are def ned by the formula:

$$\lambda_{abc} = (e_b)_{i,j}\left[(e_a)^i(e_c)^j - (e_a)^j(e_c)^i\right]. \tag{22}$$

In the above, e_a stands for each one of the tetrad basis vectors and (i, j) denote the particular components of each vector. Then, the non-vanishing λ_{abc} coeff cients, for the brane-induced background (19), are found to be:

$$\begin{aligned} \lambda_{122} &= -\frac{1}{\Sigma^2}\left(\frac{\Sigma}{2}\partial_r\Delta - r\Delta\right), \quad \lambda_{132} = \frac{i\sqrt{2}ar\sin\theta}{\Sigma\bar{\rho}}, \quad \lambda_{143} = \frac{1}{\bar{\rho}}, \\ \lambda_{213} &= -\frac{\sqrt{2}a^2\cos\theta\sin\theta}{\Sigma\bar{\rho}}, \quad \lambda_{243} = -\frac{\Delta}{2\Sigma\bar{\rho}}, \quad \lambda_{314} = -\frac{2ia\cos\theta}{|\bar{\rho}|^2}, \\ \lambda_{324} &= -\frac{ia\cos\theta\Delta}{\Sigma|\bar{\rho}|^2}, \quad \lambda_{334} = \frac{r\cos\theta + ia}{\sqrt{2}\bar{\rho}^2\sin\theta}. \end{aligned} \tag{23}$$

For each one of the above non-vanishing λ_{abc} coeff cients, there is another one following from the antisymmetry property $\lambda_{abc} = -\lambda_{cba}$.

We may then compute the so-called spin coeff cients, which are given in terms of the λ_{abc} coeff cients according to the formula $\gamma_{abc} = (\lambda_{abc} + \lambda_{cab} - \lambda_{bca})/2$. We will be interested in particular combinations of them that appear directly in the f eld equations we are looking for. These are denoted by the following symbols and are found to have the values:

$$\begin{aligned} &\kappa = \sigma = \lambda = \nu = \varepsilon = 0, \quad \rho = -\frac{1}{\bar{\rho}^*}, \quad \mu = -\frac{\Delta}{2\Sigma\bar{\rho}^*}, \quad \tau = -\frac{ia\sin\theta}{\sqrt{2}\Sigma}, \\ &\pi = \frac{ia\sin\theta}{\sqrt{2}\bar{\rho}^{*2}}, \quad \gamma = \frac{\partial_r\Delta}{4\Sigma} - \frac{\Delta}{2\Sigma\bar{\rho}^*}, \quad \beta = \frac{\cot\theta}{2\sqrt{2}\bar{\rho}}, \quad \alpha = \pi - \beta^*. \end{aligned} \tag{24}$$

We refer the interested reader to [28, 29] for the exact def nition of each one of the above quantities. Finally, in what follows we will need the Newman-Penrose operators def ned as:

$$\begin{aligned} \hat{D} &= \frac{(r^2+a^2)}{\Delta}\frac{\partial}{\partial t} + \frac{\partial}{\partial r} + \frac{a}{\Delta}\frac{\partial}{\partial\varphi}, \\ \hat{\Delta} &= \frac{(r^2+a^2)}{2\Sigma}\frac{\partial}{\partial t} - \frac{\Delta}{2\Sigma}\frac{\partial}{\partial r} + \frac{a}{2\Sigma}\frac{\partial}{\partial\varphi}, \\ \hat{\delta} &= \frac{1}{\sqrt{2}\bar{\rho}}\left(ia\sin\theta\frac{\partial}{\partial t} + \frac{\partial}{\partial\theta} + \frac{i}{\sin\theta}\frac{\partial}{\partial\varphi}\right). \end{aligned} \tag{25}$$

We are now ready to derive the f eld equations for gauge bosons, fermions and scalars propagating in the black-hole brane-induced background (19). We will be assuming that the emitted particle modes couple only minimally to the gravitational background and have no other interactions, therefore, they satisfy the corresponding free equations of motion. We will address each case separately, giving ample details in the indicative case of gauge bosons, in order to write the generalized master equation. To this end, we will employ the general f eld factorization

$$\Psi_s(t,r,\theta,\varphi) = e^{-i\omega t}\, e^{im\varphi} R_s(r)\, S^m_{s,j}(\theta). \tag{26}$$

While the radial part $R_s(r)$ will be determined in terms of the particular gravitational background, the angular part $S^m_{s,j}(\theta)$ are the so-called spin-weighted spheroidal harmonics present also in the pure 4-dimensional calculation. For particular choices of the $a\omega$ and s parameters, these reduce to particular families of angular eigenfunctions: for $s = 0$, they reduce to the spheroidal harmonics [31], for $a\omega = 0$, to the spin-weighted spherical harmonics [32] whereas, for $s = 0$ and $a\omega = 0$, they are the well-known spherical harmonics [33].

Let us start with the case of brane-localised gauge bosons, with $s = 1$. Following again the Newman-Penrose formalism, the Yang-Mills equations can be written as:

$$(\hat{D} - 2\rho)\,\Phi_1 - (\hat{\delta}^* + \pi - 2\alpha)\,\Phi_0 = 0, \tag{27}$$

$$(\hat{\delta} - 2\tau)\,\Phi_1 - (\hat{\Delta} + \mu - 2\gamma)\,\Phi_0 = 0, \tag{28}$$

$$(\hat{D} - \rho)\,\Phi_2 - (\hat{\delta}^* + 2\pi)\,\Phi_1 = 0, \tag{29}$$

$$(\hat{\delta} - \tau + 2\beta)\,\Phi_2 - (\hat{\Delta} + 2\mu)\,\Phi_1 = 0, \tag{30}$$

in terms of the particular components $\Phi_0 = F_{13}$, $\Phi_1 = (F_{12} + F_{43})/2$ and $\Phi_2 = F_{42}$ of the gauge f eld strength tensor $F_{\mu\nu}$. The above is a set of coupled differential equations with partial derivatives: as in the pure 4-dimensional case [34], they can be decoupled only for the radiative components of the gauge feld. For instance, Eqs. (27) and (28) can be rearranged to give a decoupled equation for Φ_0, namely

$$(\hat{D}-2\rho-\rho^*)(\hat{\Delta}+\mu-2\gamma)\,\Phi_0-(\hat{\delta}-2\tau)(\hat{\delta}^*+\pi-2\alpha)\,\Phi_0=0\,. \tag{31}$$

What is even more remarkable is that the above differential equation with partial derivatives can also be decoupled into a radial and an angular ordinary differential equation. To accomplish this, we assume a factorized ansatz for Φ_0 of the form of Eq. (26), and after some algebra we fnd the following equations [11]

$$\frac{1}{\sin\theta}\frac{d}{d\theta}\left(\sin\theta\,\frac{dS^m_{1,j}}{d\theta}\right)+\left[-\frac{2m\cot\theta}{\sin\theta}-\frac{m^2}{\sin^2\theta}+a^2\omega^2\cos^2\theta\right.$$
$$\left.-2a\omega\cos\theta+1-\cot^2\theta+\lambda_{1j}\right]S^m_{1,j}(\theta)=0\,, \tag{32}$$

$$\frac{1}{\Delta}\frac{d}{dr}\left(\Delta^2\frac{dR_1}{dr}\right)+\left[\frac{K^2-iK\partial_r\Delta}{\Delta}+4i\omega r+(\Delta''-2)-\Lambda_{1j}\right]R_1(r)=0\,. \tag{33}$$

In the above, λ_{1j} is a separation constant, and we have also defned the quantities

$$K=(r^2+a^2)\,\omega-am\,,\qquad \Lambda_{hj}=\lambda_{hj}+a^2\omega^2-2am\omega\,. \tag{34}$$

Note that the subscript 1 has been added to the angular and radial parts of Φ_0, as well as to the separation constants λ and Λ, to denote that these correspond to the frst radiative component of the gauge feld. A similar rearrangement of Eqs. (29) and (30) leads to a decoupled equation for the second radiative component of the gauge feld, Φ_2, which then leads to the following set of ordinary differential equations [35]

$$\frac{1}{\sin\theta}\frac{d}{d\theta}\left(\sin\theta\,\frac{dS^m_{-1,j}}{d\theta}\right)+\left[\frac{2m\cot\theta}{\sin\theta}-\frac{m^2}{\sin^2\theta}+a^2\omega^2\cos^2\theta\right.$$
$$\left.+2a\omega\cos\theta-1-\cot^2\theta+\lambda_{-1j}\right]S^m_{-1,j}(\theta)=0\,, \tag{35}$$

and

$$\frac{1}{\Delta}\frac{d}{dr}\left(\Delta^2\frac{dR_{-1}}{dr}\right)+\left[\frac{K^2+iK\partial_r\Delta}{\Delta}-4i\omega r-\Lambda_{-1j}\right]R_{-1}(r)=0\,. \tag{36}$$

By simple inspection, one concludes that the angular and radial equations for both radiative components may be written in a unique form in the following way [35]

$$\frac{1}{\sin\theta}\frac{d}{d\theta}\left(\sin\theta\,\frac{dS^m_{h,j}}{d\theta}\right)+\left[-\frac{2mh\cot\theta}{\sin\theta}-\frac{m^2}{\sin^2\theta}+a^2\omega^2\cos^2\theta\right.$$
$$\left.-2a\omega h\cos\theta+h-h^2\cot^2\theta+\lambda_{hj}\right]S^m_{h,j}(\theta)=0\,, \tag{37}$$

and

$$\Delta^{-h}\frac{d}{dr}\left(\Delta^{h+1}\frac{dR_h}{dr}\right)+\left[\frac{K^2-ihK\partial_r\Delta}{\Delta}+4ih\omega r+h(\Delta''-2)\delta_{h,s}-\Lambda_{hj}\right]R_h(r)=0\,. \tag{38}$$

In the above, the spin s is always taken to be positive, while the spin-weight $h = (-s, +s)$ is merely a parameter to distinguish the two radiative components of the given f eld.

Next, we turn to the case of brane-localised fermion f elds ($s = 1/2$). In the Newman-Penrose formalism, the Dirac equation, for a massless two-component spinor f eld, can be written as:

$$(\hat{\delta}^* - \alpha + \pi)\chi_0 = (\hat{D} - \rho)\chi_1\,, \tag{39}$$

$$(\hat{\Delta} + \mu - \gamma)\chi_0 = (\hat{\delta} + \beta - \tau)\chi_1\,. \tag{40}$$

Fortunately, the above two equations can be rearranged to give the following decoupled equation for χ_0

$$(\hat{D} - \rho - \rho^*)(\hat{\Delta} + \mu - \gamma)\chi_0 - (\hat{\delta} - \alpha^* - \tau + \pi^*)(\hat{\delta}^* - \alpha + \pi)\chi_0 = 0\,, \tag{41}$$

which further reduces to the following set of angular and radial equations [11]

$$\frac{1}{\sin\theta}\frac{d}{d\theta}\left(\sin\theta\frac{dS^m_{1/2,j}}{d\theta}\right) + \left[-\frac{m\cot\theta}{\sin\theta} - \frac{m^2}{\sin^2\theta} + a^2\omega^2\cos^2\theta \right.$$
$$\left. - a\omega\cos\theta + \frac{1}{2} - \frac{1}{4}\cot^2\theta + \lambda_{\frac{1}{2}j}\right] S^m_{1/2,j}(\theta) = 0\,, \tag{42}$$

$$\frac{1}{\sqrt{\Delta}}\frac{d}{dr}\left(\Delta^{3/2}\frac{dR_{1/2}}{dr}\right) + \left[\frac{K^2 - iK(\partial_r\Delta)/2}{\Delta} + 2i\omega r \right.$$
$$\left. + \frac{1}{2}(\Delta'' - 2) - \Lambda_{\frac{1}{2}j}\right] R_{1/2}(r) = 0. \tag{43}$$

As is obvious, the above equations can be considered as particular cases of the general equations (37)–(38) with $h = 1/2$. Similarly, the equations for the second radiative component are found to have the same form but with $h = -1/2$ as expected.

In the case f nally of a brane-localised scalar f eld, we do not need the Newman–Penrose formalism. The equation of motion of its sole component follows by evaluating the double covariant derivative $g^{\mu\nu}D_\mu D_\nu$ acting on the f eld, i.e by solving the equation

$$\frac{1}{\sqrt{-g}}\partial_\mu\left(\sqrt{-g}g^{\mu\nu}\partial_\nu\Phi\right) = 0\,. \tag{44}$$

A similar factorized ansatz (26) for Φ leads to the following pair of decoupled equations [11]

$$\frac{1}{\sin\theta}\frac{d}{d\theta}\left(\sin\theta\frac{dS^m_{0,\ell}}{d\theta}\right) + \left[-\frac{m^2}{\sin^2\theta} + a^2\omega^2\cos^2\theta + \lambda_{0\ell}\right] S^m_{0,\ell} = 0\,, \tag{45}$$

$$\frac{d}{dr}\left(\Delta\frac{dR_0}{dr}\right) + \left(\frac{K^2}{\Delta} - \Lambda_{0\ell}\right)R_0(r) = 0\,. \tag{46}$$

The above equations may be again considered as a special case of the general equations (37)–(38) but for $s = 0$. We may thus conclude that the aforementioned equations are the "master" equations, angular and radial respectively, that describe the motion of a brane-localised f eld with arbitrary spin s on the 4-dimensional induced background (19).

3.2. Decay of Spherically-Symmetric Black Holes

With the equations of motion for felds with arbitrary spin s, propagating in the four-dimensional spacetime induced on the 3-brane, being fnally at our disposal, we can now discuss the emission of Hawking radiation on the brane during the two intermediate phases in the life of the higher-dimensional black hole, namely the spin-down and Schwarzschild phase.

Although the spin-down phase precedes the Schwarzschild phase, the simplicity of the gravitational background describing the latter motivated its study from the early days of the study of Hawking radiation by higher-dimensional black holes. This motivation was further re-inforced by the belief that the Schwarzschild phase was the longest one in the life of the black hole and that most of its available energy is radiated during this phase. This belief dominated the research activity on black holes for a considerable period of time however, as we will comment later, we now believe it is a false one.

The Schwarzschild phase arises, in principle, when the black hole has lost all of its angular momentum through the emission of elementary particles via the process of Hawking radiation. During this phase, the black hole is characterized by spherical symmetry and the line-element describing the gravitational background around it is given by Eq. (6). As we are interested mainly in the radiation emitted by the higher-dimensional black hole on the brane – being ourselves restricted to live and perform experiments on the 4-dimensional hypersurface – we need to determine the line-element in which the brane-localized modes propagate. This follows, as discussed in the previous section, by fxing the values of all angular coordinates describing the extra, compact dimensions: in this case, we may easily see that this is realised by setting $\theta_i = \pi/2$, with $i = 2,\ldots,n+1$. Then, the 4-dimensional gravitational background that the brane-localised particles "see" has the form

$$ds_4^2 = -\left[1-\left(\frac{r_H}{r}\right)^{n+1}\right]dt^2 + \left[1-\left(\frac{r_H}{r}\right)^{n+1}\right]^{-1} dr^2 + r^2\, d\Omega_2^2\,. \tag{47}$$

Equivalently, the above line-element follows from the more general rotating one (19) in the limit of vanishing angular momentum, i.e. when $a \longrightarrow 0$.

The corresponding equations of motion of Standard-Model felds with spin $s = 0, 1/2, 1$ are given by a simplifed form of the master equations (37) and (38). For instance, the angular equation, in the limit $a \rightarrow 0$, takes the form

$$\frac{1}{\sin\theta}\frac{d}{d\theta}\left(\sin\theta\,\frac{dS^m_{h,j}}{d\theta}\right)+\left[-\frac{2mh\cot\theta}{\sin\theta}-\frac{m^2}{\sin^2\theta}+h-h^2\cot^2\theta+\lambda_{hj}\right]S^m_{h,j}(\theta)=0\,. \tag{48}$$

The functions $S^m_{h,j}(\theta)$ now reduce to the spin-weighted spherical harmonics [32] satisfying the above differential equation with the eigenvalue having the closed form $\lambda_{hj} = j(j+1) - s(s+1)$ – note that the eigenvalue is the same for both radiative components of a given feld, i.e. for both values of h. The radial equation is also simplifed adopting the form

$$\Delta^{-h}\frac{d}{dr}\left(\Delta^{h+1}\frac{dR_h}{dr}\right)+\left[\frac{r^2\,(\omega^2-ih\omega f')}{f}+2i\omega hr+h(\Delta''-2)\delta_{h,s}-\lambda_{hj}\right]R_h(r)=0\,, \tag{49}$$

where we have used that now $K \equiv \omega r^2$, $\Lambda_{hj} = \lambda_{hj}$, and defned $\Delta \equiv r^2 f$ where $f \equiv 1 - \left(\frac{r_H}{r}\right)^{n+1}$ is the metric function.

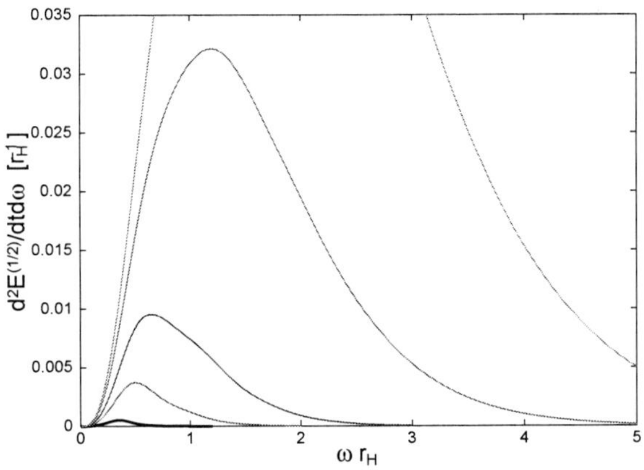

Figure 1. Energy emission rates for fermions on the brane or $n = 0, 1, 2, 4$ and 6 (from bottom to top) [11].

As already noted, the Schwarzschild phase is characterized by spherical symmetry as no preferred axis or direction exist in the given gravitational background. As a result, we expect the particles emitted by the black hole during this phase to be evenly distributed over the 4π solid angle. Solving the angular equation (48) and calculating the angular distribution of the emitted particles would have revealed exactly this behaviour. Also, since the angular eigenvalue λ_{hj}, that appears in the radial equation (49) is already known, there is again no reason to solve the angular one. Therefore, in what follows we ignore the angular equation and focus on the radial one.

Solving the radial equation, on the other hand, will allow us to determine the Absorption Probability $|\mathcal{A}(\omega)|^2$ that, together with the temperature of the black hole, determine the Hawking radiation spectra. For this we need to determine the radial function that satisf es Eq. (49) throughout the radial regime, then f nd its form at asymptotic inf nity, and from there determine the Ref ection and Absorption coeff cients. This analysis has been performed for the Schwarzschild phase both analytically [36,37] and numerically [38]. The corresponding results have also been reviewed in detail in [11], therefore, here we discuss only the main features of the radiation spectra.

For a given mass of the black hole, the geometry around its Schwarzschild phase is determined solely by the parameter n, the number of additional spacelike dimensions that exist outside our brane. As a result, the Absorption Probability, and consequently, the radiation spectra will also depend on this number. It turns out that the energy emitted per unit time and unit frequency by the black hole during the Schwarzschild phase is greatly enhanced with the number of extra dimensions. As an indicative case, in Figure 1, we depict the aforementioned differential rate for fermions for different values of the number of transverse spacelike dimensions ($n = 0, 1, 2, 4, 6$ from bottom to top) [11, 38]. We observe that the energy emission curves become taller and wider as n increases, resulting in a signif cantly larger output of energy from the black hole per unit time that is in addition distributed over an increasingly more extended range of frequencies. The same behaviour was found also for scalars and gauge bosons. In terms of numbers, the exact enhancement factors for all species of brane-localised particles are shown in Table 2 [38]- the entries here

Table 2. Total emissivities for all brane-localised species of particles [38]

n	0	1	2	3	4	5	6	7
Scalars	1.0	8.94	36.0	99.8	222	429	749	1220
Fermions	1.0	14.2	59.5	162	352	664	1140	1830
G. Bosons	1.0	27.1	144	441	1020	2000	3530	5740

give the total emissivities, i.e. the energy emission rate over the whole frequency regime, and are normalised to unity for the 4-dimensional case for easy comparison.

The entries of Table 2 reveal also another important feature: the enhancement factor, in terms of n, gets bigger the larger the spin of the emitted particle is. From studies of the emission of Hawking radiation from pure 4-dimensional black holes, it has been known that such a spherically-symmetric black hole prefers to emit mainly scalar particles, then fermions and then gauge bosons [39, 40], i.e. the spin of the emitted particle acts as a deterring factor for their emission. However, the behaviour revealed by the entries of Table 2 automatically poses the question of whether different species of particles become the dominant ones to be emitted as the dimensionality of spacetime changes. For this, the energy emission rates, for the three species of particles localised on the brane, were compared as n varied, and the relative particle emissivities were computed [38]. In Figure 2, we show the energy emission rates for scalars, fermions and gauge bosons for $n = 0$ (left plot) and $n = 6$ (right plot): from this, we conclude that indeed a 10-dimensional black hole prefers to emit f rst gauge bosons, then scalar f elds and lastly fermions. The analysis of [38] revealed also that a black hole living in a spacetime with an intermediate number of extra dimensions, i.e. $n = 2, 3$ or 4, has a "democratic" type of spectrum with all species of particles being equally likely to be emitted [38].

As a concluding remark, we may observe that the situation with the Hawking radiation in the case of the Schwarzschild phase is particularly simple. All main features of the radiation spectra are determined by the number of additional spacelike dimensions n. Therefore,

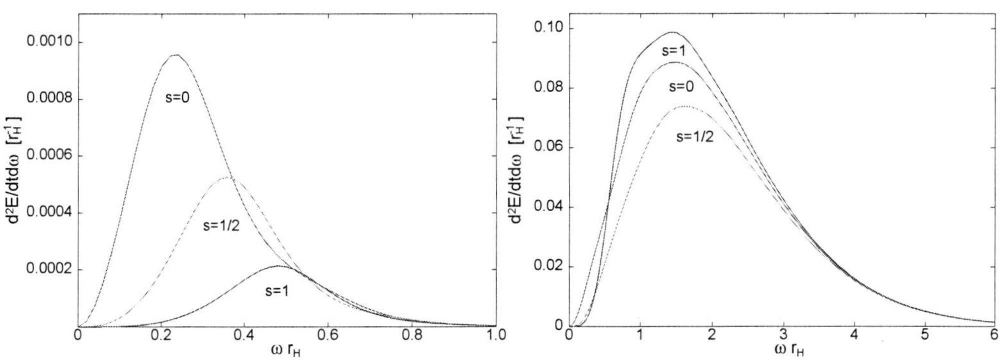

Figure 2. Energy emission rates for all species of brane-localised particles for $n = 0$ (left plot) and $n = 6$ (right plot) [38].

if the Schwarzschild phase in the life of a miniature higher-dimensional black hole, created possibly during a high-energy particle collision, could be discerned and the emitted particles detected, then the amount of energy emitted per unit time as well as the type of particles preferably emitted could help us determine the number of the extra dimensions existing in nature.

3.3. Decay of Axially-Symmetric Black Holes

We now turn to the study of the spin-down phase, the phase that precedes the Schwarzschild one and is characterized by the presence of a non-vanishing angular-momentum of the black hole. It is exactly this property that makes the line-element of the gravitational background considerably more complicated. Consequently, the corresponding f eld equations are more complex as well as technically more challenging to solve, as we will soon see.

The gravitational background on the brane in the vicinity of the higher-dimensional rotating black hole has the form of Eq. (19). The f eld equations of all Standard Model f elds propagating in this background reduce, as we have seen, to the two angular and radial master equations (37) and (38). The radial equation will provide us again with the value of the Absorption Probability necessary to determine the Hawking radiation emission spectra. Unlike the Schwarzschild case, the angular equation contains a wealth of information: f rst, it will provide the value of the separation constant λ_{hj}, that unfortunately for the spin-weighted spheroidal harmonics does not exist in a closed form, and, secondly, the angular functions themselves will determine the angular distribution of the emitted particles – note that the rotation axis of the black hole now introduces a preferred direction in space, and the angle of emission of particles measured from that axis is by itself an important observable characterizing the spin-down phase.

The set of master equations can be solved either analytically [41–43] or numerically [35, 44–47]. In the former case, an analytic formula is derived for the Absorption Probability that allows us to comprehend important properties of this quantity - unfortunately, the application of the approximation technique used relies on the assumption that both the energy of the emitted particle and the angular momentum of the black hole remain small, an assumption that inevitably restricts the validity of the result. For the derivation of the value of the Absorption Probability, under no restrictions on the energy and angular momentum, we thus turn to numerical analysis that eventually leads to the exact form of the Hawking radiation spectra. We consider both analyses useful and informative, each one in its own way, and thus we will review both here.

We start by noticing that, for $s \neq 0$, the two radiative components with $h = -s, +s$ satisfy a different radial equation due to the term proportional to $\delta_{h,s}$. In order to remove this difference and work with only one radial equation for both components, we redef ne the radial functions in the following way

$$R_s \equiv \Delta^{-s} P_s\,, \quad R_{-s} \equiv P_{-s}\,. \tag{50}$$

Then, the master radial equation on the brane takes the simplif ed form

$$\Delta^s \frac{d}{dr}\left(\Delta^{1-s}\frac{dP_h}{dr}\right) + \left(\frac{K^2 - ihK\Delta'}{\Delta} + 4ih\omega r - \tilde{\Lambda}_{hj}\right) P_h(r) = 0\,, \tag{51}$$

where the Δ''-term has now disappeared, and where we have redefned $\tilde{\Lambda}_{hj} = \Lambda_{hj} + 2s$. In order to proceed further, we need the value of the separation constant Λ_{hj}, or of the angular eigenvalue λ_{hj}. As already mentioned above, the latter does not exist in closed form as in the case where $a = 0$; it can be found either numerically, or, in the limit of small $a\omega$, expressed as a power series [34, 48–50] of the form

$$\lambda_{hj} = -s(s+1) + \sum_k f_k\,(a\omega)^k = j(j+1) - s(s+1) - \frac{2ms^2}{j(j+1)}a\omega + \cdots. \tag{52}$$

We are now ready to solve the radial equation (51) analytically [42,43]. We will apply a well-known approximation technique which amounts to solving the differential equation at two asymptotic regimes, i.e. near the horizon ($r \simeq r_H$) and far away from it ($r \gg r_H$), before matching the two solutions at an intermediate point to construct a solution valid throughout the radial regime. In the near-horizon regime, we fnd that it is particularly convenient to employ the following change of radial coordinate [42]

$$r \to f(r) \equiv \frac{\Delta(r)}{r^2+a^2} = \frac{r^2+a^2-\mu/r^{n-1}}{r^2+a^2} \Longrightarrow \frac{df}{dr} = (1-f)\,r\,\frac{A(r)}{r^2+a^2}, \tag{53}$$

where $A(r) = (n+1) + (n-1)\,a^2/r^2$. Then, Eq. (51) near the horizon ($r \simeq r_H$) can be written as

$$f(1-f)\frac{d^2P_h}{df^2} + (1-s-B_*f)\frac{dP_h}{df} + \left[\frac{K_*^2 - ihK_*A_*}{A_*^2 f(1-f)} + \frac{(4ih\omega_* - \tilde{\Lambda}_{hj})(1+a_*^2)}{A_*^2\,(1-f)}\right]P_h(r) = 0\,. \tag{54}$$

In the above, we use for convenience the dimensionless parameters $\omega_* = \omega r_H$ and $a_* = a/r_H$. Also, B_* is defned as

$$B_* \equiv 1 - s + \frac{2s + n\,(1+a_*^2)}{A_*} - \frac{4a_*^2}{A_*^2}\,. \tag{55}$$

Finally, A_* and K_* are also dimensionless quantities that follow from $A(r)$ and K when evaluated at the black-hole horizon. They are given by the expressions

$$A_* = n + 1 + (n-1)a_*^2, \quad K_* = (1+a_*^2)\omega_* - a_* m\,. \tag{56}$$

If we make the further redefnition of the radial function: $P(f) = f^{\alpha}(1-f)^{\beta}F(f)$, then we may bring Eq. (54) to the form of a hypergeometric differential equation [33]

$$f(1-f)\frac{d^2F}{df^2} + [c - (1+a+b)\,f]\frac{dF}{df} - abF = 0\,, \tag{57}$$

with

$$a = \alpha + \beta + B_* - 1\,, \quad b = \alpha + \beta\,, \quad c = 1 - s + 2\alpha\,. \tag{58}$$

The power coeff cients α and β can be determined by solving the second-order algebraic equations

$$\alpha^2 - s\alpha + \frac{K_*^2}{A_*^2} - \frac{ihK_*}{A_*} = 0 \tag{59}$$

and

$$\beta^2 + \beta\,(B_* + s - 2) + \frac{K_*^2}{A_*^2} - \frac{ihK_*}{A_*} + \frac{(4ih\omega_* - \tilde{\Lambda}_{hj})\,(1 + a_*^2)}{A_*^2} = 0\,, \tag{60}$$

respectively. The f rst of these two equations may be easily solved leading to the following solution for the parameter α

$$\alpha_\pm = \frac{s}{2} \pm \left(\frac{iK_*}{A_*} + \frac{h}{2}\right), \tag{61}$$

while the second equation for β, after a little bit of algebra, we can show that it admits the solutions

$$\beta_\pm = \frac{1}{2}\Bigg[(2 - s - B_*) \pm \sqrt{(B_* + s - 2)^2 - \frac{4K_*^2 - 4ihK_*A_*}{A_*^2} - \frac{4(4ih\omega_* - \tilde{\Lambda}_{hj})\,(1 + a_*^2)}{A_*^2}}\Bigg]. \tag{62}$$

The general solution of the hypergeometric differential equation may be written in the following way [33]

$$P_{NH}(f) = A_- f^{\alpha}\,(1 - f)^{\beta}\,F(a, b, c; f) + A_+ f^{-\alpha}\,(1 - f)^{\beta}\,F(a - c + 1, b - c + 1, 2 - c; f)\,, \tag{63}$$

however, there is a boundary condition that the above solution must respect: that no outgoing modes can exist at the horizon of the black hole. Using the fact that, in the limit $r \to r_H$, $f(r) \to 0$, and the property that $F(a, b, c; f) \to 1$ when $f \to 0$, we obtain either

$$P_{NH}(f) \simeq A_- f^{(s+h)/2}\, f^{iK_*/A_*} + A_+ f^{-(s+h)/2}\, f^{-iK_*/A_*}\,, \tag{64}$$

for $\alpha = \alpha_+$, or

$$P_{NH}(f) \simeq A_- f^{(s-h)/2}\, f^{-iK_*/A_*} + A_+ f^{-(s-h)/2}\, f^{iK_*/A_*}\,, \tag{65}$$

for $\alpha = \alpha_-$. To decide which of the above solutions is an acceptable one, we recall that, according to Teukolsky's classic analysis [30], the correct boundary condition at the horizon of the black hole, for a f eld with non-zero spin s, is

$$R_s \sim \Delta^{-s}\, e^{-ikr_*}\,, \tag{66}$$

where r_* is the well-known tortoise coordinate for the Kerr metric def ned by $dr_*/dr = (r^2 + a^2)/\Delta(r)$, and $k \equiv \omega - m\Omega$. In our case, we may introduce a tortoise-like coordinate def ned as $y = r_H(1 + a_*^2)\ln(f)/A_*$, that, in the limit $r \to r_H$, becomes identical [42] to the usual one r_*. In terms of the new coordinate, the factors $f^{\pm iK_*/A_*}$, that appear in the asymptotic solutions (64)–(65), reduce to $e^{\pm iky}$ describing an outgoing and incoming free wave, respectively. Taking into account the function redefnitions (50), Teukolsky's boundary condition translates in our case to

$$P_{+s} \sim e^{-iky} = f^{-iK_*/A_*}, \quad P_{-s} \sim f^s\, e^{-iky} = f^s\, f^{-iK_*/A_*}. \tag{67}$$

Comparing the above with the asymptotic solutions (64)–(65), we conclude that the selection $\alpha = \alpha_+$ should be rejected, and that, in addition, in the form of the solution for $\alpha = \alpha_-$, we should impose the condition $A_+ = 0$. The above brings the Near-Horizon solution to the fnal form

$$P_{NH}(f) = A_-\, f^{\alpha}\,(1-f)^{\beta}\, F(a,b,c;f). \tag{68}$$

Finally, we need to decide the sign in the expression of the β coeffcient (62). The criterion for the convergence of the hypergeometric function $F(a,b,c;f)$ demands that $Re(c-a-b) > 0$; when this is applied in our case, with the hypergeometric indices given by Eq. (58), it leads to the choice $\beta = \beta_-$.

Let us now turn to the far-feld regime, where the radial master equation (51), in the limit $r \gg r_H$, becomes

$$\frac{d^2P_h}{dr^2} + \frac{2(1-s)}{r}\frac{dP_h}{dr} + \left(\omega^2 + \frac{2ih\omega}{r} - \frac{\lambda_{hj} + 2s + a^2\omega^2}{r^2}\right) P_h = 0. \tag{69}$$

Here, it would be also useful to make the function redefnitions $P_h = e^{-i\omega r}\, r^{\frac{1}{2}(2s-1+Z)}\tilde{P}_h$, with

$$Z = \sqrt{(2s-1)^2 + 4(\lambda_{hj} + 2s + a^2\omega^2)}. \tag{70}$$

Then, Eq. (69) takes the form of a confuent hypergeometric differential equation whose solution can be expressed in terms of the Kummer functions M and U, i.e. [33]

$$P_{FF}(r) = e^{-i\omega r}\, r^{\frac{1}{2}(2s-1+Z)} \left[B_1\, M\left(\frac{1}{2} - h + \frac{Z}{2}, 1+Z, 2i\omega r\right) + B_2\, U\left(\frac{1}{2} - h + \frac{Z}{2}, 1+Z, 2i\omega r\right)\right]. \tag{71}$$

The fnal step towards the construction of the complete solution is to match the Near-Horizon and Far-Field solutions in an intermediate regime. To this end, we need to appropriately extrapolate, or "stretch" as this action is usually called, the two asymptotic solutions. Thus, the Far-Field solution should be stretched towards small values of the radial coordinate – by taking the $r \to 0$ limit in (71), we then fnd [33]

$$P_{FF}(r) \simeq B_1\, r^{\frac{1}{2}(2s-1+Z)} + B_2\, r^{\frac{1}{2}(2s-1-Z)}\, \frac{\Gamma(Z)}{\Gamma(\frac{1}{2} - h + \frac{Z}{2})}\,(2i\omega)^{-Z}. \tag{72}$$

Similarly, we need to stretch our Near-Horizon solution to values of the radial coordinate that are much larger than the horizon radius. However, in this case, we need to "shift" frst the argument of the hypergeometric function from f to $1-f$ by using the following relation [33]

$$P_{NH}(f) = A_- f^{\alpha}(1-f)^{\beta}\left[\frac{\Gamma(c)\,\Gamma(c-a-b)}{\Gamma(c-a)\,\Gamma(c-b)}F(a,b,a+b-c+1;1-f)\right.$$
$$\left. + \;(1-f)^{c-a-b}\frac{\Gamma(c)\,\Gamma(a+b-c)}{\Gamma(a)\,\Gamma(b)}F(c-a,c-b,c-a-b+1;1-f)\right]. \quad (73)$$

We may now use also the fact that the function $f(r)$ may be alternatively written as

$$f(r) = 1 - \frac{\mu}{r^{n-1}}\frac{1}{r^2+a^2} = 1 - \left(\frac{r_H}{r}\right)^{n-1}\frac{(1+a_*^2)}{(r/r_H)^2+a_*^2}. \quad (74)$$

In the above, we have used the horizon equation $\Delta(r_H) = 0$ in order to eliminate μ from the above relation. In the limit $r \gg r_H$, and for $n \geq 0$, the above expression goes to unity. Thus, the argument of the "stretched" hypergeometric function goes again to zero, and the "stretched" Near-Horizon solution takes the form

$$P_{NH}(f) \simeq A_-\,(1-f)^{\beta}\,\frac{\Gamma(c)\,\Gamma(c-a-b)}{\Gamma(c-a)\,\Gamma(c-b)}$$
$$+ \; A_-\,(1-f)^{-\beta+2-B_*-s}\,\frac{\Gamma(c)\,\Gamma(a+b-c)}{\Gamma(a)\,\Gamma(b)}. \quad (75)$$

We may simplify further the above expression by noticing that the quantity $(1-f)$, for $r \gg r_H$, can be accurately approximated by

$$1-f \simeq (1+a_*^2)\left(\frac{r_H}{r}\right)^{n+1}. \quad (76)$$

This brings Eq. (75) to a simpler power-law form

$$P_{NH}(r) \simeq A_1\, r^{-(n+1)\beta} + A_2\, r^{(n+1)(\beta+s+B_*-2)}, \quad (77)$$

with

$$A_1 = A_-\,\left[(1+a_*^2)\,r_H^{n+1}\right]^{\beta}\frac{\Gamma(c)\Gamma(c-a-b)}{\Gamma(c-a)\Gamma(c-b)}, \quad (78)$$

$$A_2 = A_-\,\left[(1+a_*^2)\,r_H^{n+1}\right]^{-(\beta+s+B_*-2)}\frac{\Gamma(c)\Gamma(a+b-c)}{\Gamma(a)\Gamma(b)}. \quad (79)$$

We observe that both stretched solutions (72) and (77) have a power-law form. Their matching would be a 'perfect' one if the powers of r were the same in both expressions: in that case, the matching would be equivalent to the identif cation of the corresponding multiplicative coeff cients. For arbitrary values of ω_* and a_*, the above is not possible. However, if we assume that $\omega_* \ll 1$ and $a_* \ll 1$, we get, from Eq. (55), $B_* \simeq 2 - s + \frac{2s-1}{n+1}$, and

$$\beta \simeq \frac{1}{2\,(n+1)}\left(1 - 2s - \sqrt{(2s-1)^2 + 4\tilde{\Lambda}_{hj}}\right). \quad (80)$$

Then, the stretched near-horizon solution (77) takes the form

$$P_{NH}(r) = A_1\, r^{\frac{1}{2}\left(2s-1+\sqrt{(2s-1)^2+4\tilde{\Lambda}_{hj}}\right)} + A_2\, r^{\frac{1}{2}\left(2s-1-\sqrt{(2s-1)^2+4\tilde{\Lambda}_{hj}}\right)}\,. \tag{81}$$

Using the def nitions (34) and (70) for the constants Λ_{hj} and Z, respectively, we see that, under the above assumption for the values of the energy of the emitted particle and the angular momentum of the black hole, the powers of r in the two stretched solutions (72) and (81) become indeed identical. Then, identifying the corresponding coeff cients in front of the same powers of r, we obtain

$$\frac{B_1}{B_2} = \frac{\Gamma(Z)}{\Gamma(\frac{1}{2}-h+\frac{Z}{2})}\,(2i\omega)^{-Z}\,\frac{A_1}{A_2}\,, \tag{82}$$

with A_1/A_2 following from Eq. (79)

$$\frac{A_1}{A_2} = \frac{\Gamma(c-a-b)\Gamma(a)\Gamma(b)}{\Gamma(c-a)\Gamma(c-b)\Gamma(a+b-c)}\,\left[(1+a_*^2)\,r_H^{n+1}\right]^{2\beta+s+B_*-2}\,. \tag{83}$$

The above conditions complete the construction of the solution for the radial function satisfying the master equation (51). By setting $s=1/2$ and 1, the corresponding solution for fermions and gauge bosons, respectively, propagating in the gravitational background induced on the brane emerges. The same formalism applies also in the case of scalar f elds: the near-horizon solution arises from Eq. (68) after setting $s=0$, while the far-f eld solution assumes a simpler form and is given in terms of Bessel functions - we refer the interested reader to [42] for more details. Finally, the above analysis applies also in the case of propagation of f elds with arbitrary spin s in the background of a non-rotating black hole [11,36,37] with the corresponding results arising from the ones presented in this section after setting $a=0$.

Our next task is to compute the Absorption Probability $|\mathcal{A}_{sjm}|^2$. For this, we expand the far-f eld solution (71) in the limit $r \longrightarrow \infty$, and obtain [33]

$$\begin{aligned} P_{FF}(r) &\simeq \left(\frac{B_1\, e^{i\pi\left(\frac{1}{2}-h+\frac{Z}{2}\right)}}{(2i\omega)^{\left(\frac{1}{2}-h+\frac{Z}{2}\right)}}\,\frac{\Gamma(1+Z)}{\Gamma\left(\frac{1}{2}+h+\frac{Z}{2}\right)} + \frac{B_2}{(2i\omega)^{\left(\frac{1}{2}-h+\frac{Z}{2}\right)}}\right)\frac{e^{-i\omega r}}{r^{1-h-s}} \\ &+ \frac{B_1}{(2i\omega)^{\left(\frac{1}{2}+h+\frac{Z}{2}\right)}}\,\frac{\Gamma(1+Z)}{\Gamma\left(\frac{1}{2}-h+\frac{Z}{2}\right)}\,\frac{e^{i\omega r}}{r^{1+h-s}} \\ &\equiv Y_h^{(in)}\,\frac{e^{-i\omega r}}{r^{1-h-s}} + Y_h^{(out)}\,\frac{e^{i\omega r}}{r^{1+h-s}}\,. \end{aligned} \tag{84}$$

At this point, we need to consider each species of brane f elds separately. Let us focus f rst on the case of scalar f elds: after setting $h=s=0$ in the above expansion, the asymptotic solution at inf nity turns out to be a superposition of an outgoing and an ingoing spherical wave. Therefore, the corresponding amplitudes may easily determine the Ref ection Probability and, from that, the Absorption Probability, namely

$$|\mathcal{A}_{0jm}|^2 = 1 - |\mathcal{R}_{0jm}|^2 = 1 - \left|\frac{Y_0^{(out)}}{Y_0^{(in)}}\right|^2\,. \tag{85}$$

For the case of fermion fields, the expanded solution at infinity does not have the form of spherical waves, therefore the above definition for the absorption probability can not be applied. It is however easy to see from the master radial equation (51) that, similarly to the 4-dimensional case [27, 29, 51, 52], the following condition holds

$$\frac{d}{dr}\left(|P_{\frac{1}{2}}|^2 - |P_{-\frac{1}{2}}|^2\right) = 0\,. \tag{86}$$

This means that the quantity inside the parentheses is a conserved quantity – as a matter of fact, this combination turns out to be proportional to the radial component of the particle current produced by the black hole. Then, we may define the absorption probability as the ratio of the flux of particles at the black hole horizon over the one at infinity, i.e.

$$|\mathcal{A}_{\frac{1}{2}jm}|^2 = \frac{F_{in}^{(H)}}{F_{in}^{(\infty)}} = 1 - \frac{F_{out}^{(\infty)}}{F_{in}^{(\infty)}}\,, \tag{87}$$

where in the second part of the equation, we have used the conservation of the total flux. The flux of fermions at infinity may be found by integrating the radial component of the conserved current over a 2-dimensional sphere at asymptotic infinity. Applying this and using Eq. (84), we find

$$|\mathcal{A}_{\frac{1}{2}jm}|^2 = 1 - \frac{|P^{(out)}_{\frac{1}{2}}|^2 - |P^{(out)}_{-\frac{1}{2}}|^2}{|P^{(in)}_{\frac{1}{2}}|^2 - |P^{(in)}_{-\frac{1}{2}}|^2} = 1 + \frac{|Y^{out}_{-\frac{1}{2}}|^2}{|Y^{(in)}_{\frac{1}{2}}|^2}\,. \tag{88}$$

If we use the explicit expressions for $Y^{(out)}_{\pm\frac{1}{2}}$ for the two radiative components with $h = \pm 1/2$, as these are defined in Eq. (84), we find that these are related through the equation

$$Y^{(out)}_{-\frac{1}{2}} = \frac{2i\omega}{\sqrt{\lambda_{1/2j} + 1 + a^2\omega^2}}\, Y^{(out)}_{\frac{1}{2}}\,. \tag{89}$$

Therefore, the final expression for the absorption probability for fermions takes the form

$$|\mathcal{A}_{\frac{1}{2}jm}|^2 = 1 - \frac{4\omega^2}{\lambda_{1/2j} + 1 + a^2\omega^2}\left|\frac{Y^{(out)}_{\frac{1}{2}}}{Y^{(in)}_{\frac{1}{2}}}\right|^2\,. \tag{90}$$

For fields with spin $s = 1$, a different technique [53] altogether must be used since there is no conserved particle current either. Through a radial function redefinition and the use of the tortoise coordinate r_*, the radial master equation is transformed to an alternative one with real, short-range potential. Then, the asymptotic solution at infinity for the gauge field is given as the sum of outgoing and incoming plane waves with constant amplitudes [53, 54], namely

$$X_{jm\omega} \sim e^{-i\omega r_*} + A^{(in)}_{jm\omega}\, e^{i\omega r_*}\,. \tag{91}$$

From this, the expression of $\mathcal{A}_{1jm}$ may easily follow as

$$|\mathcal{A}_{1jm}|^2 = 1 - |A^{(in)}_{jm\omega}|^2\,. \tag{92}$$

The constant amplitude $A^{(in)}_{jm\omega}$ in Eq. (91) is related with the constant coefficients $Y_s^{(in,out)}$ appearing in our Eq. (84), which leads finally to the result

$$|\mathcal{A}_{1jm}|^2 = 1 - \frac{16\omega^4}{B^2_{jm\omega}} \left| \frac{Y_1^{(out)}}{Y_1^{(in)}} \right|^2 . \tag{93}$$

The constant $B_{jm\omega}$ is defined as the coefficient appearing in the relation

$$Y_{-1}^{(out)} = -\frac{4\omega^2}{B_{jm\omega}} Y_1^{(out)} . \tag{94}$$

By using the explicit expressions of $Y_{\pm 1}^{(out)}$ from Eq. (84), we obtain

$$Y_{-1}^{(out)} = -\frac{4\omega^2}{\lambda_{1j} + 2 + a^2\omega^2} Y_1^{(out)} . \tag{95}$$

The above determines $B_{jm\omega} = \lambda_{1j} + 2 + a^2\omega^2$, and, leads to the final expression for the absorption probability for brane-localised fields with spin 1

$$|\mathcal{A}_{1jm}|^2 = 1 - \frac{16\omega^4}{(\lambda_{1j} + 2 + a^2\omega^2)^2} \left| \frac{Y_1^{(out)}}{Y_1^{(in)}} \right|^2 . \tag{96}$$

Having at our disposal an analytic formula for the Absorption Probability, one could investigate its asymptotic limit as $\omega \to 0$, and the corresponding behaviour of the absorption cross-section σ_{sjm}. One may easily see that, as the energy of the emitted particle goes to zero, the ratio B_1/B_2 behaves like $\omega^{-(2j+1)}$ for all species of particles. When substituted in the expression of the Absorption Probability, and after quite some algebra, this leads to $|\mathcal{A}_{0jm}|^2 \sim \omega^{2j+2}$ for scalar fields [42] and $|\mathcal{A}_{sjm}|^2 \sim \omega^{2j+1}$ for fields with spin $s = 1/2$ and 1 [43]. The absorption cross-section for each partial wave is given by the formula $\sigma_{sjm} = \pi |\mathcal{A}_{sjm}|^2/\omega^2$, and it behaves as

$$\sigma_{0jm} \sim \omega^{2j}, \quad \sigma_{\frac{1}{2}jm} \sim \omega^{2j-1}, \quad \sigma_{1jm} \sim \omega^{2j-1} . \tag{97}$$

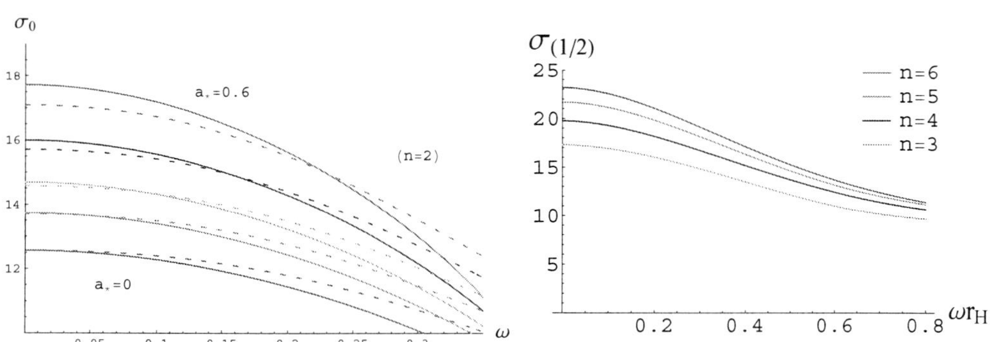

Figure 3. Absorption cross-section at the low-energy regime for scalar fields (left plot) in terms of a_*, for $n = 2$ [42], and fermions (right plot) in terms of n, for $a_* = 0.5$ [43].

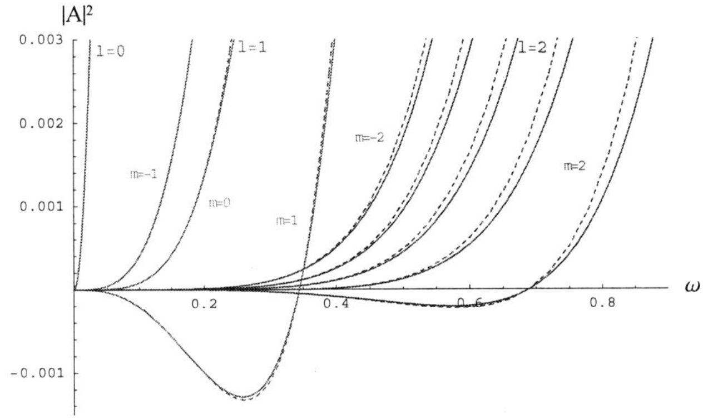

Figure 4. Absorption Probability (analytic result) of brane-localised scalar particles with $l = 0, 1, 2$ and $-l \leq m \leq l$, for $n = 6$ and $a_* = 0.4$ [42].

As a result, the absorption cross-section for the lowest scalar and fermionic modes, with $j = 0$ and $j = 1/2$ respectively, assumes a constant asymptotic value as $\omega \to 0$. On the other hand, all higher modes, as well as all gauge-boson modes, have absorption cross-sections that go to zero as the energy of the emitted particle vanishes. The behaviour of the absorption cross-section at the low-energy regime is depicted in Figure 3(a) (left plot) for scalars [42], and in Figure 3(b) for fermions (right plot) [43]. In the case of scalar f elds the universal property of the absorption cross-section being equal to the area of the horizon is recovered only in the limit of small angular-momentum of the black hole – this behaviour is conf rmed by both analytical (dashed lines) and numerical results (solid lines). The constant asymptotic value of σ_{sjm} for scalars and fermions is enhanced as either the angular-momentum of the black hole or the number n of extra dimensions increases, as depicted in Figure 3.

Finally, one may be tempted to plot the analytic formula for the Absorption Probability in an effort to check the range of validity of the derived result. Thus, in Figure 4, we depict the Absorption Probability for scalar f elds (dashed lines) [42] and in Figure 5(a) the corresponding quantity for gauge bosons [43]. In Figure 4, the exact numerical results for scalar f elds – to be discussed shortly – derived in [45] are also depicted as solid lines, and one may easily conclude that the agreement between the two sets is indeed remarkable in the low-energy regime and very satisfactory in the intermediate regime. Unfortunately, as the energy of the emitted scalar particle increases, the analytic result for the Absorption Probability breaks down quite early due to poles emerging in the expression of the Gamma functions. No such anomaly effect is observed in the case of fermions and gauge bosons which allows us to produce plots for the whole range of values of the Absorption Probability. The method for deriving the exact numerical results for fermions and gauge bosons will also be discussed shortly, and here we may present the indicative case for gauge bosons, depicted in Figure 5(b) for easy comparison. Also in this case, as well as in the case of fermion f elds, the two sets of results are in fairly good agreement especially for low values of a_*.

One could attempt to determine the Hawking radiation emission spectra by using the analytic result for the Absorption Probability. However, this would lead to spectra being

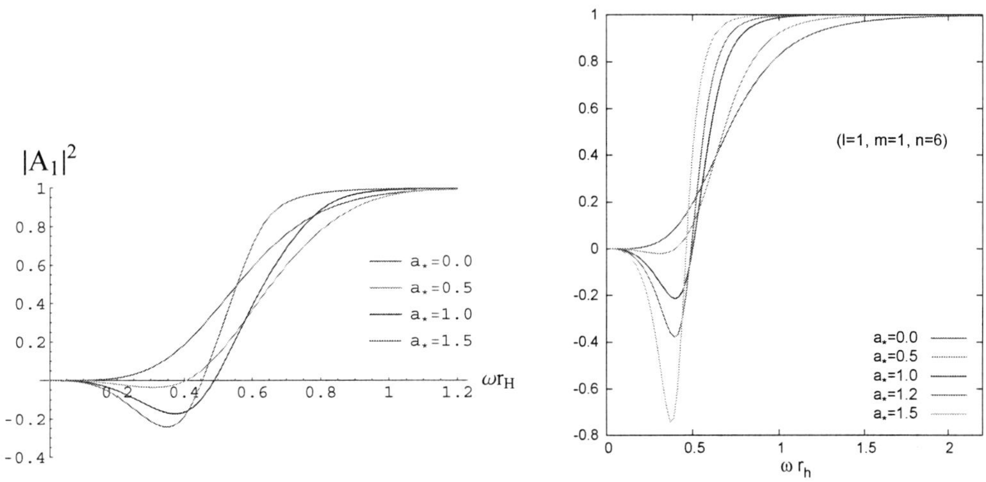

Figure 5. Analytic results [43] (left plot) and numerical results [35] (right plot) for the Absorption Probability of brane-localised gauge bosons with $j = m = 1$, for $n = 6$ and various values of a_*.

valid, in principle, only in the low-energy regime. Especially in the case of scalar f elds, the emergence of poles would cause the deviation of the derived spectra from the exact ones fairly quickly. In the case of fermions and gauge bosons, the complete emission curves, that are in good qualitative and quantitative agreement with the exact ones, can be derived, however detailed features that appear in the inter-mediate and high-energy regimes are missing. Therefore, although the approximation method can lead to very satisfactory results for the Absorption Probability and the emission spectra in the low-energy regime – and, at cases, even beyond that – one is inevitably led to the employment of numerical analysis for the derivation of the exact spectra valid at all energy and angular-momentum regimes.

As the objective now is to derive the most accurate emission spectra at all parameter regimes, one also needs an exact value of the angular eigenvalue valid for arbitrary values of ω_* and a_*. The power series expansion cannot be therefore used, but instead λ_{hj} must be found through numerical means. It turns out that, for different species of f elds, this task presents different challenges that are overcome by using different techniques. In the case of scalar f elds, the angular eigenvalue may be found [44] by using the continuation method [55], a variant of perturbation theory that allows to f nd λ_{hj}, for every value of $a\omega$, j and m, starting from its known value for $a\omega = 0$. Alternatively, the angular eigenvalue and the angular function $S^m_{h,j}$ can be simultaneously found by numerically integrating the master angular equation (37) itself. For scalars [45] and gauge bosons [35], the shooting method [56] can be used that relies on f nding the value of λ_{hj} for which the irregular part of $S^m_{h,j}$ – being in principle present due to the regular irregular points $x \equiv \cos\theta = \pm 1$ of the angular equation – vanishes. Finally, in the case of fermions [46], the spectral decomposition method [57] was used to f nd f rst the angular eigenvalue: in this method, the angular function is expanded in terms of spherical harmonics and λ_{hj} is the eigenvalue of a matrix formed by the expansion coeff cients. Subsequently, the angular equation can be solved for $S^m_{h,j}$ [46] by using a power-series expansion in terms of $(1+x)$ introduced by

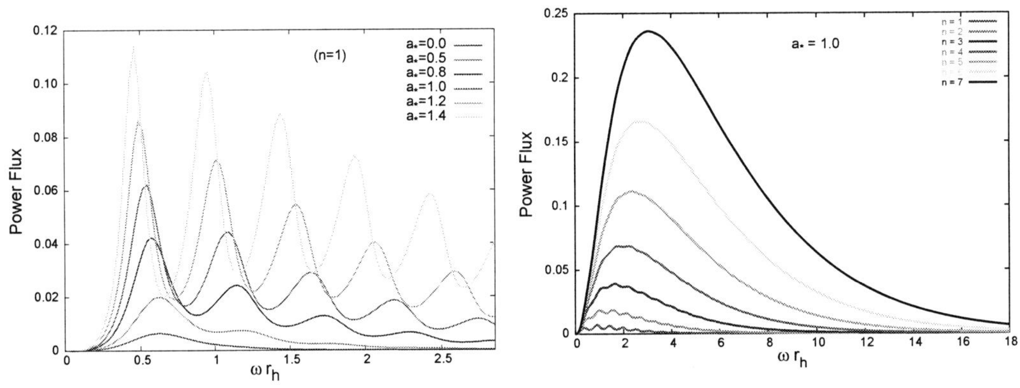

Figure 6. Energy emission rates for gauge bosons, for $n = 1$ and various values of a_* (left plot) [35], and fermions, for $a_* = 1$ and various values of n (right plot) [46].

Leaver [58], where the expansion coeff cients are determined through recurrence relations.

With the exact value of the angular eigenvalue at hand, we may now integrate numerically the radial master equation (38). The integration for all species of f elds starts from the horizon, with the boundary condition that a regular ingoing mode exists there, and proceeds to asymptotic inf nity. For the case of scalar f elds [44, 45], the amplitudes of the incoming and outgoing spherical waves at inf nity may be directly read and the Absorption Probability may be computed. For fermions [46] and gauge bosons [35], a radial function redef nition and a change of radial coordinate are f rst employed in order to ensure that the potential becomes a short-range one and the solution at inf nity is given in terms of outgoing and ingoing plane waves.

The exact values of the Absorption Probability, together with the temperature of the black hole, may then be used to derive the emission spectra for Hawking radiation on the brane. In Figure 6, we depict the energy emission rates for the indicative cases of gauge bosons, for $n = 1$ and various values of a_* (left plot) [35], and fermions, for $a_* = 1$ and various values of n (right plot) [46]. The enhancement of the differential emission rate, for all species of brane-localised f elds, as the number of additional spacelike dimensions increases is very prominent in the spin-down phase, too. We also observe that an increase in the value of the angular momentum of the black hole also leads to the overall enhancement of the energy emission rate, although of a smaller magnitude: the enhancement factor when n varies from 1 to 7 is of $O(100)$ while the one when a_* changes from 0 to 1 is of $O(10)$. A similar dependence on n and a_* is manifest in the other two emission rates that one may compute, namely the particle f ux and the rate of the emission of angular momentum – we refer the interested reader to [35,45,46] for more details on these.

As was mentioned at the beginning of this section, the non-vanishing value of the angular-momentum of the black hole introduces a preferred direction in space, that of the rotation axis. The question therefore that one can pose is whether the emitted radiation has a non-trivial distribution in space with respect to that axis. This question can be answered by computing the differential emission rate per unit time, frequency and unit of $\cos\theta$, where θ the angle measured from the rotation axis of the black hole. For this, we need the exact values of the angular function of the given brane-localised f eld that follows by integrat-

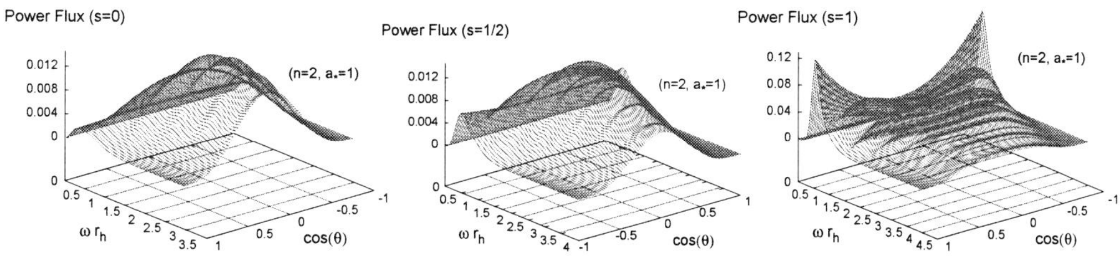

Figure 7. Angular distribution of the power spectra for scalars [45] (left plot), fermions [46] (middle plot) and gauge bosons [35] (right plot) for $n = 2$ and $a_* = 1$.

ing the angular master equation (37) by using the various techniques outlined above. In Figure 7, we present three-dimensional graphs depicting the differential energy emission rate in terms of ω_* and $\cos\theta$ for scalars, fermions and gauge bosons (from left to right) for $n = 2$ and $a_* = 1$ [35,45,46]. The angular distribution of the emitted particles is indeed non-trivial. There are two factors that determine this distribution: (i) the centrifugal force that forces the particles to be emitted on the equatorial plane; this force is independent of the spin of the particle and is enhanced as either ω_* or a_* increases, (ii) the spin-rotation coupling, i.e. the coupling between the spin of the emitted particle and the angular-momentum of the black hole; this factor tends to align the emitted particles with spin parallel or anti-parallel to the rotation axis and is more prominent the larger the value of the spin and the smaller the energy of the particle is. For the calculation of the above angular distribution, both radiative components have been taken into account in the expression of the differential emission rates (18) and, as result, the distribution is symmetric over the two hemispheres. This non-trivial angular distribution of the emitted particles is expected to be one of the most important observable effects associated with the emission of Hawking radiation from a higher-dimensional black hole during its spin-down phase.

In the previous subsection, devoted to the study of the Hawking radiation during the Schwarzschild phase, we discussed how the rate of the emission of energy or the preferred species of particles emitted could determine the sole parameter of the gravitational background, the number of extra dimensions n. During the spin-down phase, for a given black-hole mass again, the background depends both on n and the angular-momentum parameter a_* of the black hole. The similar effect that both these parameters have on the radiation spectra, i.e. enhancement although of a different magnitude, creates a degeneracy. As a result, one could not determine the value of either of these two parameters by using only the radiation spectra. We clearly need another observable that would depend rather strongly on the value of one only of these parameters and be almost insensitive to the value of the other.

It turns out that this observable is the angular distribution of the emitted radiation. If we look particularly at the low-energy channel of the non-zero spin f elds, we could see that their angular distribution is quite different. The gauge bosons, for which the spin-rotation coupling is most important, are perfectly aligned along the rotation axis independently of the angular momentum of the black hole; this is manifest at the left plot of Figure 8 [59]. Therefore, the direction of emission of gauge bosons at the low-energy channel gives also

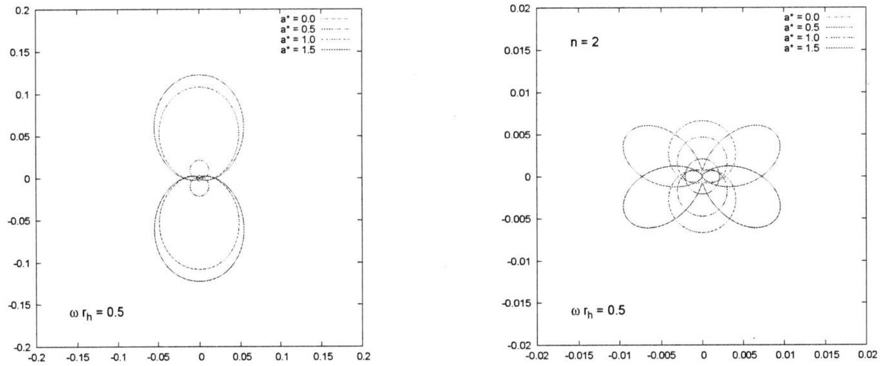

Figure 8. Polar plots with the angular distribution of gauge bosons (left plot) and fermions (right plot) in the energy channel $\omega r_h = 0.5$ emitted on the brane by a 6-dimensional, rotating black hole [59]. The upper set of curves correspond to positive helicity particles and the lower set to negative helicity particles.

the direction of the rotation axis of the black hole. Once this has been found, the angular distribution of fermions, that depends very strongly on the value of the angular momentum of the black hole, as it may be seen from the right plot of Figure 8, can give us the value of a_* [59, 60]. As desired, the above behaviour is virtually independent of the number n of extra spacelike dimensions, which may then be determined, as in the Schwarzschild phase, by comparing the predicted with the observed integrated-over-$\cos\theta$ radiation spectra.

4. Decay of Higher-Dimensional Black Holes via Emission in the Bulk

In this section, we study the emission of Hawking radiation from a higher-dimensional black hole, either rotating or spherically-symmetric, in the bulk. We will consider both species of f elds that are allowed to propagate in the bulk, namely scalars and gravitons, and present results for the corresponding radiation spectra in the cases where these have been derived. Finally, we will address the question of the energy balance between the brane and the bulk that so far remains open.

4.1. Emission of Scalar Fields in the Bulk

The easiest case to study is the emission of scalar f elds in the bulk due to the simplicity of the corresponding equation of motion. For a massless, minimally-coupled to gravity scalar f eld, this has the form

$$\frac{1}{\sqrt{-G}}\,\partial_A\left(\sqrt{-G}\,G^{AB}\partial_B\Phi\right)=0\,, \tag{98}$$

where G_{AB} is the metric tensor of the higher-dimensional spacetime. The emission of scalar f elds in the bulk has been studied during both the Schwarzschild and the spin-down phase by employing analytical [36, 61, 62] as well as numerical [38, 63, 64] methods. The equation of motion that the scalar particle satisf es can be decoupled, for both the spherically-symmetric and the rotating phase, into a radial and an angular part and the scattering problem may be solved along the same lines as on the brane. Here, we will present the main points of the theoretical framework, highlight the main steps of the analytical approach and f nally review the results that follow from the exact numerical analysis.

We will focus on the formalism that applies in the case of the rotating phase [62] since the one for the spherically-symmetric one follows readily be setting the angular-momentum parameter a equal to zero. As in the case of brane emission, we will consider a simply-rotating higher-dimensional black hole whose gravitational background is described by the line-element of Eq. (19). The equation of motion (98) can be separated by employing the following factorised ansatz

$$\Phi = e^{-i\omega t} e^{im\varphi} R(r)\, S(\theta)\, Y_{jn}(\theta_1, \ldots, \theta_{n-1}, \phi)\,, \tag{99}$$

where $Y_{jn}(\theta_1, \ldots, \theta_{n-1}, \phi)$ are the hyper-spherical harmonics on the n-sphere that satisfy the equation [65]

$$\sum_{k=1}^{n-1} \frac{1}{\prod_{i=1}^{n-1} \sin^i\theta_i}\, \partial_{\theta_k} \left[\left(\prod_{i=1}^{n-1} \sin^i\theta_i \right) \frac{\partial_{\theta_k} Y_{jn}}{\prod_{i>k}^{n-1} \sin^2\theta_i} \right] + \frac{\partial_{\phi\phi} Y_{jn}}{\prod_{i=1}^{n-1} \sin^2\theta_i} + j(j+n-1)\, Y_{jn} = 0\,. \tag{100}$$

Then, by using the fact that, for the line-element (19) the following relation holds

$$\sqrt{-G} = \Sigma \sin\theta\, r^n \cos^n\theta \prod_{i=1}^{n-1} \sin^i\theta_i\,, \tag{101}$$

equation (98) reduces to the following decoupled equations

$$\frac{1}{r^n}\, \partial_r \left(r^n \Delta \partial_r R \right) + \left(\frac{K^2}{\Delta} - \frac{j(j+n-1)a^2}{r^2} - \Lambda_{j\ell m} \right) R = 0\,, \tag{102}$$

$$\frac{1}{\sin\theta \cos^n\theta}\, \partial_\theta \left(\sin\theta \cos^n\theta\, \partial_\theta S \right) + \left(\omega^2 a^2 \cos^2\theta - \frac{m^2}{\sin^2\theta} - \frac{j(j+n-1)}{\cos^2\theta} + E_{j\ell m} \right) S = 0\,, \tag{103}$$

for the radial $R(r)$ and the angular function $S(\theta)$, respectively. The above equations f rst appeared in the literature in Ref. [66]. The quantity K appearing in the radial equation is given again by the expression $K = (r^2 + a^2)\,\omega - am$, while the constant $\Lambda_{j\ell m}$ is now def ned as $\Lambda_{j\ell m} = E_{j\ell m} + a^2\omega^2 - 2am\omega$.

For the angular eigenvalue $E_{j\ell m}(a\omega)$, there is no closed analytic form, as in the case of brane emission. It may again be computed numerically or expressed as a power series in $a\omega$ [67, 68]. In order for the power series to converge in the limit $a\omega \to 0$ by terminating at f nite order, a number of restrictions are imposed on the allowed values of the integer parameters (j, ℓ, m) specifying the emission mode: in general, m may take any integer value and j and ℓ any positive or zero integer value provided that [67]

$$\ell \geq j + |m| \quad \text{and} \quad \frac{\ell - (j + |m|)}{2} \in \{0, \mathbb{Z}^+\}\,. \tag{104}$$

For our analytic approach, we will use the power-series expression and keep the terms up to the 5th order. Under our assumption that both ω_* and a_* remain small, one may check that the analytic result agrees remarkably well with the exact numerical one. Then, the expression for $E_{j\ell m}$ has the form[2] [69]

$$\begin{aligned} E_{j\ell m} &= j(j+n+1)-(a\omega)^2\,\frac{[-1+2\ell(\ell-1)+2j(j+1)-2m^2+2n(j+\ell)+n^2]}{(2j+n-1)(2j+n+3)} \\ &+ (a\omega)^4\left\{\frac{(\ell-j+|m|)(\ell+j-|m|+n-1)}{16(2j+n-3)(2j+n-1)^2}\Big[(2+\ell-j+|m|)(\ell+j-|m|+n-3)\right. \\ &- 4(2j+n-3)\frac{[-1+2\ell(\ell-1)+2j(j+1)-2m^2+2n(j+\ell)+n^2]}{(2j+n-1)(2j+n+3)}\Big] \\ &- \frac{(\ell-j+|m|-2)(\ell+j+n-|m|+1)}{16(2j+5+n)(2j+n+3)^2}\Big[(\ell-j+|m|-4)(j+\ell+n-|m|+3) \\ &+ \left. 4(2j+n+5)\frac{[-1+2\ell(\ell-1)+2j(j+1)-2m^2+2n(j+\ell)+n^2]}{(2j+n-1)(2j+n+3)}\Big]\right\} \\ &+ o\Big((a\omega)^6\Big). \end{aligned} \tag{105}$$

The approximation method that will be applied in order to solve the radial equation analytically, and derive a formula for the Absorption Probability $|\mathcal{A}_{j\ell m}|^2$, follows closely the one already employed for the same purposes on the brane. Thus, here, we present only the steps where a modifcation appears due to the different background that the scalar particle sees. We start from the near-horizon regime ($r \simeq r_H$), and perform the same change of radial variable (53) [42,43] as in section 3.3. Then, Eq. (102) takes the form [62]

$$\begin{aligned} &f(1-f)\,\frac{d^2R}{df^2}+(1-D_*f)\,\frac{dR}{df} \\ &\qquad + \left[\frac{K_*^2}{A_*^2 f(1-f)} - \frac{\left(j(j+n-1)a_*^2+\Lambda_{j\ell m}\right)\,(1+a_*^2)}{A_*^2\,(1-f)}\right]R = 0, \end{aligned} \tag{106}$$

where A_* and K_* are given again by Eqs. (56) and

$$D_* \equiv 1 - \frac{4a_*^2}{A_*^2}. \tag{107}$$

By employing the same transformation $R_{NH}(f) = A_- f^{\alpha}(1-f)^{\beta}F(a,b,c;f)$, Eq. (106) can be brought to the form (57) of a hypergeometric differential equation, where now $a = \alpha+\beta+D_*-1$, $b = \alpha+\beta$, and $c = 1+2\alpha$. The parameters α and β are again determined by second-order constraints that may be easily solved to give the solutions

$$\alpha = \pm\frac{iK_*}{A_*}, \tag{108}$$

$$\beta = \frac{1}{2}\left[(2-D_*) \pm \sqrt{(D_*-2)^2 - 4\left[\frac{K_*^2-\left(j(j+n-1)a_*^2+\Lambda_{j\ell m}\right)(1+a_*^2)}{A_*^2}\right]}\,\right]. \tag{109}$$

[2]This expression is based on the analysis of [67] but disagrees on the sign of the second-order term. One may see that, only if the sign of that term is indeed reversed, the expression for $E_{j\ell m}$ correctly reproduces the on-brane eigenvalues, that have appeared in the literature previously, in the limit $j, n \to 0$.

The general solution of the radial equation (106) then has the form (63). The boundary condition that no outgoing modes exist near the black-hole horizon must again be applied. By using the same tortoise-like coordinate $y = r_H(1+a_*^2)\ln(f)/A_*$, in the limit $r \to r_H$, we obtain

$$R_{NH}(f) \simeq A_- f^{\pm iK_*/A_*} + A_+ f^{\mp iK_*/A_*} = A_- e^{\pm iky} + A_+ e^{\mp iky}. \qquad (110)$$

Therefore, we can set either $A_- = 0$ or $A_+ = 0$, depending on the choice for α. As the two choices are clearly equivalent, we choose $\alpha = \alpha_-$, which then demands $A_+ = 0$. This brings the near-horizon solution to the f nal form

$$R_{NH}(f) = A_- f^{\alpha}(1-f)^{\beta} F(a,b,c;f). \qquad (111)$$

Finally, the convergence criterion for the hypergeometric function, i.e. $Re(c-a-b) > 0$, should also be applied, leading us to choose again $\beta = \beta_-$.

In the far-f eld regime ($r \gg r_H$), after making the substitution $R(r) = r^{-\left(\frac{n+1}{2}\right)}\tilde{R}(r)$, Eq. (102) takes the form of a Bessel equation [33], in terms of $z = \omega r$,

$$\frac{d^2\tilde{R}}{dz^2} + \frac{1}{z}\frac{d\tilde{R}}{dz} + \left(1 - \frac{E_{j\ell m} + a^2\omega^2 + \left(\frac{n+1}{2}\right)^2}{z^2}\right)\tilde{R} = 0. \qquad (112)$$

The solution in the far-f eld regime may then be written as

$$R_{FF}(r) = \frac{B_1}{r^{\frac{n+1}{2}}} J_\nu(\omega r) + \frac{B_2}{r^{\frac{n+1}{2}}} Y_\nu(\omega r), \qquad (113)$$

with J_ν and Y_ν are the Bessel functions of the f rst and second kind, respectively, and $\nu = \sqrt{E_{j\ell m} + a^2\omega^2 + \left(\frac{n+1}{2}\right)^2}$.

The two asymptotic solutions (111) and (113) need to be stretched before being matched in an intermediate regime. When the latter is expanded for small values of r, we obtain

$$R_{FF}(r) \simeq \frac{B_1\left(\frac{\omega r}{2}\right)^{\nu}}{r^{\frac{n+1}{2}}\Gamma(\nu+1)} - \frac{B_2}{\pi r^{\frac{n+1}{2}}}\frac{\Gamma(\nu)}{\left(\frac{\omega r}{2}\right)^{\nu}}. \qquad (114)$$

The Near-Horizon solution needs f rst to be shifted according to the formula (73) and then to be expanded for large values of r. In this limit we have that $f \to 0$, the hypergeometric functions reduce to unity and f nally we f nd

$$R_{NH}(r) \simeq A_1 r^{-(n+1)\beta} + A_2 r^{(n+1)(\beta+D_*-2)}, \qquad (115)$$

where now A_1 and A_2 are def ned as

$$A_1 = A_-\left[(1+a_*^2)r_H^{n+1}\right]^{\beta}\frac{\Gamma(c)\Gamma(c-a-b)}{\Gamma(c-a)\Gamma(c-b)},$$

$$A_2 = A_-\left[(1+a_*^2)r_H^{n+1}\right]^{-(\beta+D_*-2)}\frac{\Gamma(c)\Gamma(a+b-c)}{\Gamma(a)\Gamma(b)}. \qquad (116)$$

Both solutions have again a power-law form and their matching is perfect, i.e. the powers of r in the two expressions become the same, in the small a_* and ω_* limit. More particularly, we fnd $(n+1)\beta \simeq -\ell$, $(n+1)(\beta+D_*-2) \simeq -(\ell+n+1)$, and $\nu \simeq \ell+(n+1)/2$. Then, identifying the multiplicative coeffcients in front of the same powers of r, we fnd the constraint [62]

$$B \equiv \frac{B_1}{B_2} = -\frac{(2/\omega r_H)^{2\ell+n+1}\,\nu\Gamma^2(\nu)\Gamma(\alpha+\beta+D_*-1)\Gamma(\alpha+\beta)\Gamma(2-2\beta-D_*)}{\pi(1+a_*^2)^{\frac{2\ell+n+1}{n+1}}\,\Gamma(2\beta+D_*-2)\Gamma(2+\alpha-\beta-D_*)\Gamma(1+\alpha-\beta)}. \tag{117}$$

that again guarantees the existence of a smooth, analytic solution for the radial part of the scalar feld over the whole radial regime, valid in the low-energy and low-rotation limit.

The Absorption Probability may now easily follow by expanding the Far-Field solution (113) at asymptotic infnity. We then obtain

$$\begin{aligned} R_{FF}(r) &\simeq \frac{1}{r^{\frac{n+2}{2}}\sqrt{2\pi\omega}}\left[(B_1+iB_2)\,e^{-i\left(\omega r-\frac{\pi}{2}\nu-\frac{\pi}{4}\right)}+(B_1-iB_2)\,e^{i\left(\omega r-\frac{\pi}{2}\nu-\frac{\pi}{4}\right)}\right] \\ &= A_{in}^{(\infty)}\,\frac{e^{-i\omega r}}{r^{\frac{n+2}{2}}}+A_{out}^{(\infty)}\,\frac{e^{i\omega r}}{r^{\frac{n+2}{2}}}\,. \end{aligned} \tag{118}$$

We observe that, for large distances from the black hole, the solution for the scalar feld reduces to an incoming and an outgoing spherical wave, whose value reduces faster with the distance, compared to a brane-localised scalar feld, due to the extra factor of n in the power of r in the denominators of the above expression. The greybody factor can be again easily computed from the ratio of the two amplitudes:

$$|\mathcal{A}_{j\ell m}|^2 = 1-\left|\frac{A_{out}^{(\infty)}}{A_{in}^{(\infty)}}\right|^2 = 1-\left|\frac{B_1-iB_2}{B_1+iB_2}\right|^2 = \frac{2i(B^*-B)}{BB^*+i(B^*-B)+1}\,. \tag{119}$$

The above result gives the expression of the Absorption Probability for the emission of scalar felds in the bulk by a higher-dimensional, rotating black hole. By setting $a_*=0$, the corresponding result for the emission from a spherically-symmetric black hole [36] readily follows.

The above analytic result can be put into test, regarding its validity, by comparing it with the exact numerical results that follow by integrating numerically the radial equation (102). In Figure 9, we plot the value of $|\mathcal{A}_{j\ell m}|^2$ for a bulk scalar feld, for $n=2$ and $a_*=0.4$, and a variety of modes [62]. Each curve in that graph is a combination of two lines, a dashed one representing the analytic result and a solid one standing for the exact numerical result. The agreement between the two sets of results is again remarkable even beyond the strictly low-energy regime. Once again, the emergence of poles in the expression of Gamma functions present in the analytic formula for $|\mathcal{A}_{j\ell m}|^2$ does not allow us to produce plots for the greybody factor over a more extended range of values.

A simplifed formula for the Absorption Probability may be derived by taking, as in the case of brane emission, the very-low energy limit $\omega_* \to 0$. In this limit, we fnd that $B \simeq \omega_*^{-(2\ell+n+1)}$ and, focusing on the dominant mode $j=\ell=m=0$, we fnd that $|\mathcal{A}_{000}|^2 \sim \omega_*^{n+2}$. Then, the corresponding cross-section may be evaluated by using the formula [62]

$$\sigma_{j\ell m}(\omega) = \frac{2^n}{\pi}\Gamma^2\left(\frac{n+3}{2}\right)\frac{A_H}{(\omega r_H)^{n+2}}\frac{N_j}{(1+a_*^2)}|\mathcal{A}_{j\ell m}|^2\,, \tag{120}$$

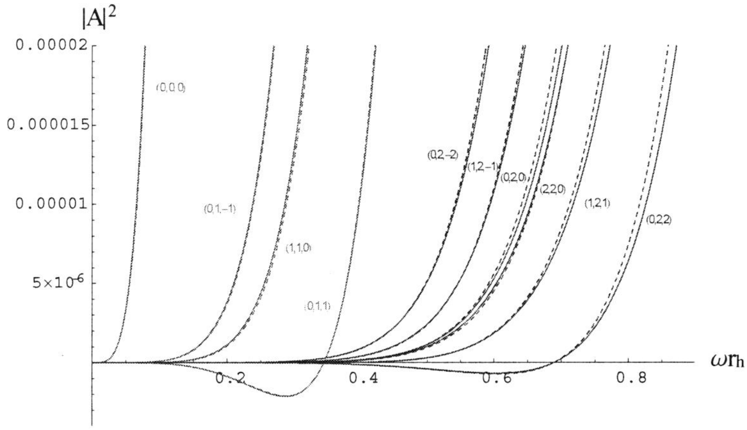

Figure 9. Absorption probabilities (analytic results) $\left|\mathcal{A}_{j\ell m}\right|^2$ for a bulk scalar f eld, for $n = 2$, $a_* = 0.4$, and various combinations of (j, ℓ, m) [62].

where

$$N_j = \frac{(2j+n-1)(j+n-2)!}{j!\,(n-1)!}, \qquad A_H = \frac{2\pi^{\frac{n+3}{2}} r_H^n (r_H^2 + a^2)}{\Gamma\left(\frac{n+3}{2}\right)} \tag{121}$$

are the multiplicity of the jth partial wave in the expansion of the wave function over the hyper-spherical harmonics on the n-sphere, and the horizon area of the $(4+n)$-dimensional rotating black hole, respectively. Substituting for the Absorption Probability, we obtain

$$\sigma_{000}(\omega) \simeq \frac{(n+1)(1+a_*^2)A_H}{A_*(2-D_*)} + \dots. \tag{122}$$

The above result allows us to examine the relation between the value of the absorption cross-section and the horizon area of the black hole for a bulk scalar f eld. For $a_* \to 0$, i.e. for a spherically-symmetric black hole, one may easily conf rm that the above reduces to A_H independently of the value of n, as was indeed found in [38]. As a_* starts increasing, the value of σ_{000} becomes gradually larger than the horizon area of the black hole, similarly to the behaviour found also in the case of a brane scalar f eld.

It is only through numerical means that the complete range of values of the Absorption Probability, for the emission of scalar f elds in the bulk, can be derived. As in the case of brane-localised scalars, the derivation of exact results demands the integration of both the radial and the angular equation, as the latter provides the angular eigenvalue $E_{j\ell m}$. This task was performed in [64] – where we refer the interested reader for more details – and the angular equation was solved numerically by using a modif ed shooting method [54, 67]. Once the value of $E_{j\ell m}$, valid for arbitrary energy of the emitted particle and angular-momentum of the black hole, was derived, the radial equation was solved following a numerical method similar to the one used for the emission of scalar f elds on the brane. By determining the amplitudes of the ingoing and outgoing spherical waves at inf nity, the Absorption Probability was found through Eq. (119). Its values for the same set of parameters, as in Figure 9, i.e. for $n = 2$ and $a_* = 0.4$, and for several emission modes, are depicted in Figure 10. One

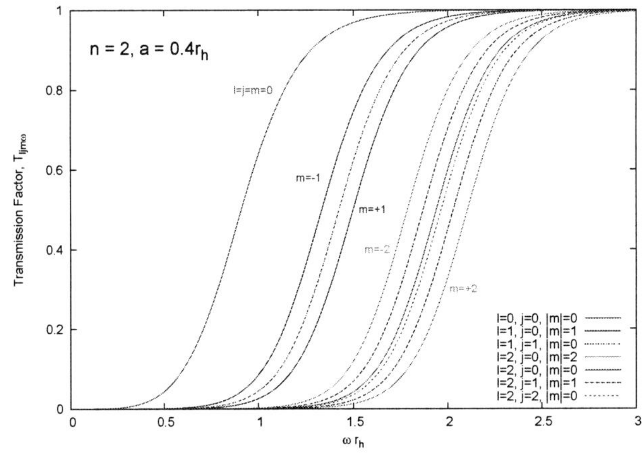

Figure 10. Absorption probabilities (numerical results) $\left|\mathcal{A}_{j\ell m}\right|^2$ for a bulk scalar f eld, for $n = 2$, $a_* = 0.4$, and various combinations of (j, ℓ, m) [46].

may easily see that the modes split f rst in ℓ, the total angular momentum number, then in m, the angular momentum number in the plane of rotation, and f nally in j, the angular momentum number in the internal n-sphere. Several of the modes in Figure 10 have $m > 0$ and they are therefore super-radiant - however, the effect is very small and cannot be seen; it is nevertheless visible in Figure 9 where the magnif cation of the low-energy regime is depicted.

Having found the Absorption Probability, we can now proceed to compute the rate of

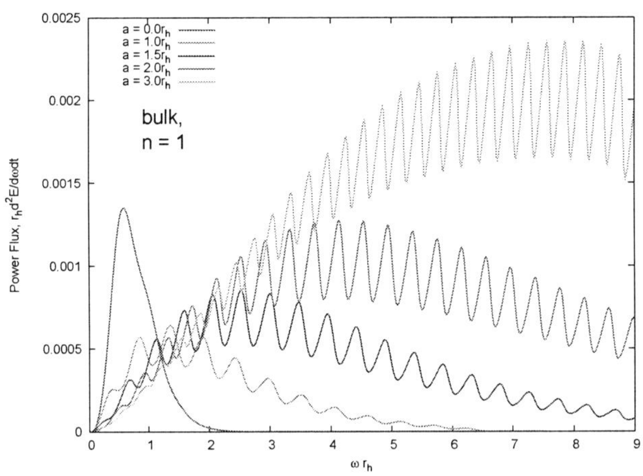

Figure 11. Energy emission rates for scalar f elds in the bulk from a black hole with $n = 1$ and variable a_* [64].

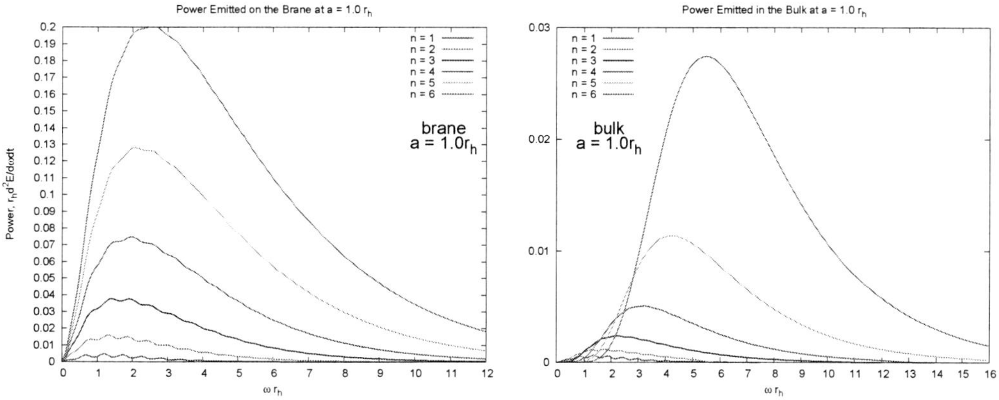

Figure 12. Energy emission rates for scalar f elds on the brane (left plot) and in the bulk (right plot) from a black hole with $a_* = 1$ and variable n [64].

energy emission of massless scalar f elds in the bulk. This is given by the expression [46,62]

$$\frac{d^2E}{dt\,d\omega} = \frac{1}{2\pi} \sum_{j,\ell,m} \frac{\omega}{\exp\left[k/T_{\rm H}\right] - 1} N_j \left|\mathcal{A}_{j\ell m}\right|^2 . \tag{123}$$

The above differs from the 4-dimensional [70] expression in the presence of an additional sum over the new angular momentum number j, and from the 5-dimensional [61] one in the introduction of the multiplicity of states N_j (121) following from the expansion of the wavefunction of the f eld in the n-dimensional sphere.

In Figure 11, we present the energy emission rate for scalar f elds in the bulk from a higher-dimensional, rotating black hole with $n = 1$ and variable a_*. We observe that the energy emission rate is suppressed in the low-energy regime with the angular momentum of the black hole but is enhanced in the intermediate and high-energy regime, leading to a signif cant overall enhancement especially for large values of a_*. A similar behaviour is observed in terms of the number n of extra dimensions, as may be clearly seen in Figure 12 (right plot). Here, the corresponding energy emission rate for emission of scalar f elds on the brane is also shown (left plot) – one may easily conclude that, for the same values of a_* and n, the emission on the brane, in the form of scalar f elds, is much more signif cant than the emission in the bulk. We will return to this point in section 5 where the question of the energy balance between the brane and the bulk will be addressed.

4.2. Emission of Gravitons in the Bulk

The second species of particles that one should study, during the emission of Hawking radiation in the bulk by a higher-dimensional black hole, is the gravitons, the degrees of freedom of the gravitational background itself. However, the progress in this case has been very slow due to the complexity of the f eld equations describing gravitational perturbations in a higher-dimensional black-hole background. For a spherically-symmetric black hole, these equations were derived only a few years ago [71] while, for the case of an axially-symmetric black hole, the derivation of the f eld equations is still in progress.

The level of diff culty of the derivation of the f eld equations of the gravitational perturbations in a given spacetime depends strongly on the structure of the corresponding line-element. One particular class of higher-dimensional spacetimes in which progress has been made in the derivation of the graviton perturbation equations is the spacetime whose line-element is described by [72]

$$ds^2 = G_{AB}\,dz^A\,dz^B = g_{\mu\nu}\,dx^\mu\,dx^\nu + S^2(x)\,d\Omega_n^2, \tag{124}$$

where $\{\mu,\nu\} = (0,1,2,3)$ and $d\Omega_n^2$ stands for the line element of an n-dimensional unit sphere S^n. The above line-element is a special $(4+n)$-dimensional case of a more general class of gravitational backgrounds where the spacetime can be written as the warped product of an m-dimensional spacetime $\mathcal{N}$ and an n-dimensional space $\mathcal{K}$ of constant curvature [73, 74]. In these type of backgrounds, gravitational perturbations can be classif ed into tensor, vector and scalar types according to their transformation properties as tensors on the constant-curvature spacetime $\mathcal{K}$.

The gravitational background around a spherically-symmetric, neutral black hole formed in a $(4+n)$-dimensional, f at spacetime is described by the line-element (6). This is a special case of the class of spacetimes described by Eq. (124) with $m=2$ and $\mathcal{K}=S^{n+2}$. Following [71], a graviton in D dimensions can then be decomposed into a symmetric traceless tensor, a vector and a scalar part. These are further expanded in terms of the spin-weighted spherical harmonics on the S^{n+2} unit sphere. Then, the radial parts of all three types of gravitational perturbations are found to satisfy a Schrodinger-like equation of the form:

$$f\frac{d}{dr}\left(f\frac{d\Phi}{dr}\right) + (\omega^2 - V)\,\Phi = 0, \tag{125}$$

where $f(r) = 1 - \left(\frac{r_H}{r}\right)^{n+1}$ is the metric function. The potential V has a different form for each type of perturbation, namely

$$V_{T,V} = \frac{f(r)}{r^2}\left[l(l+n+1) + \frac{n(n+2)}{4} - \frac{k(n+2)^2}{4}\left(\frac{r_H}{r}\right)^{n+1}\right] \tag{126}$$

for tensor- and vector-like perturbations, with $k=-1$ and $k=3$ respectively, $l=2,3,4,\ldots$ is an integer number denoting the eigenvalues of the spin-weighted spherical harmonics on S^{n+2}, and

$$V_S = \frac{f(r)}{r^2}\,\frac{qx^3 + px^2 + wx + z}{4\,[2m + (n+2)(n+3)x]^2}, \tag{127}$$

for scalar gravitational perturbations, with

$$m \equiv l(l+n+1) - n - 2, \qquad x \equiv \left(\frac{r_H}{r}\right)^{n+1} = 1 - f, \tag{128}$$

and

$$q \equiv (n+2)^4(n+3)^2, \tag{129}$$

$$p \equiv (n+2)(n+3)\left[4m\,(2n^2+5n+6) + n(n+2)(n+3)(n-2)\right], \tag{130}$$

$$w \equiv -12m\,(n+2)\,[m(n-2) + n(n+2)(n+3)], \tag{131}$$

$$z \equiv 16m^3 + 4m^2(n+2)(n+4). \tag{132}$$

The above equations for all three types of gravitational perturbations can be solved analytically in the low-energy regime [75] (and also in the intermediate-energy regime [76]) by using the same approximation method that was employed for the study of other species of f elds emitted either on the brane or in the bulk. Starting from the near-horizon regime, we treat tensor and vector perturbations together, due to the similar form of their potential, while the scalar perturbations will be dealt with separately.

Tensor and Vector Perturbations. If we make the change of variable $r \rightarrow f(r)$, the f eld equation for tensor and vector perturbations can be written in the form [75]

$$f(1-f)\frac{d^2\Phi}{df^2} + \left[1 - \frac{(2n+3)}{(n+1)}f\right]\frac{d\Phi}{df} + \left[\frac{(\omega r_H)^2}{(n+1)^2 f(1-f)} - \frac{A}{(1-f)} + \frac{k(n+2)^2}{4(n+1)^2}\right]\Phi = 0. \quad (133)$$

In the above, $k = -1$ for tensor perturbations and $k = 3$ for vector ones, and we have also def ned for convenience the quantity

$$A \equiv \frac{l(l+n+1)}{(n+1)^2} + \frac{n(n+2)}{4(n+1)^2}. \quad (134)$$

The usual f eld redef nition: $\Phi(f) = f^{\alpha}(1-f)^{\beta}F(f)$ brings the above equation to the form of a hypergeometric equation under the identif cations

$$a = \alpha + \beta + \frac{(n+2)}{2(n+1)} + G, \quad b = \alpha + \beta + \frac{(n+2)}{2(n+1)} - G, \quad c = 1 + 2\alpha, \quad (135)$$

where G is an arbitrary constant. Demanding further that the coeff cient of $F(f)$ in our equation be indeed $-ab$ yields three more additional constraints that determine the remaining unknown constants, α, β and G. Their corresponding values are then found to be:

$$G^{(V,T)} = \frac{(1+k)(n+2)}{4(n+1)}, \quad \alpha = \pm\frac{i\omega r_H}{n+1}, \quad (136)$$

and

$$\beta = \frac{1}{2(n+1)}\left\{-1 \pm \sqrt{(2l+n+1)^2 - 4\omega^2 r_H^2}\right\}. \quad (137)$$

The convergence condition of the hypergeometric function, i.e. $Re\,(c-a-b) > 0$ demands that we choose $\beta = \beta_-$. The general solution of the hypergeometric equation is of the form of Eq. (63). In the limit $r \rightarrow r_H$, or equivalently $f(r) \rightarrow 0$, we obtain

$$\Phi_{NH}(f) \simeq A_1 f^{\alpha} + A_2 f^{-\alpha} = A_1 e^{-i\omega y} + A_2 e^{i\omega y}, \quad (138)$$

where we have used the 'tortoise-like' coordinate y def ned through the relation: $y(r) = r_H \ln f(r)/(n+1)$, and randomly choose $\alpha = \alpha_-$. Then, the boundary condition of no outgoing modes in the area close to the horizon demands that we set $A_2 = 0$. This brings the Near-Horizon solution to its f nal form: $\Phi_{NH}(f) = A_1 f^{\alpha}(1-f)^{\beta}F(a,b,c;f)$ with the indices (a,b,c) given above.

Scalar perturbations. By employing the same change of variable, the field equation for scalar gravitational perturbations can be brought to the form [75]

$$f(1-f)\,\frac{d^2\Phi}{df^2}+\left[1-\frac{(2n+3)}{(n+1)}\,f\right]\frac{d\Phi}{df}$$
$$+\left[\frac{(\omega r_H)^2}{(n+1)^2 f(1-f)}-\frac{z}{16(n+1)^2 m^2(1-f)}-C\right]\Phi=0\,, \quad (139)$$

where we have defined the quantity

$$C\equiv\frac{\tilde q(1-f)^2+\tilde p(1-f)+\tilde w}{16(n+1)^2[m+(n+2)(n+3)(1-f)/2]^2}\,, \quad (140)$$

with

$$\begin{aligned}\tilde q &= q\,,\\ \tilde p &= p-\frac{z(n+2)^2(n+3)^2}{4m^2}\,,\\ \tilde w &= w-\frac{z(n+2)(n+3)}{m}\,.\end{aligned} \quad (141)$$

In this form, Eq. (139) has poles at $f=0$ and $f=1$ (or, at $r=r_H$ and $r=\infty$, respectively) while the quantity C takes on a constant value in both limits.

We then follow the same steps as before to bring Eq. (139) to the form of a hypergeometric equation. In this case, the indices (a,b,c) are given by Eqs. (135), and the powers (α,β) by Eqs. (136)–(137). The only difference arises in the value of the arbitrary constant G that, in the case of scalar perturbations, is given by

$$G^{(S)}=\frac{1}{2(n+1)}\sqrt{(n+2)^2-\frac{\tilde q+\tilde p+\tilde w}{[2m+(n+2)(n+3)]^2}}\,. \quad (142)$$

By applying the same boundary condition of no-incoming waves near the horizon, the general solution for the scalar perturbations in the near-horizon regime is given again by $\Phi_{NH}(f)=A_1 f^{\alpha}(1-f)^{\beta}F(a,b,c;f)$, with $\beta=\beta_-$ and $\alpha=\alpha_-$, as before.

In the far-field regime ($r\gg r_H$), the field equation for all types of gravitational perturbations – tensor, vector and scalar – takes the simplified form

$$\frac{d^2\Phi}{dr^2}+\left(\omega^2-\frac{4(n+1)^2 A}{r^2}\right)\Phi=0\,, \quad (143)$$

where A was defined in Eq. (134). By further setting $\Phi=\sqrt{r}R$, the new radial function R is found to satisfy a Bessel differential equation with $\nu=l+(n+1)/2$ in terms of $z=\omega r$. Therefore, the general solution of Eq. (143) can be written as

$$\Phi_{FF}(r)=B_1\sqrt{r}J_{l+(n+1)/2}(\omega r)+B_2\sqrt{r}Y_{l+(n+1)/2}(\omega r)\,, \quad (144)$$

where J_ν and Y_ν are the Bessel functions of the first and second kind, respectively, and $B_{1,2}$ arbitrary integration constants.

Expanding the above solution towards small values of the radial variable, and then shifting and stretching the Near-Horizon solution towards large value of r leads to the following power-law forms

$$\Phi_{FF}(r) \simeq B_1 \left(\frac{\omega}{2}\right)^{l+\frac{n+1}{2}} \frac{r^{l+\frac{n}{2}+1}}{\Gamma\left(l+\frac{n+3}{2}\right)} - \frac{B_2}{\pi}\left(\frac{2}{\omega}\right)^{l+\frac{n+1}{2}} \frac{\Gamma\left(l+\frac{n+1}{2}\right)}{r^{l+\frac{n}{2}}}. \quad (145)$$

$$\Phi_{NH}(f) \simeq A_1 \left(\frac{r_H}{r}\right)^{\beta(n+1)} \frac{\Gamma(c)\Gamma(c-a-b)}{\Gamma(c-a)\Gamma(c-b)} + A_1 \left(\frac{r_H}{r}\right)^{-1-\beta(n+1)} \frac{\Gamma(c)\Gamma(a+b-c)}{\Gamma(a)\Gamma(b)}. \quad (146)$$

In the limit $\omega r_H \ll 1$, the ωr_H-term under the square-root in the expression for β, Eq. (137), can be ignored and then the powers of r in the two stretched solutions become the same. Identifying the corresponding coeff cients at the front, we obtain the constraint [75]

$$B \equiv \frac{B_1}{B_2} = -\left(\frac{2}{\omega r_H}\right)^{2l+n+1} \frac{\Gamma\left(l+\frac{n+3}{2}\right)\Gamma\left(l+\frac{n+1}{2}\right)\Gamma(a)\Gamma(b)\Gamma(c-a-b)}{\pi\Gamma(c-a)\Gamma(c-b)\Gamma(a+b-c)}. \quad (147)$$

The above result applies for all three types of gravitational perturbations, tensor, vector and scalar, upon substitution of the corresponding indices (a,b,c).

Expanding further the far-f eld solution (144) in the limit $r \to \infty$, we obtain a superposition of an incoming and an outgoing plane wave

$$\Phi_{FF}(r) = \frac{1}{\sqrt{2\pi\omega}} \left\{(B_1 - iB_2)\, e^{i\left(\omega r - \frac{\pi}{2}\nu - \frac{\pi}{4}\right)} + (B_1 + iB_2)\, e^{-i\left(\omega r - \frac{\pi}{2}\nu - \frac{\pi}{4}\right)}\right\} + \cdots. \quad (148)$$

Then, the absorption probability follows from the ratio of the two amplitudes according to the well-known formula

$$|\mathcal{A}_l|^2 = 1 - |\mathcal{R}_l|^2 = 1 - \left|\frac{B-i}{B+i}\right|^2. \quad (149)$$

The expression (147) can be further simplif ed, and a compact analytical result for the Absorption Probability may thus be derived, if we expand it in the very low-energy limit $\omega r_H \to 0$. The analysis is similar to the one followed for the other species of f elds studied before and involves quite some algebra. We therefore refer the reader to [75] and here we only state the f nal result. This has the form

$$|\mathcal{A}_l|^2 = 4\pi \left(\frac{\omega r_H}{2}\right)^{2l+n+2} \frac{\Gamma\left(1+\frac{l}{n+1}-G\right)^2 \Gamma\left(1+\frac{l}{n+1}+G\right)^2}{\Gamma\left(l+\frac{n+3}{2}\right)^2 \Gamma\left(1+\frac{2l}{n+1}\right)^2}, \quad (150)$$

and it applies for all three types of gravitational perturbations upon substitution of the constant G in each case. Since the corresponding cross-section will be $\sigma_l \sim \omega_*^{-(n+2)}$ and $l \geq 2$, we conclude that the asymptotic value of σ_l for all graviton modes vanish as $\omega_* \to 0$. Coming back to the Absorption Probability, and since $\omega r_H \ll 1$, we may conclude from Eq. (150) that, as either n or l increases, $|\mathcal{A}_l|^2$ is suppressed in the very low-energy regime – this behaviour is indeed conf rmed by numerical evaluation of the above formula. It turns out that the magnitude of the Absorption Probability for tensor perturbations is by orders of magnitude suppressed compared to the vector and scalar perturbations, while the relative magnitude of the latter two is strongly dependent on the particular values of l and n. Also, one may easily see that, if we make the redefnition $\Phi^{(T)}(r) = \sqrt{r^{n+2}}\,\Phi(r)$ for the radial function of the tensor perturbations, their corresponding feld equation reduces to the one for a scalar feld propagating in the bulk. We therefore may conclude that, of all the degrees of freedom propagating in the bulk, the vector and scalar gravitational perturbations are the dominant ones in the low-energy part of the spectrum.

A point that has not been touched upon yet is the one of the multiplicities of graviton states that correspond to the same angular momentum number l. For tensor, vector, and scalar type of perturbations these are known, and can be found in [77]. They are given by the expressions

$$N_l^{(T)} = \frac{n(n+3)(l+n+2)(l-1)(2l+n+1)(l+n-1)!}{2(l+1)!\,(n+1)!}, \tag{151}$$

$$N_l^{(V)} = \frac{l(l+n+1)(2l+n+1)(l+n-1)!}{(l+1)!\,n!}, \tag{152}$$

$$N_l^{(S)} = \frac{(2l+n+1)(l+n)!}{l!\,(n+1)!}, \tag{153}$$

respectively. The multiplicities depend both on the dimensionality of spacetime and the angular-momentum number of the emitted mode, and increase as either one of these parameters increases, too. A simple numerical evaluation reveals that for a moderate value of the angular-momentum quantum number, i.e. $l = 5$, the number of states, as n varies from 1 to 6, increases already by a factor of $10^3 - 10^4$. What is particularly intriguing is the fact that it is the tensor gravitational perturbations that proliferate the most.

In order to decide about the dominance of a specifc degree of freedom emitted in the bulk it is necessary to compute the Absorption Probability, and thus the energy emission spectra, over the whole energy regime. Therefore, we inevitably turn to the numerical integration of the radial equations. This was performed in [78, 79], and the contribution of each type of gravitational perturbations to the total graviton energy emission rate, given by the expression

$$\frac{dE^{(P)}}{dt\,d\omega} = \frac{1}{2\pi}\sum_l N_l^{(P)}|\mathcal{A}_l^{(P)}|^2\,\frac{\omega}{\exp(\omega/T_H)-1}, \tag{154}$$

was computed - in the above the superscript $P = (T, V, S)$ denotes the type of perturbation. It turned out that, primarily, the tensor and, secondly, the vector type perturbations contribute the most to the total energy output of the black hole in the bulk.

As was already discussed before, during the spin-down phase of a higher-dimensional, rotating black hole, the gravitational background around it is described by the class of the Myers–Perry solutions [22]. The most general form of the solution describing a black hole characterised by $N = [(D-1)/2]$ independent angular-momentum components does not belong to the class described by Eq. (124). However, the particular case of the simply-rotating black hole (11), that has been considered throughout this review, is indeed a special case with $m = 4$, $\mathcal{K} = S^n$ and $S(x) = r\cos\theta$.

As was demonstrated in [72, 80], the tensor-type gravitational perturbations for the line element (11), which exist only for $n \geq 3$, or $D \geq 7$, can be expanded in terms of a basis of transverse and traceless harmonic tensors $\mathbb{T}^{(j,\alpha)}_{ab}$ on the unit sphere S^n according to the formula

$$\delta G_{ab} = 2S^2(x) \sum_{j,\alpha} H_T^{(j,\alpha)}(x)\, \mathbb{T}^{(j,\alpha)}_{ab}(y) . \tag{155}$$

In the above, the indices $\{a,b\}$ refer to the y coordinates along the sphere S^n, and the harmonic tensors $\mathbb{T}^{(j,\alpha)}_{ab}$ satisfy the eigenvalue equation

$$[\hat{\Delta} + j(j+n-1) - 2]\, \mathbb{T}^{(j,\alpha)}_{ab} = 0 , \tag{156}$$

where $\hat{\Delta}$ is the Laplace-Beltrami operator on S^n, and $j = 2,3,4,\ldots$ an integer number that labels the corresponding eigenvalues. Finally, α is a label to distinguish harmonic tensors with the same eigenvalue.

By working out the (μ,ν) component of Einstein's equation in vacuum, the amplitude functions $H_T^{(j,\alpha)}(x)$ are found to satisfy the following second-order hyperbolic equation [80]

$$-\Box H_T - \frac{n}{r\cos\theta}\, g^{\mu\nu}\, \partial_\mu(r\cos\theta)\, \partial_\nu H_T + \frac{j(j+n-1)}{r^2\cos^2\theta} H_T = 0, \tag{157}$$

where $\Box$ is the d'Alembertian operator for the metric $g_{\mu\nu}(x)$, and where, for simplicity, we have omitted the labels $\{j,\alpha\}$. We may now use the following factorization for the amplitude function

$$H_T(x) = e^{-i\omega t}\, e^{im\varphi}\, R(r)\, S(\theta) , \tag{158}$$

which, when substituted in the aforementioned differential equation with partial derivatives, leads to a set of radial and angular equations, namely,

$$\frac{1}{r^n}\,\partial_r\left(r^n \Delta \partial_r R\right) + \left(\frac{K^2}{\Delta} - \frac{j(j+n-1)a^2}{r^2} - \Lambda_{j\ell m}\right) R = 0 , \tag{159}$$

$$\frac{1}{\sin\theta\cos^n\theta}\,\partial_\theta\left(\sin\theta\cos^n\theta\,\partial_\theta S\right) + \left(\omega^2 a^2\cos^2\theta - \frac{m^2}{\sin^2\theta} - \frac{j(j+n-1)}{\cos^2\theta} + E_{j\ell m}\right) S = 0 . \tag{160}$$

In the above, $E_{j\ell m}$ is the separation constant of the two equations, and ℓ a new quantum number that labels the eigenvalues of the angular function $S(\theta)$.

The above two equations are identical to the ones satisfed by a scalar feld propagating in the gravitational background of a higher-dimensional, simply-rotating black hole, and they were studied in section 4.1. We have thus confrmed the fact that, in a spacetime with a line-element of the form of Eq. (124), the tensor-type gravitational perturbations are found to satisfy the same feld equations that a massless scalar feld obeys in the same background [72, 80]. The only difference between the two cases is the range of values of the angular-momentum number j: whereas, in the scalar case, it satisfes the constraint $j \geq 0$ [66], this changes to $j \geq 2$ in the case of gravitons [72].

Therefore, the whole analysis presented in section 4.1 leading to the analytic formula for the Absorption Probability, Eqs. (117) and (119), holds also for the case of tensor-type gravitons. In [81], the same set of equations for tensor-type gravitons was solved

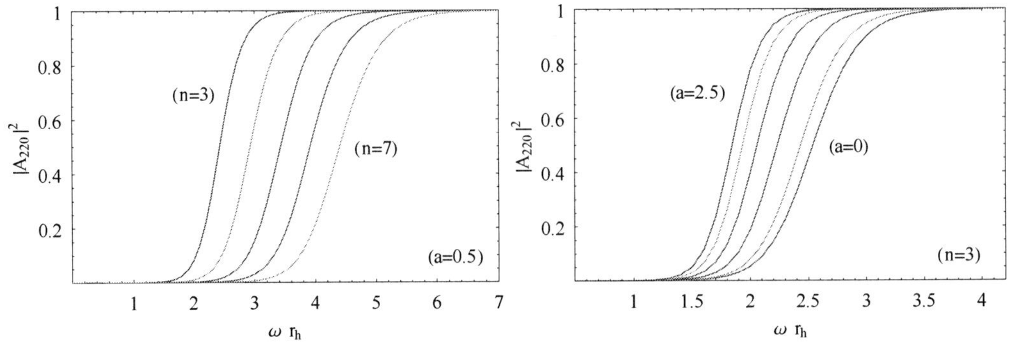

Figure 13. Absorption probabilities for the mode ($j = 2, \ell = 2, m = 0$) as **(a)** a function of $n = 3,4,5,6,7$, for $a_* = 0.5$, and **(b)** a function of $a_* = 0, 0.5, 1, 1.5, 2, 2.5$, for $n = 3$ [81].

numerically, and a comparison was performed between the analytical and numerical results. It was found once again that the two sets agree remarkably well in the low-energy regime, with the agreement extending even up to the intermediate energy regime especially for modes with low values of the angular momentum numbers (ℓ, j, m) – unlike the scalar case, no poles arise in the expression of the Gamma functions, due to the different allowed values of j and ℓ, and thus an extended plot covering all values of $|\mathcal{A}_{j\ell m}|^2$ can be constructed.

In Figure 13, we depict the exact numerical results obtained in [81] for the Absorption Probability of tensor-type gravitons emitted in the bulk by a higher-dimensional simply-rotating black hole – we consider the indicative case of the lowest mode $j = \ell = 2$ and $m = 0$. As we may see, the value of $|\mathcal{A}_{j\ell m}|^2$ is suppressed with the number of extra dimensions n (similarly to the behaviour found for scalar f elds emitted in the bulk [38, 62, 64]) but enhanced with the angular-momentum of the black hole.

The energy emission rate for the tensor-type gravitons is given by the following formula [81]

$$\frac{d^2E}{d\omega dt} = \frac{1}{2\pi} \sum_{j,\ell,m} \frac{\omega}{e^{k/T_H} - 1} N_T^j(S^n) |\mathcal{A}_{j\ell m}|^2, \tag{161}$$

where the quantity N_T^j is the multiplicity of the second-rank symmetric, traceless ($T^A{}_A = 0$) and divergence-free ($D_B T^{BA} = 0$) tensor harmonics T_{AB} that satisfy Eq. (156), or, equivalently, the multiplicity of tensor modes on S^n that are described by the same angular-momentum number j. This number is given by the expression [77, 81]

$$N_T^j(S^n) = \frac{(n+1)(n-2)(n+j)(j-1)(n+2j-1)(n+j-3)!}{2(j+1)!(n-1)!}. \tag{162}$$

Using the above formula, the Hawking radiation spectra for tensor-type gravitons in the bulk were computed in [81]. In Figure 14, we present the energy emission rate in terms both of the number of extra dimensions n and the value of the angular-momentum parameter a_* of the black hole. We observe that an increase in n leads to a signif cant enhancement of the energy-emission rate over the whole energy regime. On the other hand, the dependence on a_* is more particular: for low values of n, the energy emission rate is enhanced in the

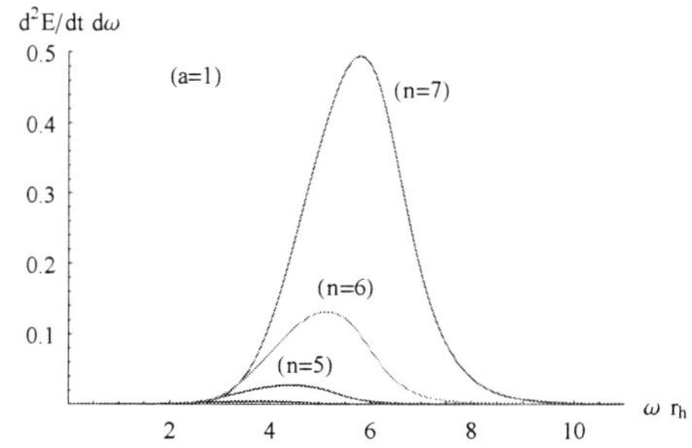

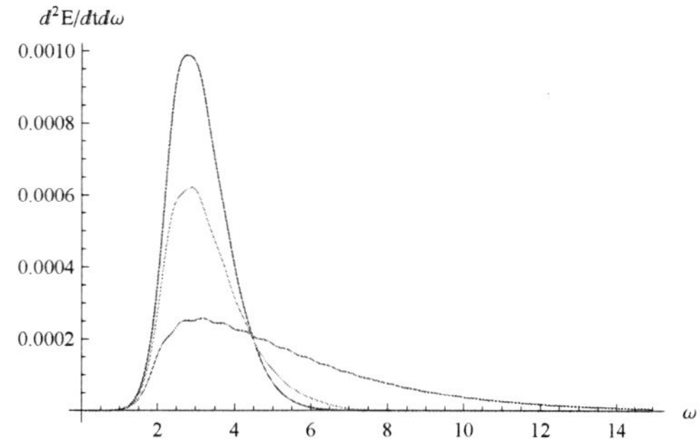

Figure 14. The energy emission rate for tensor-type gravitons in the bulk for **(a)** $a_* = 1$ and $n = 3, 4, 5, 6, 7$ (from bottom to top), and for **(b)** $n = 3$ and $a_* = 0$ (blue line, top), $a_* = 0.5$ (green line), $a_* = 1.2$ (red line, bottom) [81].

high-energy regime and suppressed in the other frequencies while for high values of n, it displays a rather mild dependence on a_*.

5. Energy Balance between the Brane and the Bulk

We f nally turn to the important question of the energy balance between the brane and the bulk. The particles emitted in the bulk by a higher-dimensional black hole, either spherically-symmetric or rotating, cannot be detected by an observer localised on the brane, therefore, the study of this emission channel may seem to have only theoretical motivation. Nevertheless, as the amount of energy that the black hole can give out is pre-determined by its mass, the percentage of energy emitted in the bulk inevitably determines the one left for the emission on the brane. As a result, in order to decide about the magnitude of any observable effects on the brane, it is imperative to determine the bulk-brane energy balance.

The f rst attempt to compute the relative bulk-over-brane emissivity, i.e. the relative energy emitted over the whole frequency regime by the black hole into the bulk over the one on the brane, was made in [38]. Performing an exact numerical analysis leading to the complete radiation spectra for scalar particles propagating in the bulk and on the brane in the vicinity of a higher-dimensional spherically-symmetric black hole, the total emissivities were computed and compared. The results are presented in Table 3, and they reveal that the bulk-over-brane ratio remains smaller than unity for all values of n from 1 to 7. However, we observe that the two channels become comparable, in terms of the energy emitted per unit time in each one of them, for large values of the number of extra dimensions. Nevertheless, one may safely conclude that the emission of scalars in the bulk is sub-dominant to the one on the brane.

Clearly, to decide about the bulk-brane energy balance for the Schwarzschild phase one needs to consider also the emission of gravitons. As discussed in the previous subsection, this effect was studied in an exact numerical way in [78]. In there, the radiation spectra for all types of gravitational perturbations were derived and the question of how the total energy emitted in the graviton bulk channel compares with the one spent on the emission of Standard-Model f elds on the brane was addressed. It was found that the energy emissivity

Table 3. Bulk-to-brane energy emissivities for different values of n [38]

n	0	1	2	3	4	5	6	7
Bulk/Brane	1.0	0.40	0.24	0.22	0.24	0.33	0.52	0.93

of gravitons in the bulk increases very fast with n, due to the increase in the number of degrees of freedom of the gravitational f eld, with the enhancement factor reaching values of $\mathcal{O}(10^6)$. Nevertheless, the overall energy carried into the bulk, compared to the one released in total in the form of Standard-Model brane f elds, is still sub-dominant. Therefore, the conclusion may be drawn that, during the Schwarzschild phase, the brane channel is indeed the dominant one, in agreement with an early work [82] where this claim was made.

As we saw in section 4.1, the emission of bulk scalars by a higher-dimensional, simply-rotating black hole was also studied via numerical means in [64]. The exact emission spectra was in fact compared to the corresponding ones for emission of scalar f elds on the brane, and found to be signif cantly smaller (see Figure 12). In comparison with the Schwarzschild phase, it was demonstrated that the total energy output of the black hole in the scalar channel in fact increases with the angular momentum of the black hole. However, the relative emissivity takes even smaller values, compared to the Schwarzschild phase, as soon as the angular-momentum parameter assumes non-vanishing values – this is clearly demonstrated in Figure 15. This is due to the fact that the temperature of the black hole decreases with a_*, while the enhancement of the greybody factor in the bulk is not large enough to compensate this decrease. As a result, the brane dominance persists even in the spin-down phase, at least in the scalar channel of emission.

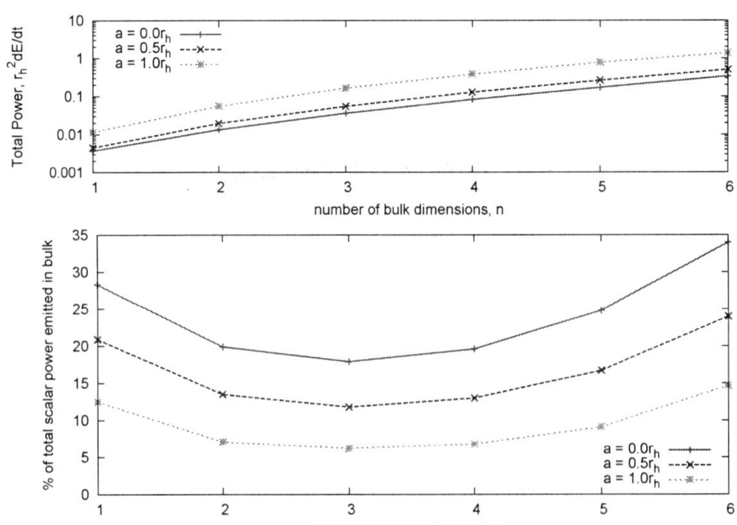

Figure 15. Total power emitted by a rotating black hole in the scalar channel (upper plot) and the % of this power emitted in the bulk (lower plot) [64].

Table 4. Total emissivity for scalars and tensor-type gravitons in the bulk during the rotating phase [81].

n	Scalars	Tensor-type gravitons	
3	0.1646	0.0013	0.8%
4	0.3808	0.0222	5.8%
5	0.7709	0.1853	24%

Nevertheless, special cases where the bulk channel may dominate over the brane one are known in the literature [83, 84]. Therefore, we still need to clarify the situation for the gravity emission in the bulk during the rotating phase. Unfortunately, only impartial conclusions may be drawn since the graviton equations are known in only some very particular cases. In [81], where the emission of tensor-type gravitons by a higher-dimensional, simply rotating black hole was studied, an estimate of the total percentage of energy going into the bulk graviton channel was made - this is given in Table 4 for the indicative case of $a_* = 1$. According to our results, for low values of n the energy carried by gravitons into the bulk is only a minute percentage of the one carried by scalars. However, as n increases, this amount becomes important and it reaches the value of 25% for $n = 5$. Unfortunately, the calculation of the total emissivity for gravitons, due to the very large number of partial modes that need to be summed, becomes extremely diff cult for larger values of n. We may thus only estimate, that, as in the case of a spherically-symmetric background, this percentage becomes even more important as n increases further. Whether the gravitons may tilt the bulk-to-brane energy balance towards the bulk during the rotating phase remains to be seen, especially after the study of vector and scalar gravitational perturbations is completed.

Finally, we would like to end by reporting results from a recent work that was also shown to be related to the bulk-brane energy balance. Up to now, the emitted f elds by the black hole were considered to be massless. However, in [69], the effect of the mass of the emitted scalar particles on the differential energy emission rates both in the bulk and on the brane, and consequently on the relative emissivity between the two channels, was studied. The same approximation method was employed in order to derive analytic results for the Absorption Probability - the analysis is similar to the one for a massless scalar f eld, therefore, here we only point out the points where differences arise.

A massive scalar f eld, with mass m_Φ, propagating in the gravitational background of a higher-dimensional, simply-rotating black hole (11) satisf es the equation of motion

$$\frac{1}{\sqrt{-G}}\,\partial_M\left(\sqrt{-G}\,G^{MN}\partial_N\Phi\right) - m_\Phi^2\Phi = 0\,. \tag{163}$$

The same factorised ansatz (99) leads to the separation of its equation of motion, even in the presence of the mass [85], into a radial and an angular part

$$\frac{1}{r^n}\partial_r\left(r^n\Delta\partial_r R\right)+\left(\frac{K^2}{\Delta}-\frac{j(j+n-1)a^2}{r^2}-\tilde{\Lambda}_{j\ell m}-m_\Phi^2 r^2\right)R=0\,, \tag{164}$$

$$\frac{1}{\sin\theta\cos^n\theta}\,\partial_\theta\left(\sin\theta\cos^n\theta\partial_\theta S\right)$$
$$+\left(\tilde{\omega}^2a^2\cos^2\theta-\frac{m^2}{\sin^2\theta}-\frac{j(j+n-1)}{\cos^2\theta}+\tilde{E}_{j\ell m}\right)S=0\,, \tag{165}$$

respectively. In the above, we define as before $\tilde{\Lambda}_{j\ell m}=\tilde{E}_{j\ell m}+a^2\omega^2-2am\omega$. We observe that the angular function $S(\theta)$ satisfies a modified higher-dimensional spheroidal harmonics equation: compared to the massless case [67], it has the energy ω replaced by the momentum $\tilde{\omega}\equiv\sqrt{\omega^2-m_\Phi^2}$. Then, the massive angular eigenvalue $\tilde{E}_{j\ell m}(a\tilde{\omega})$ is related to the massless one $E_{j\ell m}(a\omega)$, Eq. (105), by merely a shift of its argument: $a\omega\rightarrow a\tilde{\omega}$.

The analysis after this point follows very closely the one for a massless scalar field emitted in the bulk, that was studied in section 4.1. The solution in the near-horizon regime follows from a hypergeometric equation and has the form of Eq. (111) with the indices (a,b,c) and the power-coefficient α given by the same expressions as before, while β now has the form

$$\beta=\frac{1}{2}\left[(2-D_*)-\sqrt{(D_*-2)^2-4\left(\frac{K_*^2-C_*}{A_*^2}\right)}\right], \tag{166}$$

where

$$C_*=\left[j(j+n-1)a_*^2+\tilde{\Lambda}_{j\ell m}+m_\Phi^2 r_h^2\right](1+a_*^2)\,. \tag{167}$$

In the far-field regime, the solution satisfies a Bessel equation in terms of $z=\tilde{\omega}r$, and the solution has the same form, Eq. (113), as in the case of a massless scalar field, where now $\nu=\sqrt{\tilde{E}_{j\ell m}+a^2\tilde{\omega}^2+\left(\frac{n+1}{2}\right)^2}$. The two solutions are then stretched and matched in an intermediate regime, leading to the constraint (117), and the expression of the Absorption Probability according to Eq. (119). It is worth noting that the perfect matching of the two asymptotic solutions is realised not only under the assumption of low energy and low angular-momentum, but also for $m_\Phi r_H<1$; for a model with $M_*=1$ TeV and $M_{BH}=5$ TeV, we find that $r_H\simeq(4-2)\,10^{-4}$ fm for $n=1-7$, respectively [11], and the aforementioned constraint translates to

$$m_\Phi<(0.5-1)\,\text{TeV}\quad\text{for}\quad n=1-7\,. \tag{168}$$

Let us also note that, in the very low-energy regime $\tilde{\omega}\rightarrow 0$, the absorption cross-section for the lowest mode $\ell=j=m=0$ of the massive scalar field in the bulk, turns out to be $\sigma_{000}\sim A_H\,(\omega/\tilde{\omega})$. Therefore, for $m_\Phi\neq 0$, not only do we observe again deviations from the universal rule but also the absorption cross-section deviates as $\tilde{\omega}\rightarrow 0$ – this is in accordance with previous results derived in the cases of a massive field propagating in the background of a 4-dimensional Kerr [39] or of a $(4+n)$-dimensional, spherically-symmetric black hole [63].

In the case of the emission of a massive scalar field on the brane, one needs to solve again Eq. (163) but with the higher-dimensional metric tensor G_{AB} replaced by the 4-dimensional one $g_{\mu\nu}$ corresponding to the line-element (19) of the gravitational background induced on the brane by a higher-dimensional black hole. A field factorisation of the form (26) leads again to the decoupled set of equations

$$\frac{d}{dr}\left(\Delta\frac{dR}{dr}\right)+\left(\frac{K^2}{\Delta}-\tilde{\Lambda}_{jm}-m_\Phi^2 r^2\right)R=0\,, \tag{169}$$

$$\frac{1}{\sin\theta}\frac{d}{d\theta}\left(\sin\theta\,\frac{dS_j^m}{d\theta}\right)+\left(\tilde{\omega}^2 a^2\cos^2\theta-\frac{m^2}{\sin^2\theta}+\tilde{\lambda}_j\right)S_j^m=0\,. \tag{170}$$

In the above, we have defined $\tilde{\Lambda}_{jm}=\tilde{\lambda}_j+a^2\omega^2-2am\omega$. The angular function $S_j^m(\theta)$ satisfies again a modified spheroidal harmonics equation with $\omega\to\tilde{\omega}$. Then, the corresponding massive eigenvalue $\tilde{\lambda}_j(a\tilde{\omega})$ will be related to the massless one (52) through the same shift.

The near-horizon solution is again given by Eq. (68) with $a=\alpha+\beta+D_*-1$, $b=\alpha+\beta$, $c=1+2\alpha$. The power coefficient α has the same form as in the case of a massless scalar field on the brane, while β is given by the expression (166), with D_* and C_* now taking their brane values

$$D_*\equiv 1+\frac{n\,(1+a_*^2)}{A_*}-\frac{4a_*^2}{A_*^2}\,,\qquad C_*\equiv(\tilde{\Lambda}_{jm}+m_\Phi^2 r_h^2)\,(1+a_*^2)\,. \tag{171}$$

In the far-field regime ($r\gg r_H$), the radial equation (169), under the substitution $R(r)=\frac{1}{\sqrt{r}}\tilde{R}(r)$, takes again the form of a Bessel differential equation leading to the general solution

$$P_{FF}(r)=\frac{B_1}{\sqrt{r}}J_\nu(\tilde{\omega}r)+\frac{B_2}{\sqrt{r}}Y_\nu(\tilde{\omega}r)\,, \tag{172}$$

where now $\nu=\sqrt{\tilde{\lambda}_j+a^2\tilde{\omega}^2+1/4}$.

The constraint that follows from the matching of the two stretched asymptotic solutions in this case has the form

$$B\equiv\frac{B_1}{B_2}=-\frac{(2/\tilde{\omega}r_H)^{2j+1}\,\nu\,\Gamma^2(\nu)\,\Gamma(\alpha+\beta+D_*-1)\,\Gamma(\alpha+\beta)\,\Gamma(2-2\beta-D_*)}{\pi\,(1+a_*^2)^{\frac{2j+1}{n+1}}\,\Gamma(2\beta+D_*-2)\,\Gamma(2+\alpha-\beta-D_*)\,\Gamma(1+\alpha-\beta)}\,. \tag{173}$$

The above holds provided that the mass of the scalar field on the brane does not exceed an upper value: $m_\Phi<(250-500)\,\text{GeV}$ for $n=1-7$ – note that the upper value of the mass on the brane is reduced by a factor of two compared to the one in the bulk. At the very low-energy regime, we may again derive the absorption cross-section for the lowest mode $j=m=0$; this turns out again to be $\sigma_{00}\sim r_H^2(\omega/\tilde{\omega})$ which again deviates as $\tilde{\omega}\to 0$. All higher modes, both on the brane and in the bulk, have a vanishing absorption cross-section in the limit of zero $\tilde{\omega}$.

In [69], the emission of massive scalar fields was studied also in an exact numerical way both on the brane and in the bulk. In Figure 16, we depict the indicative cases of the emission of scalar fields on the brane (left plot) and in the bulk (right plot) with $m_\Phi=0, 0.4, 0.8$ by a black hole with angular-momentum parameter $a_*=1$ and for three different values of n. As expected, the energy emission rate is reduced as the mass of the emitted field increases – clearly, the emission of a massive field demands more energy to be spent by the black hole, and thus it is less likely to happen. However, it turns out that the magnitude of the effect is not the same for the two emission channels, with the brane one to be more sensitive to

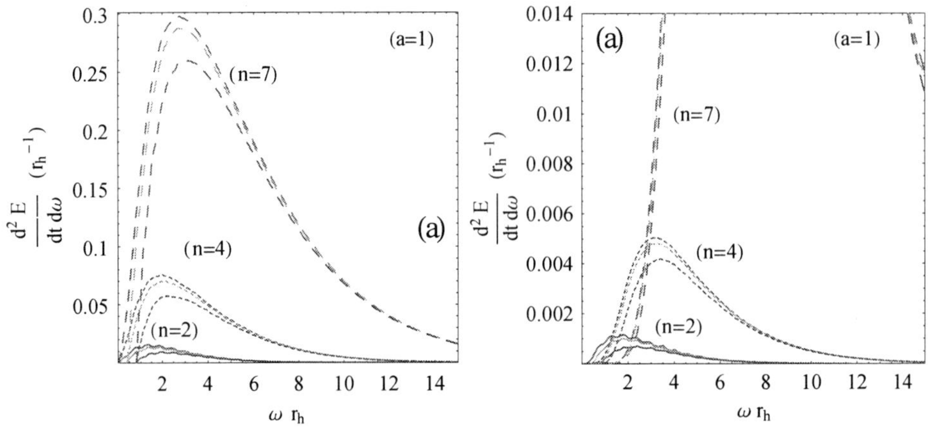

Figure 16. Energy emission rates for massive scalar f elds on the brane (left plot) and in the bulk (right plot) during the rotating phase of a black hole with $a_* = 1$ for $m_\Phi = 0, 0.4, 0.8$ (from top to bottom in each set of curves) and variable n [69].

the increase of the mass of the emitted f eld. Therefore, it is found that for a given value of the mass of the scalar f eld, e.g. for $m_\Phi = 0.8$, the bulk-to-brane relative emissivity can be increased by 34%. This enhancement holds for a black hole with low-angular-momentum ($a_* = 0.5$) leaving in a spacetime with a small number of extra dimensions ($n = 2$) – for larger values of either a_* or n the enhancement takes smaller but still signif cant values. We therefore conclude that the energy balance between the brane and the bulk is determined not only by properties of spacetime such as n or a_*, but also by properties of the emitted particle itself, such as its mass.

Conclusion

The purpose of this review was to summarise, in a concise and comprehensive way, what is currently known in the topic of the emission of Hawking radiation from higher-dimensional, either rotating or spherically-symmetric, black holes both on the brane and in the bulk. The analysis here has focused mainly on the presentation of both approximate/analytic and exact/numerical results for the spin-down phase in the life of the black hole - in this way we complete and generalise previous reviews that were devoted to the Schwarzschild phase.

We started by reviewing the provocative idea arising in the context of the new theories with Extra Dimensions, either Large or slightly Warped, that a black hole could be created during a high-energy particle collision. The creation process and the properties of the produced black holes were also discussed including the emission of Hawking radiation that, in principle, will provide the main observable effect of the creation of the black holes, and thus of the existence of extra dimensions.

We then focused on the emission of Hawking radiation on the brane, the emission channel that would be directly observable to an observer restricted to live on the brane. We studied the emission of all Standard-Model particles, namely scalars, fermions and gauge bosons, and presented analytic results, that allow us to gain a better insight of quantities such as the Absorption Probability and the absorption cross-section, as well as exact numerical results that allow us to construct the complete radiation spectra. We also discussed the feature of the angular distribution of the emitted radiation that should be a distinct observable during the spin-down phase and a valuable source of information on the higher-dimensional gravitational background.

The bulk emission channel was studied next. Here, we turned our attention first to the scalar fields, and present again both analytic and numerical results for the Schwarzschild and the spin-down phase. The emission of gravitons was discussed next, and analytical results on all types of gravitational perturbations, tensor, vector and scalar, for the Schwarzschild phase were presented. The corresponding numerical results available in the literature were also reviewed. For the spin-down phase, the results for the emission of gravitons are still incomplete: whereas both analytic and numerical results exist for tensor-type perturbations, the equations governing the vector and scalar-type gravitational perturbations are not even yet known.

We finally addressed the question of the energy balance between the brane and the bulk. For the Schwarzschild phase, the verdict is out with the emission on the brane (in the form of Standard Model fields) being the dominant one over the emission in the bulk (in the form of scalars and gravitons). During the spin-down phase, the study of the scalar channel seems to give extra support to the dominance of the brane. Nevertheless, the final conclusion has not been drawn yet as this would have to rely on the study of the graviton emission which unfortunately still misses some of its most important pieces. We have finally discussed the role of the mass of the emitted scalar particles on the differential energy emission rates both in the bulk and on the brane, and demonstrated that its effect on the relative emissivity between the two channels is, at cases, fairly important.

The intensive research activity of the creation, properties and decay process of the higher-dimensional black holes during the last ten years have created great expectations in the scientific community. Although no sightings of miniature black holes have been reported so far by particle-physics experiments, we preserve our hopes that these exotic objects will soon reveal themselves. When, and if, that happens, a window in the higher-dimensional, fundamental, hierarchy-problem-free, complete theory of all forces will have opened at last.

Acknowledgment

I am grateful to all of my collaborators for our enjoyable and fruitful collaborations over the last ten years.

References

[1] I. Antoniadis, Phys. Lett. B **246**, 377 (1990).

[2] P. Horava and E. Witten, *Nucl. Phys. B* **460**, 506 (1996); *Nucl. Phys. B* **475**, 94 (1996).

[3] J. Lykken, Phys. Rev. D **54**, 3693 (1996).

[4] N. Arkani-Hamed, S. Dimopoulos and G. Dvali, *Phys. Lett. B* **429**, 263 (1998); *Phys. Rev. D* **59**, 086004 (1999).

[5] I. Antoniadis, N. Arkani-Hamed, S. Dimopoulos and G. R. Dvali, *Phys. Lett. B* **4**36, 257 (1998).

[6] L. Randall and R. Sundrum, *Phys. Rev. Lett.* **8**3, 3370 (1999); *Phys. Rev. Lett.* **83**, 4690 (1999).

[7] K. Akama, *Lect. Notes Phys.* **176**, 267 (1982).

[8] V. A. Rubakov and M. E. Shaposhnikov, *Phys. Lett. B* **125**, 139 (1983); *Phys. Lett. B* **125**, 136 (1983).

[9] M. Visser, *Phys. Lett. B* **159**, 22 (1985).

[10] G. W. Gibbons and D. L. Wiltshire, *Nucl. Phys. B* **287**, 717 (1987).

[11] P. Kanti, *Int. J. Mod. Phys. A* **19**, 4899 (2004); *Lect. Notes Phys.* **769**, 387 (2009).

[12] P. Abreu *et al.* [DELPHI Collaboration], *Eur. Phys. J.* **C17**, 53 (2000); G. Abbiendi *et al.* [OPAL Collaboration], *Eur. Phys. J.* **C18**, 253 (2000); D. Acosta *et al.* [CDF Collaboration], *Phys. Rev. Lett.* **89**, 281801 (2002).

[13] C. D. Hoyle, U. Schmidt, B. R. Heckel, E. G. Adelberger, J. H. Gundlach, D. J. Kapner and H. E. Swanson, *Phys. Rev. Lett.* **86**, 1418 (2001); D. J. Kapner, T. S. Cook, E. G. Adelberger, J. H. Gundlach, B. R. Heckel, C. D. Hoyle and H. E. Swanson, *Phys. Rev. Lett.* **98**, 021101 (2007).

[14] T. Banks and W. Fischler, hep-th/9906038.

[15] S. B. Giddings and S. Thomas, *Phys. Rev. D* **65**, 056010 (2002).

[16] S. Dimopoulos and G. Landsberg, *Phys. Rev. D* **87**, 161602 (2001).

[17] K. S. Thorne, in *Magic without Magic*, ed. J. R. Klauder (San Fransisco, 1972).

[18] D. Ida and K. i. Nakao, *Phys. Rev. D* **66**, 064026 (2002).

[19] M. Cavaglià, *Int. J. Mod. Phys.* A **18**, 1843 (2003); G. Landsberg, *Eur. Phys. J.* C **33**, S927 (2004); C. M. Harris, hep-ph/0502005; A. Casanova and E. Spallucci, *Class. Quant. Grav.* **23**, R45 (2006); E. Winstanley, 0708.2656 [hep-th].

[20] H. Kudoh, T. Tanaka and T. Nakamura, Phys. Rev. D **68**, 024035 (2003); H. Kudoh, *Phys. Rev. D* **69**, 104019 (2004) [Erratum-ibid. D **70**, 029901 (2004)]; N. Tanahashi and T. Tanaka, *JHEP* **0803**, 041 (2008).

[21] F. R. Tangherlini, *Nuovo Cim.* **27**, 636 (1963).

[22] R. C. Myers and M. J. Perry, *Annals Phys.* **172**, 304 (1986).

[23] P. Meade and L. Randall, *JHEP* **0805**, 003 (2008).

[24] V. Khachatryan *et al.* [CMS Collaboration], *Phys. Lett. B* **697**, 434 (2011).

[25] G. Aad *et al.* [ATLAS Collaboration], *New J. Phys.* **13**, 053044 (2011).

[26] S. W. Hawking, *Commun. Math. Phys.* **43**, 199 (1975).

[27] W. G. Unruh, *Phys. Rev. D* **10**, 3194 (1974); *Phys. Rev. D* **14**, 3251 (1976).

[28] E. Newman and R. Penrose, *J. Math. Phys.* **3**, 566 (1962).

[29] S. Chandrasekhar, *The Mathematical Theory of Black Holes* (Oxford University Press, New York, 1983).

[30] S. A. Teukolsky, *Phys. Rev. Lett.* **29**, 1114 (1972); *Astrophys. J.* **185**, 635 (1973).

[31] C. Flammer, *Spheroidal Wave Functions* (Stanford University Press, Stanford, USA, 1957).

[32] J. N. Goldberg, A. J. MacFarlane, E. T. Newman, F. Rohrlich and E. C. Sudarshan, *J. Math. Phys.* **8**, 2155 (1967).

[33] M. Abramowitz and I. Stegun, *Handbook of Mathematical Functions* (Academic, New York, 1996).

[34] W. H. Press and S. A. Teukolsky, *Astrophys. J.* **185**, 649 (1973); *Astrophys. J.* **193**, 443 (1974).

[35] M. Casals, P. Kanti and E. Winstanley, JHEP **0602**, 051 (2006).

[36] P. Kanti and J. March-Russell, Phys. Rev. D **66**, 024023 (2002).

[37] P. Kanti and J. March-Russell, *Phys. Rev. D* **D67**, 104019 (2003).

[38] C. M. Harris and P. Kanti, *JHEP* **0310**, 014 (2003).

[39] D. N. Page, *Phys. Rev.* **D13**, 198 (1976); *Phys. Rev.* **D14**, 3260 (1976).

[40] N. Sanchez, *Phys. Rev.* **D16**, 937 (1977); *Phys. Rev.* **D18**, 1030 (1978); *Phys. Rev.* **D18**, 1798 (1978).

[41] D. Ida, K. y. Oda and S. C. Park, *Phys. Rev. D* **67**, 064025 (2003) [Erratum-ibid. D **69**, 049901 (2004)].

[42] S. Creek, O. Efthimiou, P. Kanti and K. Tamvakis, *Phys. Rev. D* **75**, 084043 (2007).

[43] S. Creek, O. Efthimiou, P. Kanti and K. Tamvakis, *Phys. Rev. D* **76**, 104013 (2007).

[44] C. M. Harris and P. Kanti, *Phys. Lett. B* **633**, 106 (2006).

[45] G. Duffy, C. Harris, P. Kanti and E. Winstanley, *JHEP* **0509**, 049 (2005).

[46] M. Casals, S. R. Dolan, P. Kanti and E. Winstanley, *JHEP* **0703**, 019 (2007).

[47] D. Ida, K. y. Oda and S. C. Park, *Phys. Rev. D* **71**, 124039 (2005); *Phys. Rev. D* **73**, 124022 (2006).

[48] A. A. Starobinskii and S. M. Churilov, Sov. Phys.-JETP **38**, 1 (1974).

[49] E. D. Fackerell and R. G. Crossman, J. Math. Phys. **18**, 1849 (1977).

[50] E. Seidel, Class. Quant. Grav. **6**, 1057 (1989).

[51] R. Güven, *Phys. Rev. D* **16**, 1706 (1977).

[52] D. A. Leahy and W. G. Unruh, *Phys. Rev. D* **19**, 3509 (1979).

[53] S.L. Detweiler, *Proc. Roy. Soc. Lond. A* **349**, 217 (1976)

[54] M. Casals and A. C. Ottewill, *Phys. Rev. D* **71**, 124016 (2005).

[55] E. Wasserstrom, *J. Comp. Phy.* **9**, 53 (1972).

[56] W. H. Press, S. A. Teukolsky, W. T. Vetterling and B. P. Flannery, *Numerical Recipes in Fortran* (Cambridge University Press, Cambridge, 2002).

[57] S.A. Hughes, *Phys. Rev. D* **61**, 084004 (2000).

[58] E.W. Leaver, *Proc. R. Soc. A* **402**, 285 (1985).

[59] M. Casals, S. R. Dolan, P. Kanti and E. Winstanley, *Phys. Lett. B* **680**, 365 (2009).

[60] A. Flachi, M. Sasaki and T. Tanaka, *JHEP* **0905**, 031 (2009).

[61] V. P. Frolov and D. Stojkovic, *Phys. Rev. D* **67**, 084004 (2003).

[62] S. Creek, O. Efthimiou, P. Kanti and K. Tamvakis, *Phys. Lett. B* **656**, 102 (2007).

[63] E. Jung and D.K. Park, *Nucl. Phys. B* **731** (2005) 171.

[64] M. Casals, S. R. Dolan, P. Kanti and E. Winstanley, *JHEP* **0806**, 071 (2008).

[65] C. Muller, in *Lecture Notes in Mathematics: Spherical Harmonics* (Springer-Verlag, Berlin-Heidelberg, 1966).

[66] D. Ida, Y. Uchida and Y. Morisawa, *Phys. Rev. D* **67** (2003) 084019 .

[67] E. Berti, V. Cardoso and M. Casals, *Phys. Rev. D* **73** (2006) 024013 [Erratum-ibid. D **73** (2006) 109902];

[68] V. Cardoso, G. Siopsis and S. Yoshida, *Phys. Rev. D* **71**, 024019 (2005).

[69] P. Kanti and N. Pappas, *Phys. Rev. D* **82**, 024039 (2010).

[70] A. C. Ottewill and E. Winstanley, *Phys. Rev. D* **62**, 084018 (2000).

[71] H. Kodama and A. Ishibashi, *Prog. Theor. Phys.* **110**, 701 (2003).

[72] H. Kodama, *Prog. Theor. Phys. Supple.* **172**, 11 (2008).

[73] H. Kodama, A. Ishibashi and O. Seto, *Phys. Rev. D* **62**, 064022 (2000).

[74] H. Kodama, *Lect. Notes Phys.* **769**, 427 (2009).

[75] S. Creek, O. Efthimiou, P. Kanti and K. Tamvakis, *Phys. Lett. B* **635**, 39 (2006).

[76] A. S. Cornell, W. Naylor and M. Sasaki, *JHEP* **0602**, 012 (2006).

[77] M.A. Rubin and C.R. Ordónez: *J. Math. Phys.* **25**, 2888 (1984).

[78] V. Cardoso, M. Cavaglia and L. Gualtieri, *Phys. Rev. Lett.* **96**, 071301 (2006) [Erratum-ibid. **96**, 219902 (2006)]; JHEP **0602**, 021 (2006).

[79] D. K. Park, *Phys. Lett. B* **638**, 246 (2006).

[80] H. Kodama, R. A. Konoplya and A. Zhidenko, *Phys. Rev. D* **79**, 044003 (2009).

[81] P. Kanti, H. Kodama, R. A. Konoplya, N. Pappas and A. Zhidenko, *Phys. Rev. D* **80**, 084016 (2009).

[82] R. Emparan, G. T. Horowitz and R. C. Myers, *Phys. Rev. Lett.* **85**, 499 (2000).

[83] J. Grain, A. Barrau and P. Kanti, *Phys. Rev. D* **72**, 104016 (2005).

[84] H. T. Cho, A. S. Cornell, J. Doukas and W. Naylor, *Phys. Rev. D* **77**, 016004 (2008).

[85] V. P. Frolov and D. Stojkovic, *Phys. Rev. D* **68**, 064011 (2003).

INDEX

A

B

C

N

O

P

Q

R

S

T

U

V

W

Y